Girdle Round the Earth

Two water-colours signed 'H.W.A.' (H. W. Ansell) from an album of 'Sketches on Djeddah Cable Repair January 1890'. (*Above*) signalling to ship from roof of cable house with a piece of looking glass; (*below*) a 'potential' test under difficulties.

HUGH BARTY-KING

Girdle Round the Earth

the story of Cable and Wireless
and its predecessors
to mark the group's jubilee
1929–1979

HEINEMANN : LONDON

William Heinemann Ltd
10 Upper Grosvenor Street, London W1X 9PA

LONDON MELBOURNE TORONTO
JOHANNESBURG AUCKLAND

First published 1979
© Cable and Wireless Limited 1979
434 04902 6

Printed in Great Britain by
Richard Clay (The Chaucer Press) Ltd, Bungay, Suffolk

CONTENTS

ACKNOWLEDGEMENTS

Finding out what happened – and why – in the distant past must necessarily rely on documents; in the more recent past and the present on both documents and people. The extensive research exercise which preceded the writing of this book was carried out under the guiding hand of Mr Joseph Tobin, Group General Manager (Public Relations), Cable and Wireless Limited, at Mercury House, London, and I am greatly indebted to him and the staff of his Public Relations Department, in particular Mr Peter Travers-Laney, Group Archivist, and the two graduate management trainees who were seconded to me in the initial stages, Miss Jennifer Cousins and Miss Penny David, without whose painstaking work it would have been impossible to cover the ground in such depth and in the time. Of overseas Public Relations staff I have particular reason to be grateful to Miss Alice Yao of Hong Kong and Mr Hussain Madan of Bahrain; and for her cheerful and efficient help in smoothing the mechanics of the exercise at home and overseas I thank Mrs Yvonne Filleul.

Research into early history was enriched by examination of the Campbell Stuart Papers in the library of the Royal Commonwealth Society put at my disposal by Mr Donald Simpson, librarian; and by a study of material at the India Office Library and Records, the D M S Watson Library, University College, London, and the Public Record Office.

My thanks are due to the Chairman of Cable and Wireless, Lord Glenamara, and to Mr Peter McCunn, Mr R W Cannon, Mr John Bird, Mr R A Rice and Mr W H Davies of the Court of Directors, and Mr Arthur Cooke, Company Secretary, for giving me of their time to talk of current issues and past achievements; to ex-managing directors Mr A A Willett and Mr H H Eggers; to Lord Pender, Hon. Richard Denison-Pender and Lady Wilshaw; to Mr Harold J Wilson, ex-Press Officer.

I am grateful to the many present and past members of the staff of Cable and Wireless in Britain who aided research with valuable written contributions, or were good enough to talk either to me, Miss David or Miss Cousins about their work and experiences, including Mr Clifford Angove, Mr H L N Ascough, Mr John Ash, Mr Roy Baker, Mr D C Buck, Mr Sid Chappel, Mr Peter J A Cousins, Mr H A Cox, Mr J H Evans, Mr Michael Forrer, Mr C Fox, Mr Arthur Grass, Mr Leonard Harris, Mr R W Harrisson, Mr A Hinshelwood, Mr John Hooley, Mr H Madigan, Mr Charles Martin, Mr F

Maynard, Mr D Michell, Mr D John Moore, Commander C C Muckleston, Mr John Packer, Mr M L Pitchford, Mr John Powell, Mr V G Pullen, Mr William Rae, Mr John Simpson, Mr Brian Suart, Mr R F C Thomae, Mr P F Tudor, Mr D W Weedon, Mr M Whiting, Mr N T G Williams, Mr J H Wilson, Mr Martin Woolf.

My thanks are also due to the directors, managers and staff, past and present, whom I interviewed overseas, including Mr Charles Antrobus, Mr Jack Armstrong, Mr J P Bairstow, Mr E Bissett, Mr J Bourne, Mr J Brewster, Mr Norman Brooks, Mr David Chan, Mr Paul Carrington, Mr Jimmy Cozier, Mr Galbraith Cropper, Captain W H Cross, Mr John Davenport, Mr Armald Derrick, Mr Cosimiro Duran, Mr R Ellis, Captain John Fenwick, Mr Ronald Footer, Mr Peter Forestal, Mr Courtney Frost, Mr Peter Coslett, Mr Ian Hall, Mr T Harman, Mr Michael Harvey, Mr Ronald Hobbah, Mr Ho Wing Wah, Mr Richard Keen, Mr Peter Lashley, Mr Basil Leighton, Mr Joe Leonard, Mr R de Corday Long, Mr Ling Chung Yan, Mr R Mackenzie, Mr P Malig, Mr Roderick Mann, Mr Bill Martin, Mr P Millen, Mr Terry More, Mr Jim Morrison, Mr Stanley Montoute, Mr Hubert Montplaisir, Mr Arthur Palmer, Mr Dave Parker, Mr E J W Payne, Mr Richard Pilcher, Mr Garry Peterken, Mr John Power, Mr Alfred Plater, Mr R W Rogerson, Mr Harry Saunter, Mr Frank Savoury, Mr Ahmed Omar Salaah, Mr Alec Sherman, Mr Ainslie Skeete, Mr Eric Smith, Mr Philip Sumser, Mr Edward Stoute, Mr Tony Tipple, Mr Peter Turton, Mr George Warwick, Mr Chris Watson, Mr Norman Wheatley, Mr Brian Wood, Mr R Wright, Mr Terry Vaughan; and Mr F H Weller of SITA; Mr Gordon Hanson and Mr M Finucane of Reuters; and Mr H L Pierce of The Hong Kong Bank.

I am indebted to all those who read the manuscript in draft and for their helpful comments, in particular Mr Peter McCunn, Mr R A Rice, Mr Brian Suart, Mr Arthur Cooke, Mr R E McAlister and Mr W H Davies; and for the technical checking by Mr D W Weedon, Mr John Powell, Mr Bryn James, Mr Stuart Perkins and Mr Roy George.

H. B-K
May 1979

LIST OF ILLUSTRATIONS

between pages 272 and 273

Acknowledgement is made to the following for permission to reproduce photographs: Public Archives Canada, nos 1 & 29; The Post Office Corporation, nos 4 & 5; The Marconi Company, nos 27, 28, 39, 40, 41, 48; The Royal Commonwealth Society, no 45; Lady Wilshaw, nos 44 & 60; Mary Evans Picture Library, nos 13 & 24; Australian Information Service, London, no 14; Standard Telephones & Cables, no 95; Associated Newspapers, no 122; Sport & General, no 43. All line illustrations were supplied by the Mary Evans Picture Library except for that of the Indo-European cable being laid at Fao (The Mansell Collection), the Persian Gulf form of 1868 (India Office), and the two *Punch* cartoons. Sir Edward Wilshaw's letter of engagement from the Eastern Telegraph Company of 1894 was kindly lent by Lady Wilshaw. The letter to Lord Palmerston of 1851 in the Public Record Office is reproduced by permission of the Controller of HM Stationery Office.

FOREWORD

by The Rt. Hon. The Lord Glenamara,
Chairman of Cable and Wireless Ltd.

The company in its present form was born half a century ago, though we trace our parentage back into the mid-Victorian era. In 1929 Imperial and International Communications Ltd was created from the old telegraph companies and their newer, radio based, rivals. Cables versus wireless became Cable and Wireless, although this did not become its name until 1934.

The story told in this book with remarkable thoroughness and lucidity by Mr Hugh Barty-King is a record of service dedicated to the task of enabling nation to speak unto nation by tens of thousands of men and women of many races, creeds and colours around the world. It is a record of service which has rarely been equalled in the forward march of mankind. The company is, in fact, a huge, continuing international family. The sheer unflagging continuity of its service as well as its unwavering technical excellence is an inspiration to all who see it.

We have had an exciting and illustrious past and it is important that it should be known and understood, for the present is the product of the past and the future of both. That is why this history is important – coming as it does in our Golden Jubilee year. It records the aspirations and the achievements as well as the failures and mistakes of our predecessors. The company is today what they made it. Tomorrow it will be what we make of their achievements. And the challenges already discernible, and many are still hidden in the future, are quite different from, but no less exciting than, those of the past. The new technology which is bringing together telecommunications and computers opens up new and almost limitless vistas for the skills, the innovativeness, the imagination and the tenacity which girdled the earth with telegraph cables in Queen Victoria's day.

The story recorded in these pages gives the company the confidence and the inspiration to meet the challenges which lie ahead.

GLENAMARA

May 1979

xiii

INTRODUCTION

Far-writing by electricity? It was a revolution indeed. However, for fourteen years it would only work across land. The English patent of 1837 brought Windsor, where Victoria had just begun her reign, within hailing distance of Westminster, but the sea that protected the young Queen's island kingdom confined the service given by Cooke and Wheatstone's Electric Telegraph Company to Britain. Packet steamers still had to ferry the letters and dispatches across the English Channel while scientists sought the insulating material which would free telegraphy from its land base and bring Victorian England instant communication with her overseas possessions.

Charles Wheatstone told a Committee of the House of Commons of his experiments with tarred hemp in 1840, but the breakthrough was not to come until five years later with the successful application by Michael Faraday of gutta percha. When in 1845 S W Silver & Co of Stratford, E. made a gutta percha extruding machine capable of covering wire, it seemed that Submarine Telegraphy had arrived. To two Bristol brothers, Jacob Brett and John Watkins Brett, the prospects seemed unlimited. On 16 June 1845 they registered The General Oceanic Telegraphic Company to unite Europe and America. John outlined the grandiose scheme to Sir Robert Peel, the Prime Minister, in a letter of 26 July 1845. When, for lack of funds, William Cooke had to turn down a proposition made to him by S W Silver & Co (The Gutta Percha Company) in 1846 that The Electric Telegraph Company should establish a telegraph between Dover and Calais, John Brett took the project over. In 1849 he formed The English Channel Submarine Telegraph Company which successfully laid a cable under the Straits the following year. It broke down after six days but Submarine Telegraphy had been proved to be practical

– the fault was physical not electrical. Henceforth the activity was developed by a clan of dedicated engineers and operators with its own traditions and folklore, its own technology, its own story, of which what follows is a part.

Representatives of the Prussian, Belgian, French and British governments discussed the implications of Submarine Telegraphs at an International Electric Telegraph Commission which sat in Paris at the beginning of May 1851. Britain was the only country of the four in which the telegraph service was not operated by the Government's Posts and Telegraphs Department. However, it was known that should Brett's next attempt to establish a cross-Channel telegraph service fail, the French concession lapse and The English Channel Submarine Telegraph Company be dissolved, Lord Palmerston the Foreign Secretary was giving serious consideration to obtaining an Act of Parliament to reserve to the British Government the power of laying down a telegraph line to cross the Straits of Dover, ensuring control over rates and the kind of service offered. Furthermore, he was considering the whole matter of HMG's involvement in the use of Continental Electric Telegraphs to all corners of the world, particularly India and China, as his long letter to the Board of Trade of 4 July 1851 in the Public Record Office (FO 97/197) shows.

As it was, with the aid of engineer T R Crampton, Brett's second attempt succeeded. The cable was laid to France on 25 September 1851, and at 12 noon on Thursday 13 November John Brett's private enterprise Submarine Telegraph service between England and France was opened to the public – and stayed open. It was the first commercial submarine telegraph service in the world.

In Britain the Government had the Post Office take over the country's internal telegraphs as soon as it was sure of their reliability. But it was sufficiently impressed by Brett's mastery of what was required, and intimidated, no doubt, by the difference in technology and cost, to leave the building and operating of Britain's first external telegraphs to businessmen and publicly floated companies. This is the circumstance, containing the seeds of the conflicts which were to come, on which this story is based, from which it derives its special character.

John Brett's Electric Printing Telegraph.

PART I
1851–1897

For Whose Good?

It must not be supposed however that because we can send letters to each other every day for a penny, or can talk to the end of the world in a few seconds by means of the telegraph wire, or because we can travel from place to place at fifty miles an hour for a penny a mile, that we have for those reasons alone become any better or much wiser than our forefathers. What a man says or thinks will not be wiser or better because he sends it over a telegraph wire. The man himself will not lead a better life, or even be a much wiser man, because he can travel from place to place instead of living quietly in his own home.

H O Arnold-Forster, *A History of England,* 1897

Good, the more Communicated, more abundant grows.

John Milton, *Paradise Lost,* 1667

1

Far Eastern Telegraph

1851–1872

This story starts with a bang – or rather a Loud Report.

'It was a singular coincidence' recorded the *Annual Register* for 1851

> that the day chosen for the opening of the Submarine Telegraph was that on which the Duke of Wellington attended in person to close the Harbour Sessions, and it was arranged that his Grace on leaving Dover by the 2 o'clock train for London should be saluted by a gun fired by the transmission of a current from Calais. As the train started, a signal was passed and instantly after a loud report reverberated on the water and shook the ground – a 32 pounder loaded with 10 lb of powder had been fired by the current. The report had scarcely ceased ere it was taken up from the heights, the military as usual saluting the departure of the Duke with a round of artillery.

In the fourteen years which had passed since William Cooke and Charles Wheatstone had patented the first electric telegraph system, which would only work across dry land, many had sought a reliable solution to the problem of how to make current, generated by a voltaic cell, flow for any distance through a wire submerged in water. That it took so long underlines the fact that Telegraphy and Submarine Telegraphy (and to an even greater extent Ocean Telegraphy) were technically separate operations, and accounts for the emergence of two groups of operators in a sea-bound, island country like Britain, one for inland and one for overseas instant communication.

Thus it was the events of 1851 and not 1837 which provide the *raison d'etre* of the enterprises here chronicled. In 1851 the man who founded the first of these enterprises was 36 years old. This was John Pender, born in the year of the Duke of Wellington's final dismissal of Napoleon's pretensions to the throne of France on the field of Waterloo. At this young age he had already made himself a fortune speculating on the cotton market; and in 1851 he gained a valuable asset in his second wife, heiress Emma Denison of Nottingham, who not only brought him a large dowry, great charm and social ambitions but remarkable business acumen.

John Pender's father was described as 'belonging to the Scotch Lowland middle class'. John went to the parish school in the Vale of Leven, Dumbarton, where he was born, and then to Glasgow High School, where he received a gold medal for Drawing. He returned home, and a position was arranged for him in the counting house of a local factory. The products of this factory are not known, cotton fabrics maybe, but young Pender soon mastered the business and at

17 Warwick Street
Regent Street
8th November 1851

My Lord
Submarine Telegraph between England and France
We write to acquaint your Lordship that this Telegraph will be open to the public for the transmission of messages to and from the Continent on Thursday the thirteenth day of November 1851 at twelve o'clock noon

We have the honor to remain
My Lord
Your Lordships
very obedient Servants

Davies Son Campbell & Haas

The Right Honorable
Viscount Palmerston
Foreign Office
Downing Street

21 was asked to become its managing director. It seems he moved to Glasgow some time before 1840, for on 28 November of that year he married a Glasgow girl, Marion Cearns, who tragically died the following year (16 December 1841) – giving birth to her son James? He soon established himself in Glasgow as a cotton merchant in the home and foreign trade, and then moved once again – to the centre of the cotton trade, Manchester.* Here he formed John Pender & Company which quickly became a leading distributor of the products of the looms of both Lancashire and his native Scotland. His transactions were mainly with India and were conducted on a large scale. He was a well-known figure on the Manchester Cotton Exchange, and is said to have rendered valuable service during the Cotton Famine. 'By degrees a change came over the methods by which business with India was carried on, and corresponding changes were visible in Manchester. About that time Mr Pender withdrew from it. One or two mercantile transactions of some magnitude were a little later ascribed to him by general repute, but they were not authentically known. It is understood that they had not proved very successful.' The *City News* obituarist may be right, but on the whole it seems that the importing and exporting house of John Pender & Co was highly profitable and made John Pender extremely prosperous.

During the six days in September 1850 when John Brett's first cable under the English Channel was working, the *Freeman's Journal* pertinently observed:

> The 30 mile strait of Dover being thus successfully spanned, there can no longer be any doubt but that the 60 mile strait between England and Ireland will within a very short time be spanned likewise. The extension of the telegraph to any point in Ireland is an easy matter.

It was something which had occurred to a group of businessmen in Liverpool, and the following year they formed a joint stock company to build telegraph systems and operate telegraph services on either side of the Irish Sea against the day when someone (perhaps themselves) laid an Irish Sea submarine cable. They called

---

* In the Manchester Street Directory of 1845 he is listed as an agent, his business address as 20 David Street, and his residence as Grove House, Higher Broughton. The 1847 Manchester Exchange Directory lists him among the principal merchants and manufacturers who frequented the Manchester Exchange, and gives the business as 'Commission merchants for piece goods' at 29 Dale Street. By 1852 he had Bradbury Hall, Cheshire as his private residence; the 1869 directory listed his addresses as Mount Street and Peter Street, Manchester, and Crumpsall House, Crumpsall.

Laying the Dover-Calais submarine telegraph of 1851.

it The English and Irish Magnetic Telegraph Company. It had this distinctive title because the system it used rejected the galvanic battery and substituted magneto-electric for voltaic current.

William Cooke's Electric Telegraph Company had had the field to themselves for four years, and English and Irish Magnetic were the first to challenge their British monopoly. Their wires were to radiate from Liverpool where they had their head office, and not from London. On the Irish side they built a telegraph from Dublin to Galway. Two attempts to lay a submarine cable under the Irish Sea in 1851 and 1852 failed – both by Stirling Newall. The second cable established communication between Dublin and London for three days, but it was cut by the anchor of a ship off Howth and no method was known at this time by which a broken cable could be recovered from the seabed. English and Irish Magnetic decided to attempt a lay, and were granted a charter dated 10 June 1852, incorporating the company, the prime object of which was to establish a telegraph service between the Home Office in London and The Castle, Dublin, the seat of HMG's Irish administration. Incorporation empowered the projectors, as they were called, to raise the large capital sum required. They appealed to wealthy businessmen in the North, particularly in Liverpool and Manchester. One of those who was attracted by the project, agreed to invest some of his profits in it and to join the Board, was the chairman and managing director of J Pender & Co.

By becoming a director of The English and Irish Magnetic Telegraph Company in 1852 John Pender took the first step in a career which, with the help of his new wife Emma, was to take him far from the cotton markets of Manchester and give him the new interest which became the obsession of his life.

Charles Bright was the engineer of English and Irish Magnetic, and it was he who made sure that the cable he laid for the company from Port Patrick to Donaghadee on 23 May 1853 not only worked but remained in operation for many a year. But the new means of communication was by no means confined to the West. Dr O'Shaughnessy had laid the first wires in India in 1851, and in 1854, at the instigation of the Governor-General, Lord Dalhousie, who was adamant that that vast territory should not be deprived of the benefits of the electric telegraph in spite of Hindu objections to it as unwelcome 'westernisation', an 800-mile line was opened between Calcutta and Agra. Three years later it was extended to Lahore and Peshawar. To link this internal system with the outside world was part of the scheme for which the indefatigable John Brett had obtained concessions from the French and Sardinian governments.

The world's first commercial submarine telegraph cable uncoils from the hold of the Blazer.

The first objective of Brett's Mediterranean Telegraph Company however was to link the southern coast of Europe with that of North Africa with a branch to Malta. The first section of ninety miles was to be from Genoa to Corsica where an inland telegraph was already building, and this was opened in June 1854. It should then have gone to Sardinia and finally joined the French Algerian system which had been operating since January 1854. But laying a submarine cable from Sardinia to Bône in the great depths of that part of the Mediterranean defeated them, and the scheme was abandoned. John Brett also had trouble shipping the 150 miles of cable for this project owing to the outbreak that year of the war in the Crimea where British troops were going into action for the first time since Waterloo, helping the Turks oust the Russians. Wire and instruments for the army arrived in Balaclava in December. The electric telegraph was providing the British army with communications in war for the first time in history.

That the electric telegraph worked, and with such speed, was still a marvel to the general public in Britain. It seemed astonishing that when the Emperor Nicholas I died in St Petersburg at 1 p.m. on 2 March 1855, news of his death was announced in the House of

THE NEW SIAMESE TWINS.

⁎ The submarine cable was successfully laid between Dover and Calais, the 13th November, 1851 ; and the closing prices of the Paris Bourse were known within business hours of the same day on the London Stock Exchange.

Lords that afternoon. In that year according to Dionysius Lardner
(The Electric Telegraph Popularised) the world boasted 8,000 miles of
telegraph (40,000 miles of wire) of which The Electric Telegraph Co
(capital £800,000) operated 4,500, and English and Irish Magnetic
(capital £300,000) 2,200. In the UK alone some one and a half
million pounds was invested in electric telegraphs. The system as
operated in the first sixteen years was of course fairly unsophisti-
cated – 'simplex' in fact, sending the pulses along the wire in one
direction, waiting for the line to clear, and then sending the reply
back in the opposite direction. It was not until 1853 that Doctor
Gintl, a director of Austrian State Telegraphs, first devised a
practical method of sending signals along a single wire in two
directions at the same time – 'duplex working' – adding greatly to
the speed and ease of operation. In November 1854 Carl Frisehen of
Hanover patented a duplex system in England, though Stirling
Newall had brought out a similar idea in October.

In 1854 no one who had studied the problem was seriously
considering laying a telegraph cable in a piece of sea as wide and
deep as the North Atlantic. But in that year a project was mooted to

*Landing the shore end of the submarine telegraph cable from Port Patrick in Scotland at
Donaghadee in County Down.*

link Europe with America by land line through Siberia, across the 100 mile, 300 fathom Bering Strait to Alaska and down into the United States. The Imperial Russian Government who owned Alaska at that time – they sold it to the US in 1862 – agreed to give the matter thought, and were still thinking about it when events overtook them and the plan became superfluous.

In 1853 an English emigrant to Nova Scotia, telegraph engineer F N Gisborne, had had to abandon his plan to build a telegraph from Newfoundland to New York as he had run out of capital. He had the fifty year charter he had obtained from the Newfoundland Legislature transferred to a wealthy paper merchant in New York called Cyrus Field whose imagination was fired by Gisborne's stories of the benefits such a telegraph would bring. Field, who was not an engineer and knew nothing of the problems involved, saw no reason why the wire should end at Newfoundland, and formed a company The New York, Newfoundland and London Telegraph Company, to link North America with Britain underneath the Atlantic Ocean. The rich friends whom he tried to persuade to invest in the project rejected it as far too visionary. So Field took ship to England where he had already corresponded with the man whose plans for a cross-Channel telegraph had had a similar reception three years before, John Brett.

He made more headway in London, but quickly realised that British investors would be more inclined to invest in a British enterprise. So he transferred his charter to a company which he formed in London and called The Atlantic Telegraph Company. He issued the prospectus on 6 November 1856, and one of the 345 Englishmen who risked a thousand pounds each was John Pender of Manchester, already a director, as seen, of The English and Irish Magnetic Telegraph Co, which in November 1856 merged with The British Telegraph Company and changed its name to The British and Irish Magnetic Telegraph Co. At the first general meeting of The Atlantic Telegraph Company on 9 December 1856, John Pender was elected a director. Glass, Elliot and Newalls were each instructed to manufacture 1,250 miles of armoured cable from the 7 strand copper core with its triple gutta-percha insulation made by the Gutta Percha Company. No 'return course' was needed; the sea acted as 'earth'. The signals were going to be induced by magneto-electric current, on the system used by The Magnetic. Transmission would be by releasing and stopping the flow of current with a rocking key in long and short bursts according to Samuel Morse's code.

Concessions were obtained from the administrations on either side of the Atlantic. In addition both British and US governments agreed to pay The Atlantic a subsidy of £14,000 a year up to the time when its dividend had reached 6 per cent, and £10,000 a year subsequently. This was not just by way of encouragement, it was payment to a private enterprise for a service which otherwise Government would have to have provided for sending government despatches.

The company's first attempt to lay a telegraph cable under the Atlantic in August 1857 met with failure. Sir William Brown, chairman of The Atlantic, wanted to abandon the project altogether. He advised selling the unused cable and giving shareholders their money back. John Pender supported Cyrus Field and William Thomson in insisting they try again. They still had 2,200 miles of cable; the specially fitted-out naval vessels were still standing by, and their services would be unlikely to be made available once they returned to duty. John Pender and his allies won the day. The board passed a resolution for the Telegraph Squadron to put to sea again on 17 July 1858. This time they succeeded.

On 18 August 1858 Lady Frederick Cavendish (Lucy Lyttelton) noted in her diary at Hagley Hall, Worcestershire: 'Aunt Wenlock

The 1857 Atlantic Cable is manufactured at the Glass Elliot works at Greenwich.

came and played whist with us, graciously bearing with our third-rate powers. The first message arrived in 36 minutes by the Atlantic Telegraph: "England and America are united. Glory to God in the Highest and on earth peace, good will towards men." Amen, from the bottom of my heart.'

According to George Saward,* who had been secretary of the company, communication was established on 13 August – but only just. It took sixteen hours to transmit Queen Victoria's ninety-eight word message to President Buchanan, but there was no denying the sensation it all created. 'Now the great work is complete,' wrote American authors Charles Briggs and Augustus Maverick, 'the whole earth will be belted with electric current, palpitating with human thoughts and emotions. . . . It has been the result of the great discoveries of the past century to effect a revolution in political and social life by establishing a more intimate connexion between nation and nation, with race and race. It has been found that the old system of exclusion and insulation are stagnation and death. National health can only be maintained by the free and unobstructed interchange of each with all. How potent a power then is the telegraphic destined to become in the civilisation of the world!' *(The Story of the Telegraph,* New York, 1858).

But the euphoria was short-lived. The signals, never very strong, got weaker and weaker. A fault in the insulation some 300 miles out of Valentia never completely righted itself. The battery power was raised but this only weakened the insulation further. 'Retardation', as some had suspected during the Magnetic trial, got the better of the line, and slowed down working until on 18 September 1858 'this great historical and experimental telegraph subsided into entire silence.'

News of the attempt to lay a telegraph cable across the North Atlantic and the British Government's willingness to pay a commercial company £14,000 a year for twenty-five years in order to use the service for conveying Government despatches, reached the British settlements in the West Indies as fast as steamer could bring it. In February 1857 the Governor, members of the legislature and leading merchants of Bridgetown, Barbados, gathered in the town hall to hear Captain Rassloff, a Danish civil engineer now working for the Canadian Government, outline a plan for laying a submarine

---

* *The Trans-Atlantic Submarine Telegraph;* A Brief Narrative of the principal incidents in the history of the Atlantic Telegraph Co. compiled from authentic and official documents, London, 1878 (privately printed).

telegraph cable to link North and South America. It would run, he said, through the British and Danish West Indian islands and connect them all together. The main cable would start from Florida, pass through the Bahamas and terminate at St Thomas in the Virgin Islands; a secondary cable would run through the British West Indies ending at St Lucia where it would branch into two routes, one through St Vincent, Grenada, Trinidad and the Spanish Main, and the other to Barbados, Tobago and Georgetown, British Guiana.

But before that plan could be realised the British Government's faith in ocean telegraphy was further undermined by the failure in 1858 of the project of The Red Sea and Telegraph to India Company. The Government promised this company a perpetual guarantee of 4½ per cent of its £800,000 capital provided the telegraph which it planned to operate between Suez, Aden and Karachi worked. The company had the monopoly for working the land line in Egypt between Alexandria, Cairo and Suez. It was the first attempt to link Britain with the Far East; but the cables which were made and laid by Newalls failed almost at once and were abandoned. The Government were obliged to continue the interest however, and the stock was converted into terminable annuities for fifty years from August 1858 to July 1908. The total loss to Britain

The Niagara *pays out cable at the American end in the first attempt by the Atlantic Telegraph Company to lay a line under the ocean in 1857 which met with failure.*

was put at £1,800,000, but the company's rights in the Alexandria-Suez land line came in very useful later. Apart from the financial loss it was a reverse for those who held that submarine cable was politically more acceptable than lines across the land of a foreign ruler who might at any moment become an enemy.

The failure of the Karachi cable and the Atlantic cable prompted the British Government to appoint in 1859 a Committee of Enquiry on the Construction of Submarine Telegraph Cables, with Captain Douglas Galton RE in the chair and four members each from the Board of Trade (one of them was Charles Wheatstone) and the Atlantic Telegraph Company. Its brief was to investigate the whole subject of deep-sea telegraphy with a view to ascertaining the best mode of constructing, testing, insulating and submerging cables. It was very comprehensive and everyone of note in the ocean tele-graphy world gave evidence. William Thomson (later Lord Kelvin) gave his views on the laws of induction and retardation. It had been shown that submarine cables could not be worked on the Morse system if their length in circuit was over 700 miles, so in 1858 Thomson introduced his revolutionary Mirror Galvanometer In-strument. The action of this depended on *changes* of current strength. It was not necessary to wait for the entire clearing of the line before a fresh signal could be sent.

The Committee's findings showed that ocean telegraphy was not perhaps as simple as some had been led to think, and that there was still a great deal to be learnt. To enable such technical knowledge about the electric telegraph as existed to be shared as widely as possible, in 1861 John Pender and others founded the first electrical journal in the world, *The Telegraphic Journal,* a sixpenny weekly which has been in circulation ever since, later named *The Electrical Review.*

The failures of the cables to the West and to the East, following the breakdown of John Brett's cable from Sardinia to Algeria in 1855, had an undeniably sobering effect on all concerned, not least the commercial community who would fund the routes of the future. But due to the faith of Sir Curtis Lampson, Cyrus Field, George Peabody and John Brett, The Atlantic Telegraph Company was not broken up in spite of the entire paid up capital of £460,000 being lost except for £150. A new Act of Parliament was obtained in 1858, which became operative on 1 July 1859 and gave the company power to raise £2 million.

A series of Acts of Parliament dating from 1856 had extended limited liability to every kind of enterprise (there was a consolidation Act in 1862) and this was greatly to the benefit of the telegraph companies; but the Atlantic company, finding it difficult to attract

investors, wanted in addition a government guarantee, as the Red
Sea company had had.

So Sir Curtis Lampson approached the Prime Minister and the
Chancellor of the Exchequer, and coaxed industry into petitioning
Parliament on the need for an Atlantic cable. But the Cabinet, once
bitten, were now shy of allocating public money to any new ocean
telegraph venture. They refused the Atlantic company's application
for an unconditional guarantee and also declined to become a
shareholder. They later relented however, and offered a guarantee of
8 per cent a year on the capital of £600,000, conditional on the cable
being successfully laid and a transatlantic telegraph *working.* In that
event the Government agreed to pay the company £20,000 a year for
twenty-five years as a subvention. The US Government agreed to do
the same.

Britain's rulers were by now fully alerted to the importance of ocean
telegraphy as an instrument of government, and of the apparent
need, in view of the high degree of failure in the commercial field, to
take matters into their own hands and commission the laying of
submarine cables and operate trans-ocean telegraph services themselves. In 1859 the British Government ordered a cable for the
purpose of linking Falmouth with its naval base at Gibraltar. Then
came war with China and the Government thought a better use of
the cable would be to link Rangoon with Singapore. But the Chinese
War came to an end before the cable was ready, so HMG decided to
have it laid from Malta to Alexandria – as a first link in a telegraph
to India?

The Government's inability to communicate with the Far East
had been a major handicap in its handling of the swiftly developing
Indian Mutiny which resulted in the winding up of the East India
Company after 258 years and the assumption of its powers by the
British Government (the India Office) in August 1858. So Sir
Stafford Northcote, the president of the Board of Trade, asked
Charles Bright for a Report to the Treasury on the possibility of a
line to India, and the Parliamentary Blue Book *The Establishment of
Telegraphic Communication in the Mediterranean and with India* was
published in 1859.

The first steps in connecting India with England had been taken
in 1856 when ocean telegraphy was not a practical proposition. In
that year negotiations began between the governments of England
and Turkey for building a telegraph from Constantinople eastwards
to what was then known as the Persian Gulf. The Turkish capital

The Indo-European Telegraph – landing the cable in the mud at Fao, Persian Gulf.

was already linked with the rest of Europe and England. In 1856 The European and Indian Junction Telegraph Company tried to obtain a concession in Constantinople to build a land line to Kurna at the junction of the Tigris and the Euphrates. The Turkish Government refused to grant this and in 1857 undertook to build the wire themselves but agreed to use British engineers. British army personnel were seconded to build the Ottoman Empire's telegraphs, and Lieutenant Colonel Biddulph began the work in 1859. He had retired by the time the line from Constantinople to Baghdad was opened in 1861.

This was the nearest to India so far reached by an international telegraph, and it gave the Government of India the opportunity they had been waiting for. In 1862, at the instigation of the Secretary of State for India, the Government of India* formed the Indo-European Telegraph Department. The head of it was immediately responsible to the Secretary of State, and it was not part of the Public Works Department.

The first task of this telegraph department was to construct a

---

*To be strict, it was the Government of Bombay which formed the telegraph department in 1862. The department's main administrative office moved to London in 1865, and in 1870 it was formally transferred to the Government of India.

telegraph for the Indian government from Karachi, the country's
most westerly town, to join the Turkish telegraph at Baghdad.
Determining the route of this telegraph was the responsibility of the
British colonel in the Royal Engineers, Patrick Stewart, who was
Director-General of Indian Telegraphs, and his assistants Colonel
F J Goldsmid and Major Bateman Champain. Submarine telegraphy was suspect and expensive and they were anxious to keep it
to the minimum. However, the risk of having land lines vandalised
by unconquered native tribes was considerable. So they decided to
build the first section from Karachi as a land line across Baluchistan
to Gwadur on the north coast of the Gulf of Oman (for which
permission of the Sultan of Muscat would have to be obtained).
From there the line would take to the sea, and they would have to lay
a submarine cable to Fao, at the mouth of the river Shat-el-Arab at
the head of the Persian Gulf and then another land line to Baghdad.
Stewart also planned an alternative route from Bushire, just south of
Fao, by land line across Persia northwards to Teheran to connect
with the Turkish system at Khanikin. But reaction to this suggestion
by the Persian government was slow in coming, and no action could
be taken until the two countries signed a convention.

Laying the telegraph cable in the shallow Persian Gulf needed large gangs of men.

Work began at once (in 1862) on the land line from Karachi to Gwadur. Cable ships from England were used to lay the cable from Gwadur to Fao, with two intermediate landing places at Mussendam and Bushire. All was in position and working by April 1864. The land line westwards from Fao to Baghdad was not completed until the following year, by which time there had been signed at Constantinople on 3 September 1864 a Convention between Queen Victoria and The Sultan for the Establishment of Telegraphic Communication between India and The Ottoman Territory.

On 27 January 1865 the first telegraphic communication between India and Europe was opened – a milestone in India's history. Sadly Patrick Stewart did not live to see the consummation of this great project. He died at Constantinople the week before the opening. He was succeeded as Director-General by his assistant Colonel Goldsmid.

The time for sending a message between England and India was reduced from a month to less than a week; but there were many delays, mostly due to the Ottoman telegraphs. The alternative route was eventually built by the Indo-European Telegraph Department as planned, by permission of Persia, and ran from Karachi to Gwadur, Bushire, Teheran, Tiflis and across Russia. If anything it was worse than the Fao-Baghdad-Constantinople route. Low engineering standards and non-English-speaking clerks led to mangling of texts. It cost five pounds* for every 20 words which began in your version and ended in theirs. Worried by the complaints, the British Government appointed a House of Commons Select Committee on East India Communications, which in October 1866 recommended that the Government be encouraged to establish direct communication from Bombay to Alexandria via Aden, and that any such telegraph should be administered *by a single management.*

Technically, the successful operation of a long-distance telegraph had been advanced by William Thomson's invention in 1867 of his Siphon Recorder instrument. Though he took out a patent in that year he did not perfect it until 1870 when it became standard equipment (though not in the West Indies and South America) and replaced his Mirror telegraph. The Siphon could be worked at the same speed, with the same battery power and over the same

---

*Bowing to public pressure, the rate was reduced to £2 17s for 20 words on 1 January 1869. The reduction was also designed to mollify the continental government telegraph department chiefs whom Colonel Goldsmid met at the Vienna Conference of 1868. But the new rate resulted in a loss of £20,000 the first year – carried, presumably, by the taxpayers.

distances as the Mirror, but it gave a record of signals as they were received. It was very important for the competing private telegraph companies, particularly those working through submarine cables, to be able to reduce errors in transmission to a minimum and trace them when they were made. However the Siphon was very much more expensive than the Mirror, and this delayed its introduction in the early stages.

The achievements of cable makers like Glass Elliot of Greenwich and Gutta Percha of Stratford in perfecting cables of light weight and flexibility with a high standard of insulation, physical protection and electrical efficiency, also advanced the technology. But in the final analysis successful working depended on *people.* The speed of communication, which was the only advantage which the electric telegraph had over transmission by horse and sailing ship, was always reduced whenever a message had to be handled by a series of different national administrations receiving and re-transmitting along a chain of stations. The ideal of a long distance telegraph under single control had not yet been realised, though of course there would always be delay caused by links which had to be short because of attenuation problems, whether the stations were manned by those of the same or different nationality.

Whatever its failings, at least there was some kind of telegraph between England and India, whereas messages to North America were still going by steamer, after the short-lived success of the 1858 Atlantic Cable. This was a situation, however, which John Pender and his colleagues were at last in a position to remedy.

Having won over the US and British governments, in December 1862 the directors of The Atlantic Telegraph Company used the powers they had obtained in the 1859 Act of Parliament to issue a prospectus to raise £600,000 in 8 per cent preference shares. They appointed a new Consulting Committee which included Douglas Galton, Sir Charles Wheatstone and Sir William Thomson. By May 1863 they had only raised £300,000, but they were sufficiently encouraged to put out tenders for making and laying a new Atlantic cable capable of working eight words a minute. They accepted Glass Elliot's out of eleven submitted. But improvements insisted on by the Consulting Committee increased Glass Elliot's costing to £700,000. Where to raise the rest? Glass Elliot, a comparatively small firm, could not accept greater liability but, knowing the prestige which would come from being associated with the Atlantic Cable, they were prepared to negotiate.

A weakness of previous operations had been having to employ two cable ships, each with half the cable, which met in mid-ocean where a hazardous splicing operation took place and then set off in opposite directions. But in 1864 there existed a ship large enough to take the 2,500 miles of wire – the 18,000 ton *Leviathan* built by I K Brunel in 1858, 700 ft long and 85 ft wide. When used as a passenger ship to the Far East she was re-named *Great Eastern.* The Australian route had proved unremunerative and the big ship was now in dock. On 4 April 1864, Glass Elliot came to an arrangement with The Great Eastern Steamship Company for the old *Leviathan* to rouse herself once again, and start a new life as a cable ship laying the Atlantic Cable.

But the problem of how £700,000 worth of cable was to be paid for by a company with only £300,000 remained unsolved. John Pender, who had been re-elected a director of the Atlantic Company on 17 March 1864, suggested amalgamating the Gutta Percha Company with Glass Elliot to form a new firm called The Telegraph Construction and Maintenance Company Limited (Telcon). This was duly registered on 7 April 1864 with himself as chairman and Richard Glass as managing director. It had a capital of £1 million. Telcon at once confirmed the contract with the Great Eastern company, whose managing director Daniel Gooch, joined Telcon's Board.

On 5 May the Atlantic company contracted with Telcon to make and lay the cable for £837,140, offering them £300,000 in cash and the rest in Atlantic company shares – a considerable risk for the Greenwich cable manufacturer. On 18 July, John Pender, Richard Glass and Daniel Gooch received the Prince of Wales at Sheerness to witness the loading of the cable on to the *Great Eastern.* 'All told,' reported *The Times,* 'the *Great Eastern* will leave the Thames with rather more than 25,000 tons in her, a burden almost as great as the whole fleet with which Nelson fought the battle of Trafalgar could have carried.' Captain James Anderson was put in command and the ship sailed from Valentia in Ireland on Sunday evening 23 July, paying out as she went, and sending messages back to Ireland along the cable on one of William Thomson's galvanometer instruments. The next day, after only 84 miles of cable had been laid, the line went dead, the result, it was said, of sabotage. Other faults developed, and in the first week in August the cable broke and fell into the sea. The expedition was once more abandoned. The Atlantic Telegraph Company threatened to sue Telcon, which must have been embarrassing for John Pender as he was a director of both companies. To extricate himself Pender, on behalf of Telcon, offered to make *and lay* a new cable across the Atlantic in 1866, and attempt to recover and complete the 1865 cable, all for £500,000. The Atlantic company accepted but the Attorney General ruled that their 1859 Act of Parliament would not allow them to float a new 12 per cent issue as they wanted, and that they would have to form another company to act as agents of The Atlantic Telegraph Company. So they formed The Anglo-American Telegraph Company Ltd, which had a capital of £600,000. The shareholders of the old company were too disheartened to risk their money further, and the money for the new company came mainly from the directors and associates of Telcon. Lord Brassey, the railway builder, subscribed £60,000; Daniel Gooch £20,000; John Pender £10,000. When the other Telcon directors jibbed at once again putting so much at risk, chairman Pender guaranteed the firm £250,000 of his own and his wife's money. Subscribers to the Anglo-American were entitled to expect a 25 per cent return on their capital in the event of success – and nothing for failure.

Most were by now thoroughly disillusioned that an Atlantic Telegraph could ever be made to work. But the next attempt however did not fail, and between 13 and 27 July 1866 the *Great Eastern,* with James Anderson once more in command, successfully laid Telcon's new lighter but stronger cable from Valentia to Heart's Content in Newfoundland. On 9 August the *Great Eastern* returned to the Atlantic and on 2 September managed to raise the end of the 1865 cable.*

The Anglo-American company had full control and charged £20 a message, which the American Press complained was too high. The tariff was promptly reduced but not the transmission time. In December the US Government sent a 4,000 word message from Washington to Paris which, at seven words a minute, took ten hours to transmit – and cost more than £2,000.

It is likely that without the success of the 1866 lay and the assurance, by the recovery of the 1865 cable, that the end of a cable which slipped two miles down into the ocean was not irretrievably lost, John Pender would have washed his hands of ocean telegraphy

---

*A piece of this found its way to the London head office of Cable and Wireless Ltd in 1977. It had been preserved by Charles Greenfield who died in 1887, an engineer employed by Telcon on the *Great Eastern* during the cable laying operations of 1864, 1865 and 1866, and the 1865 cable recovery expedition. He had an eighteen inch length of the 1865 cable mounted on a cross-section of the 1864 cable and encased in a glass dome. A plaque bears the inscription 'Atlantic Cable 1865 Recovered from a Depth of 1,950 fathoms after 13 Months Submersion'. Charles Greenfield was the great uncle of the husband of Mrs Marie Bell who presented the ornament to the Cable and Wireless museum on 14 December 1977.

and concentrated on his cotton business in Manchester, which now had a 'London Department', and developed the Whitworth Ordnance Company which he had helped to promote in 1863. He would probably have continued his interest in The Magnetic but only as an investor. As it was, the gamble with his £10,000 in the Anglo-American and the £250,000 guarantee to Telcon had paid off. There was no holding back now.

In December 1865 a bill was introduced in Congress granting The International Ocean Telegraph Company, formed by American James Scrymser, sole rights to operate a telegraph between Florida and the West Indies for fourteen years. At once they ordered a cable from Telcon, who laid it between Punta Rassa in Florida and Havana in Cuba via the island of Key West. The telegraph service was opened in December 1866.

International Ocean was later acquired by Western Union who, in 1865 however, were busy building the American side of the telegraph which was to link the United States with Russia by land, already referred to, via the short submarine cable across the Bering Strait. The Russians were building the line west to east across Siberia. This was a second best to a trans-atlantic cable, but when the laying of the latter was finally achieved in 1866 all work on the United States to Russia telegraph via Bering Strait ceased.

In the west, first steps were also being taken at this time to link the somewhat unstable regimes which made up South America. John Proudfoot, a Liverpool merchant, and Matthew Gray, the Glasgow engineer who was managing director of the Gutta Percha Company before it became part of Telcon, obtained concessions from the Argentine Republic (6 December 1864) and Uruguay (25 April 1865) to lay a twenty-four mile cable across the mouth of the River Plate and operate a telegraph service through connecting land lines between Buenos Aires and Montevideo, the Argentine and Uruguay capitals. For this purpose they formed The River Plate Telegraph Company (incorporated 26 September 1865) with a capital of £40,000 and its head office in Scotland. The service opened on 1 December 1866.

The pace of inter-nationalisation of the electric telegraph was accelerating, and to ensure that the mechanics of passing international traffic did not become so complicated and burdensome that they defeated their own objects, the telegraph departments of the main European governments established in 1865 an International Bureau of Telegraph Administrations with headquarters in Paris. If

FUN.—August 11, 1866.

PUCK'S GIRDLE COMPLETED.

THAT ELECTRIC SPARK *HAS* PUT A GIRDLE ROUND ABOUT THE EARTH.

An Atlantic Telegraph at last – August 1866.

the benefits of the new means of communication were to be enjoyed by the majority and to the full, there had to be some degree of standardisation of procedures, rates and telegraph practice generally.

The British Government's policy regarding involvement in the building and operation of international telegraph systems had been severely modified by the losses of the Red Sea disaster, as seen. And even when the Atlantic Cable had been successfully laid, and the earlier one recovered, its attitude continued to be cautious. It had had formally to deny it was planning to take over the Atlantic Telegraph. Then what *was* its policy, if any? The answer came from the Secretary to the Treasury who on 10 April 1867 laid before the House of Commons a minute dated 10 January which declared that while HMG were unable to subsidise any company formed to set up telegraphic connections with the East, it could 'cause surveys to be made of the proposed route and render assistance by Her Majesty's vessels in laying the cables, and also by using the good offices of the Government with foreign governments upon whose territories it may be necessary to land cables'.

Was the disclosure of this Treasury minute prompted by John Pender, unseated Member of Parliament for Totnes, by way of testing the ground before forming a first telegraph company of his own to communicate with the countries in the Far East with which John Pender & Co of Manchester did so much business? It might seem so, but in fact it was a preliminary to the Government announcing its most important change of policy since Cooke and Wheatstone took out their patent in 1837 – The Year of the Telegraph.

The Telegraph Purchase Bill presented to Parliament in 1868 was designed to enable the Postmaster–General to 'acquire'work and maintain electric telegraphs'. A grant of £8 million was to be allocated for the purchase of all the private inland telegraph companies in the UK and the telegraph interests of the private railway companies.*

As soon as lengthy and deep stretches of sea had been shown, by the success of the Atlantic Telegraph, to be no obstacle to good telegraphy, it was obvious that the Mother Country would not for

---

*It was important to distinguish Ocean Telegraphy from Submarine Telegraphy. From 1870 the British Post Office, while happy to leave the development of the former to private enterprise, regarded the latter as *their* sphere of influence, that is telegraphy between the UK (including Ireland and the Channel Islands) and the continent of Europe (the Dover-Calais link). In 1870 the British Post Office appointed a 'Submarine Superintendent and Electrician' at their Telegraph Department under R S Culley in the person of David Lumsden of Edinburgh.

long be content with second-best communication with her Indian
subjects – by aerial wires suspended on vulnerable poles across
unguarded areas of foreign territory – but actively explore the
possibilities of a secure, *submarine* route.

In May 1868 the prospectus was published for a new company
calling itself the Anglo-Mediterranean Telegraph Company. It was
a title which recalled, perhaps intentionally, the grand project of
John Watkins Brett, the pioneer of submarine telegraphy, who had
died in 1863. His Mediterranean Telegraph Company and his cables
had collapsed in the 1850s in the treacherous depths of the
Mediterranean Sea, but the technology of making and laying
submarine cable had so advanced by 1868 that the Anglo-
Mediterranean had confidently ordered from Telcon a 900 mile
cable 'suitable for laying on the deep seabed of the Mediterranean
between Malta and Alexandria'. It was a link, however, in a very
much more important chain. As *The Times* reported on 17 September
1868, 'the progress of the expedition which has left England to lay
the Anglo-Mediterranean cable cannot fail to be watched with
considerable interest, inasmuch as the completion of the line will
duplicate our means of communication with Egypt, and will also
constitute a most important adjunct to our agencies for correspondence with our Indian empire'.

Telcon completed the laying of the Malta-Alexandria cable
without mishap on 4 October 1868.

Anglo-Mediterranean took over the line Telcon were building
from Susa on the French-Italian border, down Italy, and under the
Straits of Messina to Modica in Sicily. A previously laid submarine
cable from Sicily to Malta was also taken over by Anglo-
Mediterranean. From Alexandria a line was laid overland to the
town of Suez at the head of the Red Sea, where Ferdinand de Lesseps
was building a canal to Port Said on the Mediterranean coast, which
opened in 1869. All the intermediate stations – at Turin, Malta,
Alexandria, Suez – were operated by British staff.

The British Government's intention, announced in 1868, to buy
out the shareholders of the private telegraph companies for cash and
transfer their services to the Post Office, meant that investors who
had made good dividends out of Telegraphs would be likely to re-
invest in the private companies which were free to promote external
telegraphy systems. One of them was the director and major
shareholder in The Magnetic which would soon disappear overnight
along with the others, John Pender. He intended to take the
opportunity to fulfil an ambition, born of his cotton merchandising
days, to fit together the existing links and create new ones which,

combined, would give Britain the submarine telegraph service to India, for which Government, Commerce and Press had been clamouring. But, uncharacteristically, he was too slow off the mark, and before he could complete his plans another group had formed the Anglo-Indian Telegraph Company to do that very thing. They acquired the Egyptian landing rights previously granted to the ill-fated Red Sea Company and engaged Charles Bright and Latimer Clarke as their engineers. But when they went to the City for funds there was no response. The Anglo-Indian was wound up before it began.

John Pender stepped in, formed the British-Indian Submarine Telegraph Company with himself as chairman and issued its prospectus on 29 January 1869. It was his first telegraph company.

The prospectus stated that the object was to lay a submarine cable from Suez through the Red Sea to Aden, and on to Bombay. They ordered the cable from Pender's Telcon and arranged for the *Great Eastern* to lay it. They contracted with Anglo-Mediterranean for the through working and lease of the landlines belonging to the Red Sea Company between Alexandria, Cairo and Suez.

A rival appeared in the form of the Direct English, Indian and Australian Submarine Telegraph Co whose prospectus, issued on 8 March, proclaimed the same objects as British-Indian Submarine but taking the line on to Australia. They thought better of it however; withdrew the prospectus on 21 May, and gave shareholders their money back. The field was now clear for British-Indian Submarine, who in June won concessions to lay lines connecting Bangkok in Siam, Cambodia and French Indo-China (Saigon and Cape St James) from where a submarine cable could one day be laid to Hong Kong. John Pender was getting into his stride.

That June he formed a second company, the Falmouth, Gibraltar and Malta Telegraph Company. When the prospectus was issued on 18 June, *The Times* commented, 'This is the portion alone remaining to be undertaken to complete the system of submarine lines between England and India'.

When the following year France expressed a desire to have a connection with the big new All-Sea telegraph trunk route to the Far East, the Marseilles, Algiers and Malta Telegraph Company was formed in London to lay and operate a cable which would give Marseilles and her French colonial empire in North Africa access to the line at Malta. The 800 mile cable made and laid by Telcon provided the French with a very much quicker route to Egypt, India, Cochin China, and China than the 1,300 miles of landline down Italy.

The ideal of the single company to administer the London-Bombay telegraph had not been attained, but in June 1869 British-Indian Submarine and Anglo-Mediterranean entered into an agreement for twenty-one years of cooperative working. But Pender was looking further ahead yet, and in December 1869 formed the China Submarine Telegraph Company (capital £525,000) to lay a 1,700 mile submarine cable between Singapore and Hong Kong via Saigon. Telcon were to make the cable and have it laid and working by June 1871.

In doing this he was entering a part of the world in which he found his main competitors were Scandinavians – The Great Northern, China and Japan Extension Telegraph Company.

The impact on the other islands in the Caribbean Sea of the opening in December 1866 of the International Ocean Telegraph Company's line from Havana under the Gulf of Mexico to Florida, giving communication to the US and Europe in hours instead of weeks, was considerably greater than Anglo-American's success in connecting the two sides of the North Atlantic. James Scrymser and his American engineers lost no time in offering to work the miracle for *them.* International Ocean lobbied the Colonial Office in London and on 9 January 1867 the Colonial Secretary, Lord Camanor, sent a circular to governors of the British colonies in the West Indies advising them of 'a proposal on the part of an American company, styled the International and Ocean Telegraph Co [sic] to establish telegraphic communications in the West Indian islands, one with another, and between them and the continent of America'.

In September of that year the governors received another Colonial Office despatch referring to 'the projects of the British and United States companies, styled respectively, the West Indian and American Telegraph Company and the International Ocean Telegraph Company'.

'You will therefore be at liberty, on application made by the International Ocean Telegraph Company, to propose to your Legislature a Law enabling them to land and work cables; but I have to request that you will guide yourself by the Treasury's letter of the 7th of June as to the general principles to be observed in such a Law, particularly those of not granting exclusive privileges, and of securing priority for Government messages.' Permission to land cables and erect wires over land must be subject to the usual laws on property and compensation, and to the control of any telegraph station on British ground being taken over by the local Government in time of war or insurrection.

International Ocean's initial plan was to connect the existing telegraph system in the United States (of which Florida had only become a member in 1845) with Jamaica, Barbados, Trinidad and Demerara (British Guiana) – and eventually Panama. But the company let it be known that they would include any colony ready to give them a subsidy. Money was in fact promised by the governments of St Kitts, Antigua, Dominica, St Lucia, St Vincent and Grenada. Barbados made a condition of theirs that communications should be established and working by 10 August 1872. In all, General E S Sanford and William G Fargo (founders of Wells Fargo who had the swiftest means of sending messages before being outpaced by electricity), through powers of attorney issued by International Ocean to the company's agent Baron von Hipple, managed to obtain concessions and subsidies from the more important West Indian islands amounting to £17,000 a year.

Though extensions of the Florida-Cuba telegraph to other islands in the Caribbean Sea had been foreshadowed in International Ocean's articles of association, General Smith the company's president, for reasons that are not entirely clear, considered that linking British colonies was work for a British company. It was under the auspices of International Ocean therefore that a company came into being in London with the title West India and Panama Telegraph Company Limited.

This issued its prospectus on 9 August 1869. Among the subscribers were William Andrews, manager of Siemens Indo-European, Henry Weaver, general manager of Anglo-American, and a shipowner called Henry Holmes. C W Earle was chairman of its Board of Managers which included Cyrus Field representing International Ocean.*

West India and Panama took over International Ocean's plan for the West Indies, together with the concessions and subsidies which General Smith had been granted so far (though local governments took the opportunity to tighten up their conditions and timetables). Once the telegraph was constructed, the company would operate a public service. It had no difficulty in raising a capital of £2,445,630. 'There can be no doubt' commented *The Times* on 26 August, 'that the most popular outlet now for commercial enterprise is to be found in the construction of submarine lines of telegraph.' The paper

---

*Controlling interest in International Ocean was purchased by the Western Union Telegraph Company in 1873.

stressed the importance of the new company in the light of the Panamanian government's plans to connect the West Indian system by an overland and submarine cable along the South West Pacific coast from Panama to Valparaiso across La Plata to Buenos Aires and Montevideo. Although International Ocean had only planned to lay to Panama at a later stage, the inclusion of that country in the title of the new company indicated that a connection with Colon on the north coast of Panama was an integral part of the scheme from the outset.

It was an ambitious project involving the longest submarine cable ever ordered, 4,100 miles. The Colonial Office offered a subsidy which, shareholders were told at the first general meeting of 26 November 1869, the company had accepted. It was a new challenge for Sir Charles Bright who was engaged as engineer for the project, but even his genius could not guarantee success.

He used a converted steamer the *Dacia* as his cable ship and had soon laid the submarine cable from Jamaica to Cuba and run aerial wires from Batanabo on the south coast of Cuba to Havana in the northwest. The service which put Jamaica via the Cuba-Florida line in direct touch with USA and Europe was opened for traffic on 15 September 1870. The weak link was having to cross Spanish Cuba by landline, but in 1869 a company had been formed in London for the express purpose of by-passing Cuba and laying a cable in the sea off its coast – the Cuba Submarine Telegraph Company. he laying of this 520 mile cable from Batabano to Santiago was undertaken by Sir Charles Bright in the *Dacia* which then set off south for Colon* in Panama to lay the cable to Jamaica.

After they had paid out 360 miles of cable it broke and the end fell into the sea. They spent a week with their grapnels trying to raise it without success and returned to Jamaica. Sir Charles decided to make a start on the inter-island cables and lay the first from San Juan in Puerto Rico to St Thomas in the Virgin Islands and St Kitts. It was mid-April (1871). He returned to San Juan and began laying the line to Jamaica, but after 600 miles there was another break-away from the ship and the end disappeared into the depths. The *Dacia* remained in the area three months searching for it before Sir Charles decided it was time to return to the cables which were to link the islands. The laying of these was accomplished with comparative ease – from St Kitts to Antigua, Guadeloupe, Dominica, Martinique, St Lucia, St Vincent, Barbados, Grenada and Trinidad.

---

*The Panama and South Pacific Telegraph Company was formed at this time to connect Central America with the west coast of South America.

Three weeks after the opening of the Jamaica-Cuba telegraph the Barbados government passed an Act (4 October 1870) to 'encourage and promote the establishment of communications by means of the Electric Telegraph between this island and the other West Indian Colonies and the Continents of Europe and America' which gave West India and Panama a shorter time scale than that given to International Ocean. Their subsidy of £2,500 a year for ten years would only be forthcoming so long as the work was completed in one year – specified in the Act as connecting Europe and America with St Thomas, Tortola, Barbados, Trinidad and the colony of British Guiana on the coast of South America.

In fact the *Dacia* was laying the final section from Trinidad to Georgetown in British Guiana during October 1871, but the stations were not ready. The company were not able to open the West Indian Telegraph for service on 4 October 1871 as stipulated in the Act, so they forfeited the Barbados subsidy.

The exercise in the Caribbean had taken toll of Sir Charles Bright's health and he had to return to England, in spite of the fact that he had been unable to lay the cable to Jamaica either from Colon or from Puerto Rico. The system could not open without the Panama link and this was not accomplished for another five months.*

The *Barbados Times* of 9 March 1872 carried the announcement:

> The Telegraph Company have given notice that the cable is laid and in working order all along the line from Havana to Demerara and through the States to Newfoundland and from there to the UK; and the communication being completed, messages can be forwarded from this island to any part of the world.

The service opened on 7 March. One of the first messages sent from Barbados acquainted the owners that their fishing vessel had been wrecked on the Crane rocks.

West India and Panama not only ran a telegraph service, but was itself a user of the system. It sent out a daily news bulletin with reports from Europe and the USA, with ruling market prices in London and New York. The first bulletins were put out in 1872, but they did not become official and regular until 1875 when it was one of the services provided by the company in return for the £2,500 annual subsidy (denied in 1871) given from 1 January 1875 for ten years.

With the excitement of the opening, the disappointments of the five months delay were forgotten but it was soon evident that the

---

*The Puerto Rico to Jamaica cable was finally laid in 1873.

quality of the service fell far short of expectations. As early as 6 July 1872 the *Barbados Times* was voicing public dissatisfaction:

> Complaints are becoming rife in these islands about the inconveni-
> ence which has been occasioned to the mercantile community and
> others by growing carelessness or incompetence on the part of the
> Telegraph employees. Owing to the exhorbitant rate of the tariff,
> messages must be necessarily condensed as much as possible to avoid
> incurring a heavy expense.

A member of the public would spend ten or twelve dollars on a telegram and then either receive no reply at all or one which was so unintelligible that he had to send a second message in order to obtain a satisfactory explanation of the first.

There had been as many complaints about the two telegraph lines to East India as there had been about those to the West Indies, and sensing this the German firm of Siemens offered to obtain the necessary foreign concessions to build for the Indian Government a complete route from the English Channel across Europe and Russia to Teheran to link with the existing line to India via Bushire and the Persian Gulf. This would give India an external route of its own, independent of any other country's telegraph. The Indian Government had so far set its face against employing private commercial companies to operate its internal telegraphs and was reluctant to do so for a scheme of the kind Siemens were now submitting. But in view of the greater usage of the internal telegraphs than anticipated, they lifted their objections and accepted Siemen's offer. The Germans formed a company in 1868 which confusingly they chose to call The Indo-European Telegraph Company. The work began at once and much of it was done by British sappers of the Royal Engineers under Major Bateman Champain. The line, which ran from Teheran through Lower Russia, under the Black Sea for a stretch and across Germany, was completed in 1870. The Government of India now had a third overland telegraph to Europe of their own, worked by 'educated and disciplined clerks' instead of illiterate, unreliable Turks. These operators who received and retransmitted the messages from station to station half across the world were employees, however, not of their Indo-European Telegraph Department but of the commercial Indo-European Telegraph Company.

'In 1870,' wrote Patrick Stewart's successor as Chief Director of the Government Indo-European Telegraph, now Colonel Sir Frederick John Goldsmid (in *Telegraph and Travel*, 1874) 'the aspect of

Eastern telegraphy entirely changed. The Indo-European Telegraph Company, projected in 1868, with a name closely assimilated to that of the Government Department, completed work on 30 January 1870 and opened on 31 January. In a month or two it became thoroughly organised, and messages were transmitted by it in fewer hours than they had taken days before, and with an accuracy of reproduction and precision of order to which no semblance of an approach had ever previously been made.'

Goldsmid's pleasure was that of a technician who delighted in seeing it all *work*. Not all those who used the system reacted quite so effusively.

> The establishment of a direct telegraph line between England and India in 1870 was an event far-reaching in importance. The delay in communication was a great advantage to the Government of India insofar as it of necessity left the initiation of policy in urgent matters to its own hands and enabled it to confront the Secretary of State with accomplished facts. But all this was bound to change when the Secretary of State had to be kept constantly informed of the course of events in India, and was in a position to issue immediate orders. Henceforth the Secretary of State exercised a far more effective control over the administration of India than was the case before, and the Viceroy really tended to be a mere 'agent' of the Secretary of State. (R C Majundar, H C Raychaudhuri and Kalikunkar Datta, *An Advanced History of India*, 1960).

As seen, the sections of yet another Indo-European telegraph were now falling into place. A cable had been laid by British-Indian Submarine from Bombay to Suez; the old landline from Suez to Alexandria had been leased by Anglo-Mediterranean; a submarine cable had been laid from Alexandria to Malta (to link with the landline from Sicily) by the Anglo-Mediterranean. It now remained for the final section to be laid from Malta to England. There were thoughts of taking a cable from Malta to Marseilles and leasing an overland line for this last section of the Australian Telegraph from the French Government to Calais.* In the event it was decided to lay a submarine cable all the way to the English coast via Gibraltar. But where should it land?

The name of the company which John Pender had formed in June 1869 to undertake the laying of this cable would seem to indicate the first choice – 'Falmouth, Gibraltar and Malta Telegraph Company'. Falmouth was on the south coast of Cornwall to the extreme west of

---

*See John Pender's remarks at a meeting of the Eastern Telegraph Company of 9 October 1876.

the British Isles which, with its fine natural harbour, had been a busy station of the Mail Packet Service since 1688. What more suitable place for the terminal of the new means of communication 'without the aid of mail bags' which would so greatly reduce the need of a mail packet service? The Electric Telegraph Company had opened an office in Arwenack Street, Falmouth, as long ago as 1857, so it would have been easy to link the line from out of the sea to the inland telegraph, now run by the Post Office, which ran to London. Or perhaps it would be better to land it at a more southerly point, near the Lizard for instance? Cliff top watchers on Pen Olver still identified passing grain carriers through telescopes and hoisted huge boards carrying the letters I C U ('I see you') to tell their masters that their arrival had been noted. The news was then transmitted to the appropriate shipping offices over the inland telegraph.

The Scilly Islands Telegraph Company had landed their cable from St Mary's in 1869 at quiet Sennen Cove beside Lands End. Why waste 25 miles by landing as far up-Channel as Falmouth? A survey was made of all the coves and bays between Lands End and the Lizard, and they chose a deep, secluded, sandy bay not three miles from Lands End called Porthcurno.*

Falmouth, Gibraltar and Malta ordered their cable from Telcon, whose two cable ships, *Scanderia* and *Edinburgh,* began laying the cable from Malta on 14 May 1870. The first section of 1,200 miles was landed at Gibraltar. The Portuguese Government, Britain's oldest ally, had asked that the cable should land at Lisbon (Carcavelos) on its way to England, and this was done by the *Scanderia.* The final section was laid by the *Hibernia,* chartered from the Anchor Line, which had laid part of the Bombay-Suez cable in the Red Sea. Connection to Malta was immediately effected. It had all been accomplished in twenty-five days.

Thus, Falmouth, Gibraltar and Malta held the key position as the British terminal of this lengthy international telegraph. The company not only obtained an exclusive concession from the Post Office to use the wire between Penzance and London, but the right to set up a telegraph office in London where messages could be taken from the public direct, and not through the Post Office. The company also won a concession from the Portuguese Government to establish and work a telegraph between Portugal, England, Gibraltar and Malta.

The telegraph between London, Porthcurno and Bombay was under the administration not of a single company as the Select

---

*Today's spelling; though it has previously been spelt Porthcurnew and Porthcurnow.

Landing the cable from Malta and Gibraltar at Porthcurno in Cornwall in 1870.

Committee on East India Communications had hoped, but three, all of which, however, had a common leader in John Pender. It was Pender, therefore, who was the host, nominally as chairman of British Indian Submarine, of the party given on 23 June 1870 to celebrate the completion of the Porthcurno-Bombay telegraph, and it was he who took the first message on Sir William Thomson's Siphon Recorder which also made its commercial debut that afternoon. Standing and listening beside him stood HRH the Prince of Wales. Emma had seen to that.

From 24 June on, the line was hardly ever at rest – and users of it came from a far wider field than the UK. But of course the success of the fourth Bombay-London telegraph – the Red Sea Line as they called it – brought alarm and despondency to the ranks of the Indo-European Telegraph Department of the Indian Government. Between 1865 and 1870 its earnings had been £92,000 a year and its expenditure £87,000, thus realising a small profit. With the efficient Siemens line in operation, traffic over the land routes at once began to increase.

The £2 17s rate (wrote Sir Frederick Goldsmid in retrospect in 1874) would doubtless now have yielded a satisfactory profit, to both

Indo-European Company and to Indo-European Department; but the Red Sea Line projected in 1869 was at this time also opened, and a great proportion of the messages at once diverted to it. The total traffic increased, but divided between three routes it was insufficient to pay any of them.

In 1871 the rates on the Siemens Teheran route and the Red Sea route were raised to £4 10s; but the Turkish Line of the Indo-European Department, being subsidised by state revenues, kept to the £2 17s rate and made a commercial loss.

The Red Sea Line would have been even more uneconomical without the aid of men like Edward Bull whom the *West Briton* described as the 'indefatigable manager and principal electrician' of the Falmouth, Gibraltar and Malta Telegraph Company's station at Porthcurno, and the men he trained to operate the service once the excitement of the laying and testing of the cables had been completed. To transmit at ten words a minute on a circuit employing a 'Speaker' (Mirror) or a Siphon Recorder, accurately and without being asked for repetitions, needed a degree of concentration and manual dexterity, which operators in these early days rarely possessed. But then Telegraphy, however romantic it may have seemed to the Press, did not attract young men of any great calibre – and the pay was low, the hours long and, in the case of the remote Porthcurno, tedious. Of the sixteen young men who formed the first staff of the Falmouth, Gibraltar and Malta's station at Porthcurno, only a nucleus would have been trained operators. The rest were 'probationers' picking up the skill for no pay as they went along, under the eye of the Superintendent, Edward Bull. There was no formal school. Eighteen-year-old Henry Clarke arrived at Porth-curno on 24 November 1870, and spent December, January, February and March learning how to work the Mirror, Morse Inker and Recorder instruments before being posted to the staff and starting to earn a wage of £52 a year – a pound a week. There was a large wastage; many who came were unable to write; some found the techniques beyond them; some had temperaments wildly unsuited; others made themselves unsuitable by over-drinking.

Historians have long regarded the period 1869–70 as the years when the British Government – Gladstone and the Liberals – came closest to dismembering the British Empire. There was a noisy 'Little England' movement; the Manchester Radicals talked of 'Separatism'. In 1852 Disraeli had told Lord Malmesbury 'these wretched colonies will all be independent too in a few years, and are a millstone round our necks'. In 1866 he asked Lord Derby 'what is the use of these colonial deadweights which we do not govern?' One school of thought wished to reconstruct the empire as a centralised unit, advocating throughout the 1870s the idea of an imperial federation. Gladstone's concept on the other hand was of a self-governing empire whose connection with the mother country was based on Freedom and Voluntaryism. The crux came in 1870 with the appointment of a Select Committee to examine the Government's colonial policy. Gladstone denied there was anything 'new' about it. He told the Committee the colonies must be taught to be self-reliant.

In May 1871 Arthur Knatchbull-Hugessen, the Liberal spokesman for the Colonial Office in the Commons, stated categorically the government had no wish to dismember the empire.

> Self-reliance does not mean separate existence. A colony may be great and self-reliant, and still maintain an intimate connection with the mother country. The government wish to retain them bound to this country by ties of kindred and affection.

Intimacy is difficult at arms' length, particularly when the arms are 12,500 miles long. For Egypt, India – Canada even – there was some semblance of reality in talking of Hands Across the Sea; but it was stretching the metaphor to apply it to Australia. John Pender was a Liberal (in spite of his dislike of Mr Gladstone). Without ascribing to him sophisticated political motives which interfered with his main driving force which was to make money, he leaned towards the Liberal view of empire and was happy to find himself in a position in which he was able to advance it by providing the people of Australia and their local rulers with a means of expressing their feelings of affection for Britain more forcibly than currently available, because more instantly.

As India was Britain's source of raw cotton, so the colonies in Australia were Britain's principal source of raw wool. Each colony was self-governing and independent of the others. There was no legal union except the Crown. All the time the legislative council of New South Wales led the campaign for an even greater degree of self-government. In its 'Declaration, Protest and Remonstrance' of 1851 it demanded an end to imperial control of taxation, land policy, and revenue, a dilution of the Crown veto and a constitution 'similar in outline to that of Canada'. It also displayed an early pugnacity towards Britain which was not to desert it in the years to come and was to have a crucial effect on this story.

Britain conceded the demands of New South Wales, and in 1856–7 democratic constitutions were introduced into the colonies in Australia, which by 1859 were providing the mother country with half her wool imports, worth £40,000,000 a year. But the census of 1861 showed that the total population was still only just over a million. So, compared with India, Australia was new and small. New Zealand, annexed by Britain in 1840 after 2,000 settlers had been there for some twenty years, was even newer and smaller. The islands did not receive responsible government until the mid-1850s.

The initiative regarding a telegraph came from the Antipodes. The only way the Australian colonies could communicate with distant Europe was through the mail bag placed on the monthly P & O mail steamer, a service which was heavily subsidised by all the colonial governments in Australia. The speed with which new Australia was developing and small Australia was growing, demanded quicker ways of asking urgent questions and receiving pertinent answers than that. The first to appeal to John Pender to bring them the benefits of nineteenth century ocean telegraphy were the governments of Queensland and Southern Australia, and he decided to accept the latter's proposal. His cables were of course already more than half-way there.

Within six months of assembling in front of Porthcurno with the rest of the Telcon fleet which had laid the cable from Malta in June 1870, the *Scanderia* was steaming back through the Mediterranean, down the new Suez Canal, into the Red Sea and crossing the southern tip of India on its way to Madras on the east coast, to set in motion the first stage of creating a submarine telegraph line between England and Australia. With her was the *William Cory*. Between them they had collected 1,800 miles of Telcon cable from Greenwich at the bidding of John Pender's latest firm, The British India Extension Telegraph Company, which he had formed in October 1869 with a capital of £460,000. The Indian Government had a telegraph from Bombay to Madras by land, and British India Extension's orders to the two Telcon ships were to lay across the Bay of Bengal to Penang on the west coast of Malaya and down to the island of Singapore at the end of the peninsula. They completed the line some time in December; telegraph stations were established at Singapore, Penang and Madras; and the line opened for traffic on 5 January 1871.

For the last section Pender had to float yet another company, the British Australian Telegraph Company with a capital of £660,000 (January 1870). In November 1870 Telcon's 3,100 ton steamer *Hibernia* laid a cable from Singapore to Djakarta, (then called Batavia) in Java, from where there was a Dutch landline to

Banjoewangie at the south east corner of the island. The last link was laid in the reverse direction.

The *Hibernia* returned to England to collect more cable (made in Greenock) and left on 3 August 1871 accompanied by the 2,315 ton steamer *Edinburgh,* used as a store ship and collier, and the 600 ton passenger steamer *Investigator* with the commander of the expedition on board, Captain Robert Halpin and his engineers and electricians. Crew and cable-laying staff numbered 300 men. By now they were well prepared against the tedium of such voyages – the two big ships sailed round the Cape and took eighty-eight days to reach Port Darwin, the nearest anchorage to Java in the Northern Territory, Australia. They brought good furniture with them, extensive libraries, pianos and billiard tables. There was no reason to rough it.

John Pender had agreed with the Southern Australian government to terminate his submarine cable, which now stretched to Porthcurno, at this northerly point in Australia on the condition that they built an Overland Telegraph Line to connect with it from Port Augusta at the head of Spencer Gulf just north of Southern Australia's capital Adelaide. The overland distance was 1,973 miles, and the work was divided into a southern, central and northern section each in the hands of a different team. The line from Port Darwin to Adelaide, and from there to the principal cities in the south and east, was completed by 23 June 1872 when the first message was sent through the whole 12,500 miles from London to Adelaide. But the next day a fault appeared in the submarine cable between Port Darwin and Banjoewangie and the line went dead. The location of the trouble was not found and rectified until 21 October, the date normally given as the opening of the All-Sea Australia to England Telegraph.

By the time the London-Adelaide telegraph opened for business, the main section of the line from London to Djakarta (Batavia) on Java had come under new management.

On 2 May 1872 shareholders of British Indian Submarine were called to an Extraordinary General Meeting with John Pender in the chair to discuss the merits of amalgamating with Anglo-Mediterranean, Falmouth, Gibraltar and Malta, and Marseilles, Algiers and Malta. They gave their approval and the proposal was adopted unanimously. Pender presided over similar meetings of the three companies concerned in quick succession, and on 1 June he registered The Eastern Telegraph Company which embraced all four.

John Pender was chairman. He was 57. He was now a Member of
Parliament again. He had offered himself as Liberal candidate for
Linlithgowshire in 1868 but he was defeated. At the beginning of
1872 however he was elected for Wick Burghs (and held the seat
until the general election of 1885). His other telegraph operating
interests at this time were British Australian, China Submarine and
British India Extension, and of course he also had a big investment
in the construction company Telcon. He was certainly now living in
the London house he had bought in Arlington Street and the country
estate at Footscray, Kent, which he rented from the Vansittart
family. His son James Pender by his first wife was now 31, but he
took no interest in his father's pursuits and lived the life of a country
gentleman. Emma had given birth to a son in 1855 who styled
himself John Denison-Pender. He was to become associated with his
father's telegraph empire, but in 1872 he was a schoolboy of 17.

The Eastern Telegraph Company had ten directors besides John
Pender: Lord N H Hay, Julius Beer, Thomas Brassey MP, George
Elliott MP, Baron E d'Erlanger, Colonel Glover RE, Rt Hon W N
Massey MP, G C Nicol, Lord Alfred Paget and Sir C Wingfield MP.
There were two Extraordinary Directors: His Excellency M Drouyn
d'Lhuys (chairman of Marseilles, Algiers and Malta) and the
veteran American telegraphist Cyrus Field. Pender chose as managing director Sir James Anderson, the one-time captain of the *Great
Eastern* who was released by the Cunard company to take up the
appointment. The company secretary was George Draper. Its head
office was at 66 Old Broad Street in the City of London, and over it
flew its house flag, an adaptation of the flag of the East India
Company with the red and white horizontal stripes and union jack in
the top right corner, but with ETC instead of EIC in the bottom left-
hand corner.*

The company had a capital of £3,800,000 in 380,000 £10 shares of
which 339,700 were fully paid up. One £10 share in Eastern
Telegraph was given for £100 worth of Anglo-Mediterranean shares,
for £200 Falmouth, Gibraltar and Malta shares, for £120 Marseilles,
Algiers and Malta shares, and for £120 British Indian Submarine
shares.

Eastern Telegraph's fixed assets consisted of many thousands of
miles of submerged cable, numerous telegraph stations, and expensive equipment but, most important of all, *people,* who in 1872 began
to build up the tradition of loyalty and expertise in the specialised

---

*Roderick Mann, retired Cable and Wireless manager, has one at his home in
Bayona outside Vigo, Spain; and there are probably many others in other parts of
the world.

field of Overseas Telegraphy, the preservation of which was always the most telling argument against fragmentation when technical, economic and political reasons appeared to be overriding.

Duplicate

No. 4. PERSIAN GULF CABLE, 1868-9.

STATEMENT OF CABLE REPAIRED AT KEYHAM.

Coil.	Length spliced up to last return.	Length spliced up since last return.	Total length spliced up to date.	Resistance of Conductor Per N.M.	Resistance of Dielectric Per N.M.	Number of splices in total length.	REMARKS.
	N.M.	N.M.	N.M.	Ohms.	Megohms.		
No. 1.	22·83	2·86	25·69	5.527	4160	155	
No. 2.	28·06	6·07	34·13	5.602	4694	171	about 45 fathoms to go on to this complete at this mile to this work for sea use.
TOTALS	50·89	8·93	59·82			326	

Greatest length between splices 18690 *feet.*

Least „ „ „ 306 „

Total Amount of Cable Condemned 6·22 *N. Miles.* { 2·44 miles of this in laying about up for splicing for sea use. }

Dated this *Twelfth* day of *June* 1869.

F. C. Webb

W. H. Taylor

2

South American Links

1872–1878

It was as chairman of The Eastern Telegraph Company, which now operated two-thirds of the line, that John Pender sat on the left of the Colonial Secretary, Lord Kimberley, who presided over the Australian Telegraph Banquet held in the Cannon Street Hotel, London, on the first anniversary of the completion of the Port Darwin to Porthcurno cable by the connection being made at Banjoewangie – 15 November 1872.

The Prime Minister, Mr Gladstone, was unable to attend but sent a letter of regret which Lord Kimberley read out. 'I should have been very glad, had it been possible, to be present on this occasion when the strong attachment which now binds, as it has ever bound, the Australian colonies to the Mother Country is cemented by the completion of so important and so useful a work.' Proposing the toast 'Telegraphic Enterprise' the Colonial Secretary said they were sometimes told the British had lost their spring forward, and Britain was not the nation it was. He denied that. 'The latest, if not the greatest of all human inventions, is the telegraph (hear, hear). Is not the English nation at least as forward as any other nation in bringing this invention to perfection and practical use, acting in unison with our kinsmen in the United States throughout whose vast dominions a system of telegraphy has been established?'

It was pleasing that Government should applaud the achievements of Private Enterprise, but was such admiration the first step to acquisition? At least one shareholder of Eastern Telegraph not only thought it might be but considered it should be. To William Ford, the presence of the Prime Minister at Cyrus Field's Thanksgiving Dinner, and the Colonial Secretary chairing the Australian Telegraph Banquet, meant only one thing: 'there is at the present time an intention upon the part of the authorities to obtain this Submarine Telegraph System from the proprietors [shareholders]'. Before the company decided to duplicate the Lisbon cable they should ask the Chancellor of the Exchequer whether the Government contemplated taking over the submarine telegraphs. If he said, Yes, they should not go ahead. A deputation to the Chancellor might lead to the appointment of a committee 'to consider whether it is not absolutely necessary *for the benefit of our colonies.* [our italics] that these telegraphs should be in the hands of the Government'. They were now only being used by the big commercial houses who could afford the rates. In Ford's opinion, every man ought to have the opportunity of sending a message at a low rate to the uttermost part of the earth. 'You can never have cheap tariffs unless the telegraphs are under the management of the Government, because naturally our duty is to make the highest possible return on our money. We are associated together for profit; and therefore so long as you have these companies so long must you pay construction and maintenance companies* those enormous sums which enable them to pay 25 and 30 per cent to their shareholders. While the existing condition of things remains, so long must you be content to pay promoters and to fill the pockets of those men who now finance the capital of companies. Therefore, I say, so long as all these things continue to exist you can never have cheap tariffs. Now, I would appeal to our directors to do all in their power to place the Submarine Companies in the hands of the Government, because in their hands I am certain we should have speedy and cheap communication with all parts of the earth.'

John Pender was patently taken off his guard, but he made a statesmanlike reply. 'As an individual, apart from my connection with the direction of the company, I heartily endorse every word that Mr Ford has said. But as chairman I am bound to take rather a different view from what he has put before you.' Sending a deputation to the Chancellor would take time to organise and might lose them their concession by overrunning their eight months. But that apart, you did not always make the best bargain when you went and priced yourselves. They were strengthening themselves in the work they were doing, 'and when the Government does come to us with any proposal for the acquisition of our property, of course they must be prepared to pay well for it'. As directors they had to look to giving shareholders a good dividend, and maintaining the system in a highly effective state. 'Let us go on doing our work, and doing it well

---

*A hit at John Pender's interest in Telcon.

I trust, and in that way I feel confident we shall be in a better position to make a good bargain with the Government.'

Thus, in the year of its formation The Eastern Telegraph Company was contemplating in all seriousness the possibility of its assets and operations being acquired by the British Government – a state of affairs which in the coming years it learnt to live with.

The shareholders gave their consent to the duplication of the line from Porthcurno to Portugal's capital via Vigo, and the 623 mile cable was laid and opened the following year. In Spain the cable from Cornwall was landed on a sandy beach called Al Cabre in the village of Bouzas a few miles to the south of Vigo, and then taken down the coast to Carcavelos on the mouth of the river Tagus from where a landline connected it with the office in Lisbon. Clearance of the cable ship for customs on arrival at Vigo was undertaken by Estanislao Duran e Hijos, the firm which under Casimiro Duran Gomez acts as agent for the company today.

When the four companies were merged to form The Eastern Telegraph Company their names had become well known in the foreign countries in which they operated, and for that reason for some time they were retained. The concessions which gave them the right to operate telegraphs in these places were in the names of the original companies. But the branch at Vigo was the first to be set up by The Eastern Telegraph Company, and as such has a special place in this story. It was to become a very big and important branch and to house the sixty or so Porthcurno-trained bachelor operators the company eventually built a large residence on the outskirts of Vigo (now in the centre) called La Finca which still stands today in its fine garden with the date palms and tennis court in Avenida Olivie, together with the separate manager's house, The Chalet. All the fittings and furniture, the carpets and curtains were sent out from England, and La Finca became a great social centre. It was the sporting young Englishmen of The Eastern Telegraph Company who, it is claimed, introduced the game of Association Football – soccer – which they had learnt to play at the public schools which had invented it. Many of them married Spanish girls, and the daughters of the English managers married handsome young Spaniards whom they met at La Finca receptions. Their descendants still live in Vigo today.

However proficient they were with their feet on the football fields of Porthcurno during their training as operators, it would appear that in the 1870s they still had difficulty in making their hands wield a pen. 'I am extremely disappointed to find that many of the staff at Porthcurno cable station are indifferent writers, especially among the probationers' wrote James Anderson, the old sea captain who was now managing director of Eastern Telegraph, to Edward Bull, the superintendent.

Three days after Christmas, one year, James Anderson told Bull to read out the following statement from him to his assembled staff:

> I have heard with great regret upon 3 occasions within the last 2 months that some of our staff at Porthcurnow are conspicuous for the use of bad language and drinking. Now I am by no means willing to learn that I cannot control a number of young Gentlemen at a station in England, any or all of whom I have the power of replacing at a very short notice.

Bull was obviously finding it hard to keep his unruly staff in order, not nly out of office hours but on the job as well. He took to issuing his own memos:

> Owing to errors we shall lose Reuters business again. This is most serious. In future all Reuters messages and those for Boltons agencies must be repeated.

> Mr Stacey at Bombay reports . . . that the errors continue to be very bad and that he has lost one very valuable customer in consequence of the mutilation of his messages. It appears that the errors are made chiefly in figures and it is an evidence that very great want of care exists that errors should be made in figures, because by rule, these figures are, or should be, repeated.

And it was not only how they wrote but what they wrote.

> This day 19 cases of this year have been brought before me of messages having the wrong station 'to' upon them. This is so serious an error and so needless that I insist upon such negligence being at once stopped. I will do my best to trace out these 19 errors and bring them home to the men who committed them. This inexcusable carelessness loses us many good customers and brings disgrace upon us . . . There are such gross mistakes as Glasgow for Manchester, Alexandria for Malta, London for Manchester, Calcutta for Colombo, Glasgow for London.

> The number of resignations and discharges for misconduct and incapacity has been so great since the Company commenced that I have not had an opportunity of giving those who are well disposed and deserving, of all the advantages I anticipated and wished. It has been impossible to foresee so many resignations and dismissals and I have not been in a position, in addition to replacing these men, to prepare a sufficient number of clerks to relieve those who might otherwise have been promoted.

The M.D. is much disappointed to learn that many of the staff habitually neglect to attend Divine Service on Sundays. This should not be; a man who neglects his (first) religious duties, will end by neglecting his (second) duties to his neighbours and employers. The M.D. therefore expects that every man will go to some place of Divine Worship regularly every Sunday when off duty. A return must be made to H.Q. every week of those who attend Church for the information of the M.D.

John Pender's main purpose in laying a second line from Porthcurno to Lisbon was to anticipate the traffic which he knew would soon be coming to Portugal from that country's erstwhile colony, Brazil, which had thrown off the yoke of the Portuguese government in 1822. A member of the Portuguese royal house was ruling as constitutional emperor.

South America was what some have called Britain's 'informal empire'. It was largely due to British influence that Portugal was persuaded to agree to terms for the independence of Brazil in 1826. As a result of Lord Ponsonby's mediation in 1828, recognition of La Plata and Uruguay followed. The new republics received large loans from the City of London. Chile received a million pounds in 1822 while fighting for its independence. Lord Palmerston told Parliament a few years later that some £150 million had been lent to South America. British adventurers and mercenaries helped them on land and sea in their fight to expel the Spanish. Earlier, English merchants had made a more peaceable contribution to the development of South America; even before the declaration of Argentinian independence there was a thriving English trading community in Buenos Aires.

The government of Brazil had set up its own Telegraph Department to run its internal telegraphs in 1863, and by 1872 most of the other states had followed suit. But as yet none of them had any telegraphic link with each other, except Argentine and Uruguay who were joined by the 25 mile cable which Proudfoot and Gray had laid across the mouth of the River Plate in 1866, as seen. By 1872 the technology of submarine telegraphy had improved to such an extent that a telegraph to link any of the southern republics with Brazil's capital 1,100 miles up the coast became a practical possibility.

An earlier attempt by Arturo Marcoartu to lay a cable from Montevideo to Rio in 1867 had failed. But in April 1872 he joined forces with a Londoner, James Reid, to have the lapsed concession from the Uruguayan government revived and transferred to a company which he and Reid formed for the purpose, the Montevidean and Brazilian Telegraph Company. They raised £135,000 and ordered a submarine cable from Henley's works in North Woolwich.

Three months later (July 1872) Andres Lamas, Uruguayan minister to Brazil, and his brother Pedro obtained a forty year concession from the governments of Brazil, Uruguay and Argentina to lay a cable from Buenos Aires (opposite Montevideo across the River Plate) to Rio, and formed a local company incorporated in Rio with capital raised entirely in South America called Companhia Telegrafica Platino-Brazileira, known variously as the River Plate and Brazil Telegraph Company or 'The Platino'. On 12 September 1872 the Lamas brothers transferred their concession to this company. A submarine cable was ordered from Siemens Brothers.

That summer – 26 July 1872 – Buenos Aires had been linked with Valparaiso by a spectacular 800 mile landline across the Andes. Chile being joined with Argentina by telegraph was an occasion for national rejoicing. Chilean and Argentinian flags flew on all public buildings in Buenos Aires.

A significant year in the history of Ocean Telegraphy, 1872, closed on a note of uncertainty yet great promise. The foundations of John Pender's telecommunications empire had been firmly established in The Eastern Telegraph Company, a grouping which the first six months' trading would seem to have justified. The revenue of the four companies to 30 September 1871 was £149,050, but in the same period in 1872 it was £175,840 – an increase of 19 per cent despite the rate for messages to India being reduced from £4 10s to £4. Shareholder William Ford had been right. Only the bigger traders had been able to pay the high rates, and to avoid them small businesses went to houses which regularly sent telegrams to India and persuaded them to include their messages inside their messages. This 'packing', as it was known, was resented by John Pender; 'trading upon our revenue' he called it. World trade was entering on one of its periodic depressions in the 1870s, and traffic with India had fallen off in the first six months of 1872. Pender wanted to introduce a one-word tariff, but the Indian Government ('not so advanced as we are') were against it. He offered to come to an arrangement with the rival operators, Siemens Indo-European Company, who had not paid a dividend for some years, but they declined. From the Indian Government's point of view Pender's company held a trump card. 'One reason why the Eastern line obtains the bulk of the traffic,' wrote Frederick Goldsmid in *Telegraph and Travel*, 'is that being in the hands of a single company its actions are freer, and its interests can better be pushed than those of the other lines which, belonging partly to a company and partly to different governments, cannot move in any matter without previous consultation of the directorates and consequent loss of time. But the chief reason of the advantage possessed by the Eastern Company undoubtedly is that it is intimately associated with the Eastern companies beyond India, and this association secures to it a virtual monopoly of the trans-Indian messages.'

Pender's plans to amalgamate even further were resisted by those whose investments were confined to Eastern Telegraph and regarded other ocean telegraph companies as commercial competitors whose success would reduce the Eastern's share of the cake. Ford criticised Pender's current negotiations with the Great Northern company. Many directors of the Eastern were also directors of British and Australian and China Submarine, but that applied to few shareholders. The problem of reconciling strictly commercial interests with the ideal of a world telegraph system organised on wholly international lines, for which John Pender had expressed a liking in the presence of his party chief Mr Gladstone at that Thanksgiving Day dinner given by Cyrus Field, had never been more acute. It was a conflict of interests which was to remain for another three-quarters of a century.

More urgent than deciding who had the right to run a cable up the east coast of South America was the need to find the money and expertise to lay one under the South Atlantic and connect the continent for the first time with Europe. As a writer in the *Brazilian Mail* pointed out (20 July 1872), a merchant in London could send a message and receive a reply to and from India in less than twenty-four hours, but he had to wait six to nine weeks for a reply to his letter to South America.

At the end of 1872 some English promoters formed The Great Western Telegraph Company to link New York with England and the West Indies. They planned to lay a cable from Lands End to Bermuda, from where two lines would radiate, one to New York and the other to St Thomas in the Virgin Islands from where, stated the prospectus, 'it would require but comparatively short a cable to continue telegraphic communication to Brazil'. They ordered a cable from Hoopers Telegraph Works which had been formed in 1870.

The purpose of the Falmouth, Gibraltar and Malta company in laying its cable to Lisbon had always been one day to extend it to

1–2 Brunel's 18,900-ton *Leviathan*, built in 1858, failed to make money as a passenger ship to the Far East for which she was re-named *Great Eastern* (*above*). But being the only ship large enough to take the 2,500 miles of cable in holds reconstructed as tanks (*see below*) required to link Ireland with Newfoundland in one trip, she was commissioned to lay the successful 1866 Atlantic Telegraph which John Pender helped to finance. Here she is at Heart's Content, Newfoundland.

3 The cable-winding machinery on the deck of the *Great Eastern* (*above*) is not too dissimilar from that used today.

4 A contemporary drawing of operators at the British end of the Atlantic Telegraph in the makeshift hut at Valentia in Ireland.

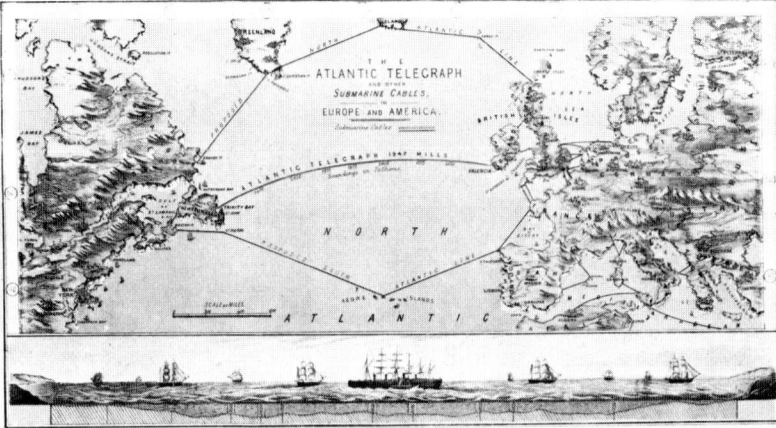

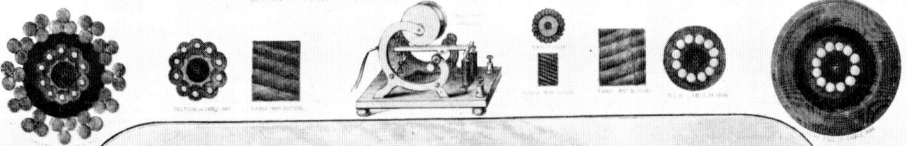

5 A broadsheet published on 14 July 1866 to mark the finally successful attempt to lay a telegraph cable under the North Atlantic. The sectional drawing of the *Great Eastern* shows how the cable was stowed in three huge tanks. At the foot is a map of 'Proposed Ocean Telegraphs and Overland Routes Round the World' across the Russian Empire to Australia and across the Bering Strait to North America.

SOUVENIR
OF THE
INAUGURAL FÊTE
AT MR PENDER'S 18 ARLINGTON STREET
TO CELEBRATE THE OPENING

OF

DIRECT SUBMARINE TELEGRAPHIC
COMMUNICATION TO
INDIA
JUNE 23RD 1870

6–9 When John Pender (*above*) had completed his direct submarine telegraph between London, Porthcurno and Bombay – the Red Sea Line – he gave a party at his London house in Arlington Street off Piccadilly on 23 June 1870, attended by the Prince of Wales, at which he received the first message from India on Sir William Thomson's Siphon Recorder (*left inset*). (*Left*) the cover of the souvenir booklet printed for the occasion; (*below*) an *Illustrated London News* drawing of the reception – note the grapnel and coil of cable hanging from the gallery (centre).

10 Typical of the cable ships of her day, the c.s. *Edinburgh* (*above*), 2,200 tons gross, built in 1855, had 40,200 cubic feet of cable coiling space. She was one of the ships which laid the cable from Malta to Porthcurno via Carcavelos in 1870 for the Falmouth, Gibraltar and Malta Telegraph Company; and the following year was laying the cable from Banjoewangie (Java) to Port Darwin (Australia). She was last used as a cable-layer in 1880.

11 Landing John Pender's Red Sea Line Telegraph cable from the *Great Eastern* at Aden on 2 March 1870 – an illustration from J. C. Parkinson's *The Ocean Telegraph to India* published in 1870.

12 Sir James Anderson (*above*) was Commander of the *Great Eastern* when she laid the Atlantic Cable of 1866, and successfully survived transfer from the sea to the boardroom when John Pender made him Managing Director of his Eastern Telegraph Company.

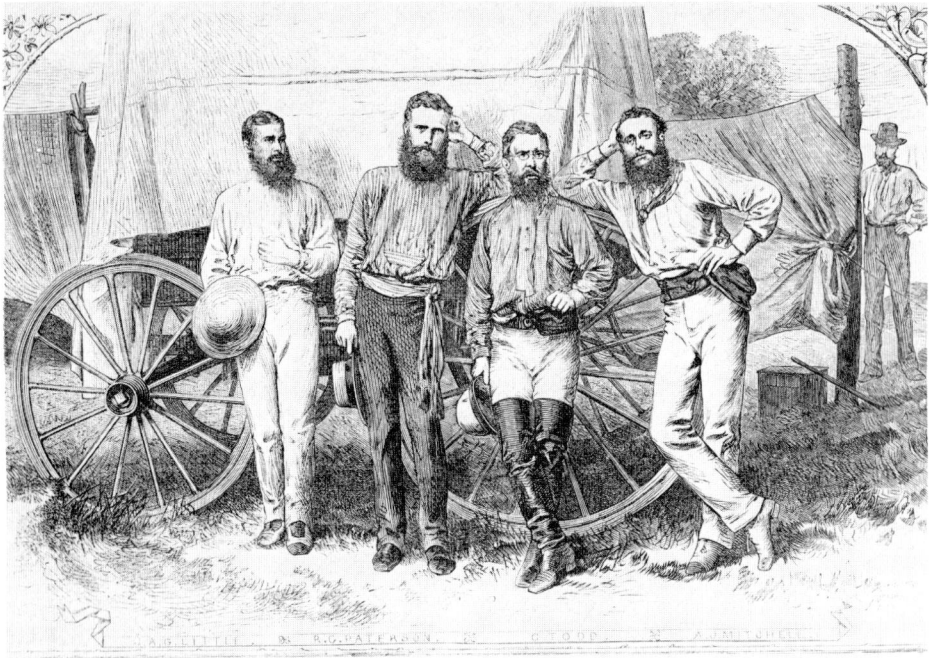

13–14 The Australian engineers who each undertook to lay one of the three sections of the 1,973-mile overland telegraph from Port Darwin to Adelaide with which John Pender's British Australian Company was to connect. But Robert Paterson (*second from left*), who had the northern section, miscalculated the difficulty of winning a bonus for completing the work by 1 January 1872, and was six months late. (*Below*) A group of those attending the ceremony at the planting of the first pole of the Australian Overland Telegraph on the northern section near Darwin on 15 September 1870.

Brazil; and their representative, Jules Despecher, had come to an arrangement with Iringu de Souza, Baron de Maua, who had won a concession for a Brazil-Portugal cable from the King of Portugal dated 16 August 1872. By that time Falmouth, Gibraltar and Malta had become part of The Eastern Telegraph Company, and in January 1873 several investors with associations with that company formed The Brazilian Submarine Telegraph Company with the object of pursuing the project of a Brazil-Portugal telegraph in earnest.

Viscount Monck was chairman, and among the directors were Sir James Anderson, managing director of the Eastern, and Sir Daniel Gooch, chairman of the Great Western Railway. Monck announced their plans at a first general meeting on 7 May 1873. They had obtained concessions from the governments of Portugal and Brazil, and the latter's concession also gave them the right to continue the cable down the coast. Their cable would be the longest yet. He was well aware, he said, that the unfortunate contingencies which had happened to sister enterprises had created a certain distrust in very long sea cables. But the line would be divided into sections and no one section would be as long as the North Atlantic cable.

Brazilian Submarine placed an order with Telcon who before 1873 was out had laid the first 622 miles from Carcavelos (Lisbon) to Madeira. A fault developed and the cable had to be cut in 2,400 fathoms. The ends were not recovered till the following year when sixty miles were hauled in and a repair made. They then laid the 117 mile Madeira-Cape Verde Islands section, followed by the 1,808 mile St Vincent to Pernambuco (Recife) section on the north-east corner of Brazil. The telegraph between Lisbon and Pernambuco was opened on 23 June 1874 – and the shareholders of the Eastern were happy that the foresight of their directors had anticipated the Pernambuco-London traffic which would come across their now duplicated Carcavelos-Porthcurno line.

Cottages were being built around the Cornish bay, which was the British terminal, to house the increasing staff needed to handle the new traffic. Their performance on the football field was no doubt of a high order, but beside the hearth their behaviour it seemed was of a lower standard. On 11 May 1874 Edward Bull posted the following appeal on the Porthcurno noticeboard:

> Complaints are made to me of the state of the fire grate in the office, made by night duty clerks spitting on it. I have had two spittoons placed in the office, therefore there should not be any occasion for further complaints.

Laying and operating coastal cable to link the ports of Brazil from Para (Belem) to Rio Grande de Sul was undertaken by a new group formed on 26 April 1873 to take over the Great Eastern company and acquire the Platino company's concession between Rio and Montevideo, called The Western and Brazilian Telegraph Company which worked in close co-operation with Brazilian Submarine.

The whole line from Lisbon to the River Plate – the Silver River – was inaugurated by the Argentine government on 15 August 1874, but it was not until 1875 that a full telegraph service was available to the public. It was a moving moment for President Sarmiento in the last two days of his presidency. 'We this day,' he said, 'attend an inauguration which, were the earth and water alive to feeling, they would understand the sentiments of the greatest emotion, flying upon the wings of electricity. We who have been born in the present age may consider ourselves blessed by the prodigies which from its commencement have been almost daily realised; and history appears to mark out in slow and solemn steps the glorious future presented to our view. Four ages ago three miserable ships crossed with fear and trembling the profound ocean which separates us from the old world. At the commencement of the present century a Spanish ship was the bearer of news from Europe of twelve months previous date communicating that the then king of this country, Fernando VII was the captive of Napoleon. Thirty years ago the learned Andres Bello, a native of Venezuela living in Chile awaited an answer to letters addressed to the correspondents of his own country three years previously.'

Western and Brazilian were connected to the systems to the north and to the west by companies formed for the purpose. The Central American Telegraph Company contracted with Hoopers to make and lay a 1,000 mile cable from Para to Demerara (British Guiana), the most southerly point of the West India and Panama system which linked the West Indian islands, and now had an outlet through Florida to the United States, providing via the Atlantic Cable* a second link to Europe. The cable had a T-piece – the first of its kind – midway between Para and Demerara running into Cayenne, the

---

*The 1865 cable continued working until 1877 but Telcon did not lay another to take its place until 1880 taking only 12 days, a record, and using the shore ends of the 1866 Cable which had broken down altogether in 1872. Anglo-American had Telcon lay two more cables in 1873 and 1874. The laying of the 1874 Atlantic Cable was the last operation of the *Great Eastern* as a cable ship. In 1873 The Direct United States Cable Company was formed to compete with Anglo-American, with the idea of laying a cable direct to the shores of America without first landing at Newfoundland. They eventually decided however that the working of a single line of that length would be too slow, and they had Siemens lay it for

capital of French Guiana. But, as with the Pernambuco to Para section, faults soon developed along the Central America telegraph which finally failed altogether. It was abandoned in 1876 and the company wound up.

In 1876 Telcon formed the West Coast of America Telegraph Company to run the telegraph they had built linking Peru with Chile, the Lima-Valparaiso line. The company was reconstructed the following year and given closer ties with John Pender's Eastern company.

Though the South American republics were now independent of Spain, their ties with the ex-Mother Country were still strong, and these were strengthened in 1874 when Telcon laid a cable from Barcelona to Marseilles for John Pender's private company, the Direct Spanish Telegraph Company formed in 1871 to connect England (Kuggar, Kennack Sands on the east side of the Lizard) by submarine cable with the north coast of Spain (Bilbao).

It had been John Pender's original plan, as seen, to start the Eastern Telegraph system not at Porthcurno but Marseilles. However negotiations with the French Government for permission to operate a landline over France came to nothing. Then, much to Pender's disgust, the French changed their minds (or were better persuaded) and granted a concession to another Englishman, Hon Philip Stanhope, a concession hunter whose sole object was to sell it to the highest bidder. He spent four years trying to find a purchaser. Pender realised that the line built by an operator who had the French concession (even though it was only for twelve years) would be a serious threat to the Eastern Telegraph system, and in 1876 he paid Stanhope £100,000 for it and leased a landline from the French Government for £5,000 a year. Pender received Eastern Telegraph's 4,000 shareholders' consent to this expenditure at a meeting on 9 October 1876 at which he said he had been working on trying to obtain the French concession for seven years, but 'Mr Stanhope probably got it more or less through favour'. He normally set his face against paying money to a third party for concessions; negotiations should always be between buyer and seller, but this transaction had

---

them to Halifax, Nova Scotia where it was connected to the United States by landline. It was opened in 1875. The previous year yet another trans-atlantic cable had been laid by La Compagnie Française du Télégraphe de Paris à New York, starting at Brest and landing at Porthcurno en route. This cable of the 'PQ' company, as it was known, was concreted into the cliff face at Porthcurno and terminated in a black hut where it remained till 1919. PQ stood for 'Pouyer-Quertier' the name of one of their directors. See John Packer's Note 'The Old PQ' at Appendix A, in which he states that the first Brest-Cornwall cable was not laid until 1880.

become very complicated. He did not know what Stanhope had paid the French Government; so what it cost the company not to have obtained the concession direct is not known. But there would have been no difficulty in raising the amount, for London's reputation as the financial centre of the world was very real.

The principal non-British telegraph operator in Britain's informal empire in South America was, as seen, the Companhia Telegrafica Platino-Brazileira incorporated in Rio de Janeiro in July 1872 in the expectation of raising its capital from Brazilian investors. But the full amount was never forthcoming and in 1873 the Platino issued a prospectus in England, inviting English investors to subscribe. There was a considerable response, and may thought the company should move its headquarters from Rio to London. The Brazilian Government baulked at this but relented in March 1877. No action was taken on the royal decree however until the following year when the Platino was wound up and re-formed in England as The London Platino Brazilian Telegraph Company. The Montevidean company was liquidated.

Ford's point that though there was a group of promoters common to the Boards of the main ocean telegraph companies there was no pool of telegraph investors was not lost on John Pender who at this time was seeking, as seen, to enlarge his Eastern Telegraph group even further. Though speed had been greatly improved by the use of duplex working on submarine cables for the first time in 1873, there were too many occasions when a line broke down altogether and the income of an investor in that company would dry up. The British India Extension telegraph had recently been silent for a fortnight. There were policy factors, too, which might reduce one company's earnings, while not affecting another's. In 1873 the Eastern Telegraph handled 11,258 twenty-word messages which earned them £63,000. But in 1874 the tariff was altered from twenty to ten words minimum. They handled more messages – 13,319 – but income was down to £54,000. Rates were fixed at international conferences of governments.

To remedy this and give promoters a combined capital for extending and duplicating lines, in 1873 Pender formed The Globe Telegraph and Trust Company. A share in Globe included a proportion of shares in other telegraph companies, but mainly Pender's Eastern Telegraph group and Anglo-American which ran the North Atlantic Telegraph and had Pender as a director and shareholder. An investor in the Globe, said its promoters, would avoid the risks of the shares of any one company unduly fluctuating – an early form of Unit Trust. But it was much more than this as the preamble to its articles of association demonstrated. The principal object was

> the acquisition and amalgamation in one Company of the principal lines of submarine telegraph and the land lines used in connection therewith. It is proposed, in the first instance, to acquire the shares and other securities of the companies owning them, issuing the shares of this Company in exchange for them, or to raise funds for the purpose of paying for them, and as opportunity offers, acquiring and absorbing the businesses and properties of the companies themselves.

In addition the memorandum of association introduced the feature of the manufacturing interest into the objects of the company. But so comprehensive a scheme did not meet with public support and its role became purely that of a trust company.

Pender did not need the ploy of a trust company in order to amalgamate his own companies, and in May 1873 he liquidated British Australian and transferred its business and property, together with that of China Submarine and British Indian Extension, into a new company he formed for the purpose which he called The Eastern Extension Australasia and China Telegraph Company. It had a capital of £3,000,000; Pender was chairman and Rt Hon W H Massey MP, vice-chairman. Sir James Anderson, Lord William Hay, and Viscount Monck were once again among its directors.

So the telegraph from Britain to Australia and China was now in the hands of two companies and one government Posts and Telegraphs Department. The Eastern Telegraph ran the service from Britain (London and Porthcurno) to Bombay; the Indian Government's Telegraphs the landline from Bombay to Madras; and Eastern Extension the service from Madras to Port Darwin, and on to Saigon, Hong Kong and Shanghai. On 1 February 1876, Eastern Extension opened a telegraph service between Botany Bay, Sydney, New South Wales to Blind Bay, Nelson, New Zealand. Telcon had laid the 1,300 mile cable with the *Hibernia* and *Edinburgh* two months ahead of schedule. At the dinner held at the Penders' London home, 18 Arlington Street, to celebrate the opening of this telegraph, John Pender was a principal speaker; and going into the drawing room 'Herbert of the Colonial Office' came up to him and congratulated him. In the course of the conversation that followed, according to Emma Pender, Herbert assured the chairman of the Eastern Telegraph and Eastern Extension that if he should ever want any business transacted 'in his office' connected with telegraphs, he would take charge of it himself. Pender took the opportunity of asking him about Barbados in the West Indies, and stated that he was against the confederation then being mooted, an opinion with which Herbert agreed.

Laying and opening new lines was comparatively easy. Maintaining an efficient and profitable service was the problem. There were now five cable making and laying companies all looking for contracts who had no interest in the service that was provided along the cables once they had laid them. In the view of some, much duplication of submarine cables took place merely in order to keep the construction firms busy and their shareholders remunerated. At an Extraordinary Meeting of the Eastern on 16 December 1875 a member said the only reason he could see for spending another £700,000 on duplicating the Suez-Bombay cable was that the Eastern Telegraph's directors were also Telcon's. At the same meeting a Mr Stewart complained that the directors were not earning shareholders anything like a sufficient revenue. 'There is no doubt,' he declared, 'there has been a great deal of jobbery in the inception and conduct of the business of the Eastern Telegraph Company.'

There was 'great interruption'.

'I must protest against that remark,' said John Pender from the chair. 'Do you mean to tell me that gentlemen on this side of the table have been stock jobbing for their own private interest?'

Stewart said he did not know who might have been jobbing (uproar and cries of 'sit down'). Pender asked him if he charged the Board with having been jobbing with a view to personal aggrandisement. If the shareholders approved of his imputation then the directors were not worthy to sit there. Stewart said he made no imputation. 'If you state distinctly that there has been no arrangement with the Construction Company . . .' began Stewart to hisses and cries of 'withdraw'. He withdrew his remark, but demanded that the meeting be adjourned for six months. Pender said only one director of the Eastern Telegraph was a director of Telcon, [Lord Alfred Paget] and the question of where to order a new cable was always referred to a committee of which he was not a member. In fact, however, every cable required by a Pender company was made and laid by Telcon.

Whatever the merit of Stewart's jobbery accusation, there was no doubt that the Eastern Telegraph's earnings fell well below their competitors'. The demand for overseas telegraph services was not growing as fast as the number of new firms entering the field. There was a threat that the traffic to Egypt would be halved by the entry of a company which obtained a concession from the French Government to lay a landline across France and then to Alexandria by sea. Eastern Telegraph were only passing twenty-three messages a day between Britain and Australia, and if a second company started to compete there would be no profit for either. The ingenuity of users in condensing messages and a wide use of codes had further reduced traffic, and so income; though James Anderson thought the road to greater earning power lay in increasing the facilities and increasing the number of *messages,* no matter how condensed or encoded. And, of course, the errors of the operators, particularly in transmitting figures (for which a letter code was invented at Porthcurno) showed no improvement and discouraged wider use. The telegraphs of both Eastern Telegraph and Eastern Extension beyond Egypt had broken down in December 1875, and as there were no cable repair ships in the area there was six weeks' delay before the nearest reached the trouble spot. This was the most cogent reason for duplicating the Suez-Bombay and Madras-Straits Settlements cables.

The accusation of jobbery at the Eastern Telegraph meeting must have been disconcerting for a businessman who, outside the world of ocean telegraphy, moved in high political circles and claimed, with some reason, to have the ear of government ministers.

On 28 November 1875 Emma Pender wrote a letter* from 18 Arlington Street to Sir William des Voeux, the professional diplomat who had married her daughter Marion, and was then the Governor of the West Indies island of St Lucia, in which she described the part played by her husband in the drama of the Khedive and the Suez Canal shares.

> 28th. The news you will find in the papers of the purchase by our government of the Khedive's Suez Canal shares will astonish you as it has done the country generally. We knew that the French were in some difficulties as to finding the £4,000,000 and Mr Pender urged, through Lady Derby,† that we should go in for them. Then on Tuesday the Rothschilds telegraphed to the Foreign Office that they could be had. And on Thursday morning the bargain was concluded. I understand Mr Pender's letter, but without his name being given,

---

*The typed transcripts of the correspondence of Emma Pender with her daughter Marion and her son-in-law William des Voeux are in the possession of Hon Richard Denison-Pender, and extracts from them are reproduced here and elsewhere in this book by his kind permission.

† In the general election of February 1874, Mr Gladstone's Liberal government was defeated, and Benjamin Disraeli formed a Tory administration with the Earl of Derby as Foreign Secretary.

was read at the Cabinet council and at least did a great deal towards influencing the decision. People guess this, but we must not mention the part he has had in it. On every account do not name it. I was amused by Levy of the *Telegraph* coming to try to get it out of me. Mr Reeve said 'I don't ask you to confess it, but I feel sure you have both had a finger in the pie'.

While commercial managements settled the mechanics and economics of telegraphy, governments of nations which had become transit areas for international telegraphs met in St Petersburg to discuss ethics and politics.

Though there had been meetings in Paris in 1865, in Vienna in 1867 and Rome in 1872, the International Telegraph Conference of 1875 was the first of any real significance. At St Petersburg the representatives of government P & T departments laid down the principle that everybody in the world had a right to correspond by means of international telegraphs, and the contents of their telegrams must be kept secret. Governments reserved the right to decide which lines in their territory would carry through traffic, which they could suspend whenever they saw fit. They agreed that the rates for messages sent along the whole of an international line could not be altered without the consent of all interested governments. Of most significance was the decision to standardise tariffs on the word-rate rather than the message-rate with its built-in minimum tariff which Reuters got round by 'packing' several cablegrams into one. Code words were to be no more than ten characters.

John Pender as a commercial operator could not be a delegate, and throughout his life he expressed his resentment at having to accept far-reaching decisions of this sort, vitally affecting his enterprises, made by people to whom he could not put his case officially. Though barred from the conference room he could of course lobby delegates in the corridors and dining rooms, and this he certainly did with characteristic vigour, but rarely to any great effect.

He was an international figure with world-wide interests and liberal views about the desirability of making cheap telegraphic communication universally available commensurate with paying reasonable dividends. But he could make little progress in achieving this ideal so long as the key policy decision of determining rates was taken by others. One way of mitigating this restriction, in his view, was to concentrate management into as few hands as possible. But there were many who saw his 'idealism' as paranoia and were quick to paint the creator of the Globe Trust as a self-seeking monopolist. From Pender's point of view it seemed good business in 1876 to try and merge Anglo-American with the new Atlantic telegraph company Direct United States. Anglo-American were now operating the French PQ cable too, but in spite of this their gross earnings were only 7 per cent on their capital (£480,910 on £7,000,000), whereas since they opened their service in 1875, Direct United States were earning 11 per cent (£143,610 on £1,300,000). Direct United States were already receiving two-thirds of the receipts of the two companies put together. This was too great a threat to Anglo-American, and Pender sought to eliminate the competition by amalgamating the two companies within the Globe Trust. When Henry Labouchere MP, a shareholder in Direct United States, heard of this, he mounted a bitter personal attack on John Pender in *Truth,* the magazine he had just founded to expose what he considered was scandal.

'It may perhaps be supposed that, from his numerous chairmanships, Mr Pender has made submarine telegraphy a speciality,' wrote Labouchere in *Truth.* 'On the contrary, he was, and unless he has withdrawn from business, still is, a commission agent in "yarns" and "grey cloths". The market value of the securities of the Cable Companies of which he is the leading spirit are as follows: Eastern, 30 discount; Eastern Extension, 27 discount; Brazilian Submarine, 35 discount; Globe 55 discount. I should be glad to know what amount of money he has himself invested in these companies, and what may be the amount of salary which he receives as their chairman per annum for managing property in a manner which has produced such disastrous results to investors?'*

Litigation followed to prevent the forced marriage of Direct United States and Anglo-American.

The High Court were to decide, wrote Labouchere, whether a minority of shareholders in a solvent and flourishing public company could be liquidated at the dictates of a majority who had suddenly become the adherents of a competitive undertaking 'so waterlogged as to be obliged to destroy opposition by colourable devices and questionable raids upon its competitors'. He implied the move was designed, among other things, to bring more work to Telcon. The

---

*At the general meeting of 15 July 1873 it was stated that directors of Eastern Telegraph were paid £5,000 a year; Eastern Extension directors £4,000; and Anglo-American £3,000. As William Ford pointed out, by extending the Eastern cable to India and Australia the directors received £9,000 a year. Sir Daniel Gooch of the Great Western Railway, which had a capital ten times that of Eastern Telegraph, received £5,000. Eastern Telegraph's articles of association gave directors £7,100 a year plus additional remuneration for every 1 per cent above 10 per cent, but this was lowered to £5,000 with 1 per cent above every 15 per cent (only as far as 20 per cent).

Memorandum of Association of the Direct United States, which had the support of the US and Canadian governments, stated that no 'working agreement' would ever be made with Anglo-American – a clause that Pender later described as 'iniquitous, if not sacrilegious', adding that in any event his proposal was for a 'friendly alliance'.

By virtue of their artificial monopoly, accused Labouchere, they clung to high rates which lost them business. Why did they not grant rebates to big users, as the Direct company did? The shareholders of Anglo-American were drowning and Pender (whom Labouchere claimed to like) could not be blamed for trying to save them. Behind the actions of the 'Cable King' was an implied philanthropy which Labouchere found hypocritical. 'We are resolved to resist to the last any attempt to convert our flourishing and independent company into the handmaiden of that debauched and dropsical Old Man of the Seas, the Anglo Company.' (11 January 1877).

The Times found the Globe concept unacceptable too. 'Nothing could well be devised more destructive to joint-stock enterprise than this machinery of "trusts" by which the controlling power over companies may be acquired and exercised in defiance of every private right by a group of speculators. That this movement to get possession of the Direct Company is the enterprise of a few, working partly through the Globe Trust, is beyond question; for the movement did not begin before last September, and we have a list of the shareholders of the Direct Company in our possession, which shows that the bulk of the shares held by the fusionists has been acquired since then. Their interest in the welfare of this company is therefore of a remarkably recent origin – it in fact dates from the time when the competition of the Direct Company may be said to have begun to tell seriously on the receipts of its rival, and therefore on the value of the enormous stake which the Globe Trust has in the stock of that rival.' (6 February 1877).

Jobbery or not, Pender's tactics succeeded, and on 26 June a resolution for the liquidation and reconstruction of Direct United States was passed by 3,725 votes to 419; and on 17 July Pender, Gooch, Anderson and the rest formed a second Direct United States Cable Company to replace that registered in 1873. This new company pooled resources with Anglo-American, the French PQ Company and the American Telegraph and Telephone Company (ATT). Anglo-American took 49 per cent, ATT 23 per cent, Direct 16 per cent, and the French company 12 per cent.

John Pender spent that Easter at Porthcurno. It should have been a refreshing change from the metropolis and the strain of the courtroom, but he found anything but peace. Emma supplies the detail.

> There is a telegraph station near to the Lands End, put down in a secluded bay, and no doubt tryingly lonely in these days when men of every station in society require so much amusement and excitement. The Company have furnished it with a billiard table, Library etc. and now it is a kind of school for preparing young men for the telegraph positions abroad. But just now the whole of this community are at loggerheads, each with all the others. Sir Jas. Anderson went down, but did no good amongst them, so he came back and said nothing. Matters went worse and he sent a clerk to try to soothe them, but no good effect followed, and at last he had to confess the whole story. So your Father has gone down and the probability is he may disperse the whole colony. I expect his having to stay till Tuesday. The weather is bad and this is perhaps as good a way for spending his Easter as any he could find. (Letter to Marion, 23 March 1877.)

The trouble at Porthcurno stemmed from Edward Bull's inability to maintain discipline. Dallying with the servant girls and slack attendance at the office had become the rule. And then there were those 'escapades', such as the one when the gravestones at St Levan changed places during the night with those in Sennen churchyard. So in June they appointed a matron. In a letter to G O Spratt, the 23-year-old assistant superintendent, H W Ansell, the company secretary, told him 'the very presence of a matron should ensure order and decency in the house'. On 27 June, Ansell instructed Spratt that Bull's salary would be paid to the end of the month. Then 'he must clear out as soon as possible and convenient, and the sooner the more agreeable for all concerned'. Two days later he wrote: 'Mr Saunders will be down to Porthcurno, probably by Monday's night mail, to represent the Managing Director and take temporary charge with a view to reorganising matters in accordance with his wishes, especially in respect to learners' classes which I trust that you have already lost no time in starting on a better footing than under the late regime'.

Much of the indiscipline had been due to the fact that the senior staff were housed in scattered farms and cottages. 'I am often greatly inconvenienced by the absence of these seniors from illness either through colds or the extreme weather coming in on night duty through pouring rain and having to walk two or three miles,' complained the superintendent. Irritation caused too by Anderson's continued insistence on Sunday worship. 'The Church Return is rather a sore point,' wrote William Ash, the new superintendent who had served with the Magnetic company. 'Those who like to go have a certain repugnance to their religious acts being brought forward in an official manner, and those who go only because their name shall appear, of course get very little good.'

It was a refreshing, commonsense attitude that soon earned Old Man Ash the respect both of his superiors and those he had to instruct. He lasted thirty-four years. His son and grandson followed him in the Eastern Telegraph and its successors.

In his foreword to Dan Cleaver's *History of PK* written in 1953, G G Wren, who was born in 1862 and as an 8-year-old boy saw the first cable being landed on Porthcurno beach, described the conditions of 1877. There were no baths, no gas, no bicycles. It was, of course, well before the days of motor cars. 'Those who wanted to go to Penzance had to hire a horse and trap from Farmer Ellis and pay 10s 6d for it or walk. I walked.' In July the following year he became one of the first six operators to be sent from Porthcurno to Singapore.

John Pender's trip to Porthcurno can have been anything but relaxing. Normally, whenever he could get away, he went to Homburg for the waters – he was suffering from numbness in his left arm. He was now 62 and, as his wife reminded him, he could less easily shrug off such pains than twenty years before. He did a considerable amount of travelling abroad, apart from visits to his Scottish constituency.

In September 1877 Emma was reporting that her husband was making 'his long meditated voyage to the telegraph stations in the Mediterranean' and this was presumably taken in the *John Pender*. He was to tell his shareholders of this tour three years later. He had interviews, he said, with state telegraph directors and foreign ministers – 'in fact all the ruling powers'. 'I have always been under the impression that we ought as far as possible to give our system of telegraphy something of an international character.' On all sides he had had an expression of perfect confidence in the company, not only for secrecy but for accuracy and speed. 'When any important message is required to be sent it is sent over our system in preference to any others.' As an Englishman he was proud of the company's establishments. 'Our young men would do credit to any service.' Telegraphy had got into the position that it was an important factor in politics and in commercial and social life, and there were very few men who now put their hands to a letter when they could telegraph.

In October (1877) he was in Constantinople. It was from here in 1876 that Telcon had laid a cable for the Black Sea Telegraph Company (formed 1874) to Odessa on the north shore, where it met the line of Pender's competitor the Siemens Indo-European company on its way through Russia and Germany to Britain.

It was a dangerous spot to be in, for in April that year Russia had declared war on Turkey and invaded Roumania. For the solution of the 'Eastern Question' the powers had resorted to force. HMG sent a note to the Czar warning him against any attempt to blockade Suez or occupy Egypt. In July the British Cabinet resolved that it would declare war on Russia in the event of her seizing Constantinople, the gateway between Asia and Europe. Britain's £30 million a year trade with the Middle East, 5 per cent of her total foreign trade, demanded a stable Turkey and above all not dominated by an ambitious, imperialist Russia. Early in 1878 the British Fleet sailed for the Black Sea, and Lord Derby, with whom the Penders were so friendly, resigned in protest and was succeeded as Foreign Secretary by Lord Salisbury. By a secret Anglo-Turkish agreement to check Russian advance in Asia Minor, Britain promised to defend Turkey against further attack, and in return was allowed to occupy Cyprus.

Eastern Telegraph had no station or branch in Constantinople, and John Pender's presence there while the Russo-Turkish War was at its height was in order to acquire one by buying out the Black Sea Telegraph Company at a time when the price would probably be low. He would also be looking to the day when settlement of the Eastern Question would restore confidence in the reliability of the overland lines between Britain and India via Turkey and Russia (the Siemens Indo-European line had been out of service from May to August 1877) and lead to a greater demand for international telegraphy than ever before.

In Constantinople he negotiated the purchase of the Black Sea company (capital £130,000) for £74,000 which, by giving Eastern Telegraph an office in the Turkish capital, won them what Pender hoped would be control of all the international telegraphs in the Mediterranean. His plans were capped by deciding to lay a cable from Larnaca on the now British island of Cyprus to Alexandria.

On 13 July 1878, Emma Pender wrote to her daughter Marion:

> Harry [her elder son] started for Cyprus in the train of Sir Garnet Wolseley this morning. Know that this Island is a new acquisition, given to England by a treaty with Turkey, and is intended by us to be used as a fortress to keep Russia in check! Sir Garnet was appointed governor and the Eastern Telegraph Company wanted someone to go out to look for the best place for laying a cable, building a house, etc, etc. In fact, as Harry said of his commission, he went out as a prophet to announce to them where new cities would be raised over the land, where the earth would pour out riches, where the seas would cast up

treasure. It is only a few days since we were startled by the news of the
treaty, and Harry had not twenty-four hours in which to make his
preparations and gather up his instructions. I miss Harry sadly
already. We were reading together and he has an organ in the hall
which wailed for hours sometimes and which I loved to listen to. I do
not think I could bear to hear it now. Make much of the few years
your child will be with you. There is no certainty after the first
separation for school.

Young Henry Denison-Pender, who, as assistant to Sir James
Anderson the managing director, was, it seemed, being groomed to
take over the management of the Eastern Telegraph group, accompanied his 63-year-old father to Rome where, as chairman, John
Pender was making an inspection of company property. In the
Italian capital Henry contracted a disease (typhoid?) from which, to
the distress of his parents and all who had worked with him, he very
soon died. His sister Marion in the West Indies will have remembered her mother's last letter with its prophetic comment – 'there is
no certainty after the first separation for school'. His brother John
will have realised that the succession, if he wanted it, would now be
his.*

The time was not ripe for buying out Siemens Indo-European
company but John Pender did the next best thing. He made a Joint
Purse Agreement with them which came into effect on 1 September
1877. It was on the basis that British Indian Submarine had 60 per
cent of the traffic to India and 80 per cent of that beyond – to China,
Java and Australia. The Indian Government were asked to join the
combination but, less concerned than the two commercial companies with making a profit in the light of their subsidy, they
declined. But the ganging up of his rivals made Major Champain
anxious, and in the months that followed he was pressing Sir James
Anderson to modify his terms.

The compromise of a tripartite Joint Purse Agreement – Pender,
Siemens, and the Indian Government – was finally reached in the
summer of 1878. Pender's patient sparring with the Indian civil
servants prepared him for the long drawn out fight with the
increasingly telegraph-conscious 'colonies' whose concern with
low rates was to dominate the scene from now on and lead to the first
challenge of John Pender's position as the unquestioned world leader
in Ocean Telegraphy. His need to impress on governments the case

---

*He was elected to the board of Eastern Telegraph on 2 February 1882.

for leaving long-distance telegraphs to the commercial companies had now become urgent. He told his shareholders that summer that he had seen to it that the governments with whom he had had to deal realised that though Eastern Telegraph's headquarters were in London they were virtually an international company; that through their way of management, having their ships on the stations and their cables ready at all times for extending or repairing the system, their interests were better in the hands of the Eastern Telegraph company than in any others'. They could do the work not only more cheaply but more quickly and more efficiently. The company, he assured investors, was gaining these countries' confidence. He hoped soon to announce that a cable would be connected from Aden to Zanzibar, Mauritius and the Cape of Good Hope. And Eastern Extension were negotiating for a second cable to Australia.

Was this in answer to the news that had reached him in Old Broad Street, maybe through the local agent he had appointed in Australia in 1876, of the discussions that had taken place at a conference in Sydney, capital of New South Wales, in January 1877 – or was he ignorant of the plans being formulated by an opposition which was to prove more formidable than any he had previously encountered?

For at this conference, to which the Press are unlikely to have been admitted, the governments in Australia and New Zealand were debating whether *they themselves* would have a submarine cable laid from Australasia to North America, and operate a telegraph service over it. They passed nine resolutions on the subject. One authorised the New Zealand government to ascertain if the United States administration would pay a subsidy to secure construction of a cable from New Zealand to USA. A copy of the resolution was sent to the Colonial Office, and to Sir Julius Vogel, Agent-General of New Zealand in London.

In August 1877, Sir Julius Vogel and Hon Archibald Michie, Agent-General for Victoria, made a report in which they referred to four possible routes for a cable to England via Asia. They also hinted there might be a route from New Zealand to San Francisco via Honolulu, but that it was impracticable in view of the great depths.

The Sydney Conference of 1877 saw Australia making the first move in a campaign to challenge the assumption that governments would not, and could not, undertake the laying of submarine cable or operate an ocean telegraph service. But an even greater threat to any major telegraph project, whether John Pender's, the Australian Government's or anyone else's, was waiting to make an entry on the communications scene.

3

Consolidation

1879–1891

At the International Telegraph Conference which met (for the first time) in London in the summer of 1879 there was no official reference to the instrument which a Canadian immigrant from Scotland had demonstrated to the Society of Telegraph Engineers in London in October 1877. But then, when Colonel Reynolds, Alexander Graham Bell's London representative, had offered to give a demonstration of his 'telephone' to the British Post Office, he had been told by R S Culley, its Engineer-in-Chief, that he already had details of it and that in any case in his view the possible use of the telephone was very limited.

For forty years the world had become used to talking about the miracle of the telegraph which enabled people to speak to one another across great distances. But of course it was only far-*writing*. Far-speaking was only possible from 1877 and it was many years before Telephony took its place beside Telegraphy as a regular means of instant communication.* It came earlier to small communities than to more extended populations – a Barbados Telephone Company, for instance, was formed in 1884. But for the most part Government Posts and Telegraph Departments all over the world were in no mood to have their investments in people and equipment thrown away as quickly as that. After the rebuff from the Post Office in 1879, Bell's firm, The Telephone Company, established the first telephone exchange with eight subscribers at Coleman Street. Though William Preece, the Chief Electrician, had counselled otherwise, Culley was confident that his decision had been

---

*In 1880 the High Court ruled that the telephone was a telegraph within the meaning of the Telegraph Acts of 1869 and 1873.

right, and the development of the Telephone caused little stir in St Martins-le-Grand or Old Broad Street.

The representatives of the Posts and Telegraphs Departments who met in London that June discussed International *Telegraphy,* though doubtless in the corridors of the conference hall hypothetical questions were put and answered about a medium as yet untried over long distances (let alone under the seaz, though Thomas Edison's carbon transmitter showed that sound could be sent very much further than with Bell's original equipment. Maybe one day?

For the present, the main debating point remained overseas telegram rates, and if conference were to adopt the latest proposals of the British Post Office the financial basis of the Cable Companies looked like being undermined. On 15 January 1878 the chief executives of the five ocean telegraph companies wrote to C H B Patey to protest at the BPO's intention to submit a plan to reduce the tariff for international European telegrams, which they claimed was equal to a diminution on present rates of between 30 and 80 per cent. They also objected in emotional terms to the proposal that different rates should be charged for the same routes.

> The proposal of the Department (they wrote) involves the destruction, more or less rapid and certain, of British enterprise and capital hitherto liberally devoted to submarine telegraphy, and the possible extinction of the £22,284,806, or thereabouts, invested in these undertakings. It is proposed that the European tariffs should be so inordinately reduced that either the Companies, or the other lines in competition with them, must succumb. The destruction of the Companies is secured because their competitors would be Government lines of telegraph with unlimited purses at their back. The propositions of the Department are thus based upon the competition of Government lines, at rates which it must be admitted will involve a loss, with the lines of private Companies.

The effect of the Post Office's policy would be

> to destroy cable property, at least cables of any considerable length. The tariffs are far too low to be remunerative for long cables, even to pay expenses in some cases, much less to provide a reserve fund; consequently the long costly cables, expensive to maintain, must give way, and between this country and foreign countries only the shortest cables would be preserved.

There was no demand for such a reduction, and even if there was, it ought not to be granted at the expense of the taxpayers and of the large undertakings put in peril by it. After longer or shorter struggles with insolvency by the Cable Companies, it would ultimately bring about a monopoly of telegraphic communication in the hands of the Post Office and its partner the Submarine Telegraph Company.

James Anderson, who signed for Eastern Telegraph, and the others who signed for the Indo-European, Direct Spanish, Anglo-American and German Union, ended their letter with a grave warning.

> The Companies desire to impress upon your Administration that this question is of vital consequence. It involves the Companies' existence, and they must certainly succumb if the propositions of the Department were carried out . . . It would indeed be an ungrateful result that the first telegraphic conference held on British soil should legislate for the destruction of British enterprise.

James Anderson had meetings with Patey early in the New Year and got him to agree not to table any tariff reduction without first consulting the cable companies. 'I submit it would be unseemly if the Companies should be found at variance with your Department when the Conference meets in London. We really represent the capital and organisation which unite Great Britain with the external world, and upon every principle, commercial and political, may fairly claim not only the support of your Department, but the amplest assurance that none of our interests shall be adversely affected for the sake of any new idea which is not of a vital character in respect to international telegraphy.' (Letter dated 24 January 1879; India Office Records, 1/PWD/1/182.)

Anderson was also at cross purposes with the Indian Government Telegraph Department. In February he received a letter from Colonel Champain complaining that Eastern Telegraph were acting against the spirit of the Joint Purse Agreement by opening offices for the soliciting of business in Liverpool and Manchester, connected to London by a private line leased from the Post Office. Anderson was obviously angry, and penned a private letter to Champain before replying in an official manner.

> I would like to say that I regret very much to have received such a letter written in a captious and vexed tone unlike you, and I venture to think a good deal unnecessary . . .
>
> I have done nothing with the object of getting Indian traffic over our line. We have gone on a steady course, never solicited a single Indian or trans-Indian message; nor have we obtained a single facility from the Post Office that is not available to all comers; nor do we give a single facility to the public other than in accord with the Indo-European. My special conviction is that these private companies are

only strong as against rivals and new combinations in proportion as they occupy the ground and get as near as possible to those who employ us [overseas telegram senders]. We want their money and we desire a reputation for good work and such rapid despatch as no rival can hope to excel . . .

It is no part of our wish to pose as an over-reaching company. Yet we must occupy the ground whenever and wherever we can do so with a probability of securing and improving our interests.

Improve a commercial company's interests or improve the world's chances of having a telegraph service cheap enough for universal use? The point was put clearly enough in a long leading article in *The Times* of 14 June 1879, the week the international conference opened.

It will at once be manifest that a telegraphic administration admits of being conducted upon two widely different principles. The object of one direction may be either to make a profit by the transmission of messages at remunerative rates, or to cheapen rates with only a secondary regard to profit, for the sake of promoting intercourse by increased facilities and in order to secure certain indirect benefits which may be expected to ensue.

It is the theme of this story. In the view of the leader writer, international telegraphy, for which there was negligible social use, afforded comparatively little scope for the enlargement of business by the temptation of cheapness. It was for mercantile and diplomatic use; the single word and not the ten or twenty word message should become the universal international unit.

It is clear that both shareholders and Governments will eventually be compelled to admit the principle that telegraphy should be paid for, as a rule, by those who send messages rather than by those who do not, and this is the point upon which the granting of increased international facilities must chiefly turn. On the present occasion it is understood that our own Post Office authorities have brought forward some proposals of great apparent liberality, but that these, even prior to any set discussion of them, have been almost universally condemned by the majority of the delegates.

The Post Office had once overdone its liberality to the employer of that leader-writer. The Press sent messages and helped pay for the telegraphic system but, as the Fourth Estate, successfully claimed to be treated as a special case on a par with Governments, and to be charged a lower word rate in view of the regularity and volume of their usage. When the Post Office took over the telegraphs in 1870 they reduced the Press rate to an extent that made that part of their business totally uneconomic. Patey told the Select Committee on the Revenue Department's Estimates of 1888 that the Press rate lost the Post Office nearly £200,000 a year. On the continent the papers, and news agencies like Reuters, set up networks of private wires on which they had the exclusive use for, say, three hours a night four days a week. But in spite of concessionary rates this was costly too.

It had been hoped that human error would be largely overcome by mechanisation. But the Siphon Recorder had not had the effect which Eastern Telegraph had anticipated.* Gaining the confidence of Public, Mercantile Community, Government and Press in the *reliability* of the telegraph should have been a prime objective of the international conference in London, but the dual management of the world's telegraphs by Governments and Cable Companies inhibited a wide-ranging review of *all* the problems involved.

The two 'sides' came together however at the dinner given by their Joint Reception Committee at the Freemasons Tavern on 17 June, over which someone had diplomatically suggested that John Pender should be invited to preside. The chairman of Eastern Telegraph and its associated companies in making the speech of the evening took the opportunity of telling the guests a few home truths. Telegraphy, he said, though already of vast importance, was still in its infancy, and its growth would depend on, in a great degree, the wisdom of the decisions of those who took part in that conference. The commerce of the world was now conducted by telegraphy. Transactions which formerly took months were completed in a few minutes. There were a million and a quarter miles of landlines costing £40 million, a vast network operated by governments. The submarine companies whose home was in England were not entitled to vote at these conferences nor were they admitted to their deliberations, but so far they had had no occasion to criticise any of their decisions. He trusted it would be the same in the future. He hoped they would not hastily recommend any reduction of tariffs to a point that would be barely remunerative. 'Unless the governments are prepared to take over the companies' lines, they must bear in

---

* 19 Sept 1873 – Memo from Ansell to all Superintendents:
'Of course you are aware that these recorder instruments are very costly and the object in introducing them was 3-fold: 1st, in expectation that through them we should be able to economise staff; 2nd, that by the record we should very materially lessen the number of errors; 3rd, that by the reduction in the number of errors, we should considerably lessen the number of repetitions required. Now as far as the results yet attained, (with the exception of Gibraltar), we do not see very much . . . effected in the staff. Again the errors are very little reduced and the repetitions asked for by the public are almost as numerous as they were before the introduction of the recorder.'

mind that investors are entitled to a fair return for the capital invested (hear, hear).' Fifteen years ago there were little more than 2,000 miles of submarine cable; now there were 66,000 representing a capital of £25 million, the bulk of which had been found in England. 'All this has not been done with the aid of governments, but by a handful of men of enterprise.'

There was a plan afoot, he said, to lay a cable to Cape Colony in South Africa. If it had been in existence two years before, the Zulu War might have been avoided; ten millions of money and much blood would have been saved. It was not the British Government's fault, but a House of Commons which was so often penny wise and pound foolish.

Having appealed for recognition of the problems of private enterprise, he ended his oration wearing the cap of the liberal idealist. 'Personally I am a strong advocate for the telegraph being international. I hope the day is not too distant when governments throughout the world will see their way to the absorption of the submarine telegraph companies so as to make one united system of telegraphs.' What Mr Patey and Colonel Champain thought of that aspiration is not on record, but doubtless they had the grace to congratulate him on his speech when they greeted him at the garden party which John and Emma Pender gave at Foots Cray Place in July.

John Pender was well aware that his ideal – absorption of the submarine telegraphic system by the world's governments – could never be achieved 'at a stroke', and in 1879, mindful of the British Government's willingness to help finance the Atlantic Telegraph of twenty years before and the cause of their disengagement from further financial involvement, he set in motion a compromise scheme by which Private Enterprise invited Government to become its partner. This was for the Cape Cable to which he had briefly referred in his dinner speech.

Telcon had a bigger interest in this operation than Eastern Telegraph. They contracted to make and lay the 3,900 mile cable for £950,000 but by an agreement dated 9 May 1879 the British Government were persuaded to grant a subsidy subject to the approval of the House of Commons. The government of Cape Colony agreed to contribute £15,000, and the Natal Government £5,000; another £5,000 came from the Portuguese Government. Eastern Telegraph considered it 'inexpedient' that it should take the entire responsibility for operating the service and formed The

Eastern and South African Telegraph Company for the purpose with
a share capital of £400,000 and mortgage debentures for £600,000.
HM Treasury nominated an official director on its Board.*

Eastern Telegraph already had a submarine cable at Aden and
the new plan involved laying one along the east coast of Africa from
Aden to Durban in Natal which the Government had connected by
landline to Cape Town the year before, the sea being too deep for a
submarine cable. It landed *en route* at the island of Zanzibar, at
Mozambique and Delagoa Bay. It was designed to operate at
fourteen words a minute. The rate from Durban to Aden would be 5s
a word; Aden to London 3s 9d a word; Government messages would
go at half rate, and the Treasury agreed to pay an operating subsidy
of £17,500 a year for twenty years. The Sultan of Zanzibar said he
was unable to contribute financially, but wrote to the British Consul
offering assistance by granting Eastern and South African exclusive
landing rights for fifty years.

The work was completed in 1880 when Eastern Telegraph bought
Telcon's £200,000 stake in Eastern and South African, making their
total holding £390,000. It was estimated that the Cape Cable would
carry twenty messages a day, but in spite of the trouble in Egypt and
the general trade depression, the daily traffic was double that figure.

Many would have thought that Eastern Telegraph and its
associated companies had grown fast enough in eight years and that
there was a real danger of it 'over-reaching' itself. Eastern Telegraph, Eastern Extension and Eastern and South African together
owned and operated 30,184 miles of submarine cable at the
beginning of 1880, and their combined capital was £9,186,000. They
were paying dividends of 5 per cent, but those who had had shares
from the beginning were getting 7½ per cent. For long Eastern
Telegraph shares had stood at a discount of between 30 and 40 per
cent, giving a return of 7¾ per cent, but in 1880 they were at a 10 per
cent discount. There was a world depression and there was no better
barometer of trade than the international telegraphs. John Pender
was the first to admit that investment in ocean telegraphy was
speculative, if only because of the long periods during which, at this
time, cables went silent. A main expense was the use of cable ships to
repair broken cables – 1½ per cent of their capital every year – which
would have had to be met whoever were the operators, governments
or companies. Such annual expenditure would be reduced, of course,

---

* In the list of directors printed in the annual report of Eastern Telegraph of 5 July
1881 there appears for the first time
   OFFICIAL DIRECTOR   }
   Appointed by H.M. Treasury } Charles W. Stronge Esq., CB

in proportion to the extent that submarine cable routes were duplicated.

The collaboration between Government and Companies over the Cape Cable had set a good example, but the Egyptian Government did not follow it in renewing their concession to Eastern Telegraph which expired in 1879. They took over the management of the cables themselves. Then the French Government laid their own cable to their colony of Algeria and charged the same rate from Paris to Algiers as to any other part of France. John Pender was appalled. No one could compete with that. 'The Government of France has not treated us as we think they ought to,' he told his shareholders in January 1880. 'They have appointed a new Minister of Telegraphs of great energy with a determination to make telegrams exceedingly cheap. As he gains experience of our system we hope he will take a broader view and deal fairly and more liberally with us, for he must be aware that if private companies are to exist there must be dividends. Private enterprise has been the great pioneer of tele-graphs, while governments have only in view the reduction of rates to the public, drawing as they do upon their long purses to make up the difference.'

Private enterprise had also been the pioneer of telephones, and by 1880 those in the instant communication business had had time to consider the role telephony was likely to play *vis-à-vis* telegraphy. John Pender was one of the first to recognise its significance and in 1880 became chairman of The Oriental Telephone Company. 'The telegraph companies over which I preside,' he told an Eastern Telegraph meeting in July 1881, 'have important interests in India, and the telephone system I believe must eventually be a feeder to the telegraph system, as the former supplies a want for short distances which cannot so well be supplied by the telegraph. If the Oriental Telephone System had fallen into the hands of the people opposed to telegraphs it might have caused us some trouble. It was for that reason, not from any desire on my part for additional labour, that I associated my name with the Oriental Telephone Company, believ-ing that I should thus benefit the interests I represent in the telegraph companies, and at the same time promote the growth of the telephone system'.

But investors in international telegraphs had more to fear from government competition than any that might arise at this stage from the telephone. Plans for the biggest non-private enterprise sub-marine cable of them all which, as seen, had already been envisaged at the Sydney Conference of 1877 were taken a stage further with the determination in 1879 of the terminal of the Canadian Pacific

Railway and thus of the east to west overland telegraph. In a letter dated 11 June 1879, Sandford Fleming, Chief Engineer of the CPR, called the attention of the implications of these events to F N Gisborne, the early pioneer who was now Superintendent of the Telegraph and Signal Service in Ottawa.

> If these connections are made (he wrote) we shall have a complete overland telegraph from the Atlantic to the Pacific coast. It appears to me to follow that, as a question of imperial importance, the British possessions to the west of the Pacific Ocean should be connected by submarine cable with the Canadian line. Great Britain will thus be brought into direct communication with all the greater colonies and dependencies without passing through foreign countries.

Such plotting in high places showed that many* were ready to think of ocean telegraphy without Private Enterprise in spite of the recent precedents which demonstrated the extent to which Government was in its debt. During the Afghan campaigns of 1878, 1879 and 1880 the British and Indian Governments relied heavily on the 'commercial' telegraph systems. In 1881 the British Government conducted all the negotiations with the Boer leaders in the Transvaal through the telegraph, settling the dispute without resort to arms. Again in 1882 the assistance which the cable companies were able to give Government in the tense situation which built up in the Middle East proved their indispensability.

In his attempt to modernise Egypt, nominally a tributary of the Ottoman Empire which was even more primitive, Ismail Pasha, its ruler or Khedive, had gone too fast. By 1876 he owed the mainly British and French firms who had undertaken his costly irrigation, agricultural, road and telegraph schemes, and the bankers from whom he had borrowed at reckless rates of interest, some £93,000,000. The sale of his Suez Canal shares had been an effort to stave off bankruptcy, and it had failed. As a result, an international Caisse de la Dette Publique was established in Cairo to watch over the interests of foreign investors and handle repayments to foreign bankers. When this proved ineffective the Khedive was forced to submit to a government which included a French and a British minister who in an attempt to return Egypt to solvency imposed strict Financial Control. The Khedive obstructed every measure and the Sultan of Turkey deposed him in favour of his son Tewfik. But the Egyptian landowners, peasants and a rag-bag army under one

---

* A leading campaigner for bringing the submarine cable system under State control was J Henniker Heaton MP who wanted universal penny postage.

Arabi Pasha would have none of him, and at the end of 1881 forced him to dismiss his European advisers and end the fiscal reforms. The British and French fleets sailed for Alexandria, but such a show of naval force had no effect whatever. There were serious riots in Alexandria during 11 and 12 June 1882. Anyone British of course was a target for attack, and the staff of the Eastern Telegraph Company had lucky escapes. An eye-witness account of one of the telegraph operators in Alexandria during the riots tells how

> six of us left the quarters about 3 in the afternoon, and went for a row round the men-of-war and returned about 6. We were going to land when we accidentally met some officers of the ships who warned us not to go on shore, so we were miraculously saved from being probably all killed. We went and stopped on board HMS *Invulnerable* where we were all well treated by the officers. We came ashore on Monday afternoon and were escorted to the office by an escort of soldiers and two officers with fixed bayonets.

In this letter home the writer said the amount of work occasioned by the situation was 'something fearful'. They all had to do double duties besides taking their turn as sentries for two hours at a time. They had blockaded the quarters 'and intend to defend it as long as we can'.

On 9 July when Arabi Pasha refused to surrender his shore forts, the British Navy bombarded Alexandria; on the 12th the Eastern Telegraph's telegraph station was burnt down and one of their clerks, a Frenchman called Ternan, murdered. That same day, in the words of another letter-writer,

> Rainier, one of the clerks with three outsiders started to come down to advise us not to come ashore [they had once again set out in rowing boats to visit the fleet in Alexandria harbour]. The three outsiders were all killed on the way, and Rainier, by a wonderful miracle, escaped with severe scalp wounds on the head and a damaged nose. Our washerwoman who had a fortnight's washing of ours had her house broken into and nearly everything stolen, so we are in the lurch.

While the town was being bombed, Eastern Telegraph men took the Alexandria end of the cable and put it on board their cable ship the ss *Chiltern* and telegraphed to London an eye-witness, minute by minute account of the destruction of Arabi's forts as it took place.

In spite of temporarily dirty linen, morale managed to remain high, as recorded by the First Lord of the Admiralty who visited the stations shortly afterwards and told John Pender, 'I never was more proud of my country or an institution connected with it, than on my

visit to your offices in the East and in the Levant . . . The young men
are gentlemen. I saw the place as compared with other offices there,
and it was clean – so tidy and businesslike'. Old Man Ash had
obviously cured them of the old Porthcurno habit of spitting into the
fire grate, but it seemed that escapades had changed from group and
anti-social to lone and self-destructive. In 1879 Probationer Baglehole (father of the archivist?) fell to his death from the cliffs near
Land's End while trying to reach a bird's nest. But there was
nothing they could learn in Cornwall to stem the effects of cholera
which played havoc with the company's staffs in Egypt at this time.
To help them recover and keep their spirits up they were given extra
allowances, and many were brought home to convalesce at the
company's expense. Maintaining full staffs in such circumstances,
and providing operators for John Pender's expanding empire, put a
great strain on Porthcurno. In the mid-1880s there were more than
thirty probationers learning the job and as many trained staff
operating the terminal whose responsibilities had been increased by
having to maintain the through connections of the cable laid in 1878
to the Scilly Islands owned and operated by the Post Office. Though
of course this was work for 'engineers' who were rather looked down
on by 'operators' – an attitude which persisted for some time.

Lord Alfred Paget, a director of Eastern Telegraph, also visited
the Eastern Telegraph's Mediterranean stations in his yacht. At
Chios he found another wrecked telegraph office, but the result of a
natural disaster, the earthquake which had shaken the building to
pieces. But Lord Alfred was able to report that though the station
was in ruins, the operators had erected a wooden hut over the
terminal and were working the line with customary efficiency. Later
Pender toured his branches in the same area in the cable ship *Volta*
with a party that included Sir James Anderson and the famous
Captain Shaw of the London Fire Brigade.

But such peaceful inspection was no longer possible when Arabi
declared a holy war against the British. Much against anti-
imperialist Gladstone's inclinations, the Cabinet decided to send an
expeditionary force to Cairo. Operations began on 18 August and
the Egyptian army was defeated at Tel-el-Kebir on 13 September.

John Pender felt the Egyptian War, and the service he had
rendered the Government during it, had done a great deal to
increase the respect which politicians held for Eastern Telegraph.
Pender considered the company had made great sacrifices in the
Egyptian War. It had cost it £62,000. But it was worth it. 'It has
given us a very important position not only with the British
Government but also with other governments. At this very moment

[July 1883] another nation – a friendly nation – is engaged in war in the Far East. What do they do? They know the benefit we conferred on our own government by the telegraphic communication we brought to bear in assisting them, and the French Government have therefore come to us; and I believe at the present moment there is a convention signed by that government which will enable the Eastern Extension Company to carry the telegraph system from Saigon to Tonkin in French Indo-China'.

He claimed, too, that the Eastern Telegraph was an important factor in the success of the Suez Canal. Before there was a telegraph, ships carried their own letters with orders for the cargoes they were to pick up. But now a telegram was sent to their destination setting up the cargo to be collected before it sailed.

He believed that the Egyptian War would not have been as short-lived as it was had it not been for the facilities which the Commander-in-Chief had through the telegraph, enabling him to maintain communication with the Admiral and other branches of the expedition 'in such a way as to continue the campaign as a compact whole. I believe the telegraph did signal service at the time and I am glad to say in doing this service I believe it was appreciated to a certain extent'.

Governments have always been reluctant to risk public money on unproven ideas – and with some reason – and for some years after their impetuous backing of the Atlantic Telegraph, HM Treasury refused to finance submarine telegraphs further than across the English Channel and North Sea. When the Post Office took over the internal lines in 1870 they leased the cable they acquired from Cooke's Electric and International company to the Submarine Telegraph Company which already had a cable to Belgium and France. The service to Europe provided by the Submarine Telegraph company was as much a Post Office service as the telegraph to Ireland, the Scillies and the Channel Islands, and the Cable Companies were happy that HMG should continue to confine its interest in submarine telegraphy to this one company. But by the 1880s instant communication by ocean telegraphy, whose practicality private enterprise and finance had now proven, had become too important a factor in foreign policy – and, when diplomacy broke down, military strategy – for a government to remain aloof. Nowhere was this more apparent than on the continent of Africa where Portugal, Germany and France were already establishing settlements. It was not a matter of 'interfering' with a profitable commercial activity for the increased revenue it could bring the public purse, or influencing overseas trade. It was merely that the determination of a new route was too sensitive a decision to be dictated by the advantages it would bring to 'proprietors' concerned with a maximum return on their investment.

When in 1885 HMG decided to take the lead as regards another cable to Africa, John Pender could not but follow, and with as good grace as he could muster. He explained the new situation to his shareholders as 'following the flag', an emotive phrase rightly calculated to win support. 'We have always considered it important that we should follow the English flag. Wherever the English flag floats trade is likely to be developed.'

The Mother Country, declared the Government, needed a link with the west coast of Africa, and invited tenders for the construction of such a system. There was already a line from London and Porthcurno to St Vincent in the Cape Verde Islands off the western bulge of Africa. It formed the final section of Brazilian Submarine's cable from Pernambuco. Eastern Telegraph tendered jointly with Brazilian Submarine and were awarded the contract. They formed The African Direct Telegraph Company of which they contributed a third of the capital. The British Government agreed to subsidise it with £19,000 a year for twenty years (representing the minimum annual traffic it would guarantee to bring it at Government rates). African Direct had Telcon lay a cable from St Vincent to the nearest point on the African coast, Bathurst in the Gambia, and then to the British territories to the south: Freetown in Sierra Leone, Accra in the Gold Coast and Lagos and Bonny in Nigeria. The laying was completed, and a telegraph service with Porthcurno-trained operators was in working order by September 1886.

At about the same time Telcon promoted The West African Telegraph Company to lay a cable from Dakar in Senegal to Bathurst, and connecting with French, Portuguese, Spanish and German territories down the coast as far as Loanda in Angola. In 1889 West African joined the Eastern Telegraph Group, and made a single system with African Direct and Eastern and South African, providing a duplicated submarine route from London to the Cape.

Collaboration was shown to work. The British Government were happy, and so were Eastern Telegraph investors. The Brazilian Government, however, were not as ready to collaborate with Western and Brazilian and Brazilian Submarine. When the former were refused a grant to lay a cable to Africa, they suggested the government should buy the company. When Brazil refused to do this, the company proposed a joint purse agreement which they also turned down.

In 1885 Baron Capanema, Director-General of Brazilian Government Telegraphs, came to London to sort matters out with the English directors of the two South American companies. His government had built competitive land lines and his intransigence stemmed from resentment that Western and Brazilian had reduced the rates of their submarine telegraph to those of the government's system. Capanema rejected all schemes for an alliance between Government and Private Enterprise and returned to Rio with his sense of grievance still burning. But Britain's link with her informal empire in South America had been reinforced in 1884 when Brazilian Submarine laid a second cable to St Vincent and Portugal from where cables already led to Porthcurno.

In that year too a Royal Commission appointed by HMG had recommended a link via all British territories to the islands of Britain's formal colonial empire that lay to the north-east of the South American continent, the West Indies. But the greatest pressure to strengthen the telegraphic communications of the British Empire came at this time not from the imperial government but from colonial governments – on the other side of the world, in Australasia.

Up to 1883, when they formed a Federal Council, the three colonies of New South Wales (the original), and breakaway Queensland and Victoria, had been aggressively separate, with railways of different gauges and tariff walls. But the expansion of Russia to the North Pacific, the appearance of France and Germany in Australasian seas and the fear that the latter might seize New Guinea, had gone far towards discarding provincialism and creating a sense of common purpose, though Australians' realisation of their common kinship was the main factor. Inspired by a great statesman, Sir Henry Parkes, the colonies acquired a sense of nationhood and came to appreciate the benefit of belonging to a united Empire.

Unity of the British Empire meant different things to different people. At one extreme were the Imperial Federalists who worked for the ideal of a Parliamentary Federation of the Empire. But for the majority of immigrants the idea of a legislative federation binding them to the United Kingdom was anathema. Close co-operation and formal empire unity was as far as most of them were prepared to go. Their ideal was Alliance with the Mother Country, not Federation.

But whatever the political bond, commerce demanded cheap and reliable instant communication with the financial and business centre of the world which was London, and the system of the British Australian Telegraph Company was neither. The rate of 10s 8d a word was prohibitive, and the odds that all sections of the circuitous route via Java, Singapore, across India by land line, up the Red Sea, through the Mediterranean and the Bay of Biscay to Porthcurno, would all be working at the time you sent your message were very small.

In 1883 at another Australasian Conference in Sydney the Postmaster-General of New South Wales tabled a scheme for laying a submarine cable from New Caledonia (to which there was already a cable to the Australian continent?) to Fiji, Honolulu and San Francisco. It was a project, he said, in which all the colonial governments concerned could participate by subscribing to the £3 million capital of the company they would form to build and operate the telegraph. The proposed rate was 2s 6d a word. When John Pender heard of this he feigned indifference. Charge what you like, he told them, so long as you guarantee Eastern Telegraph's present level of earnings. But when a commercial rival appeared on the scene in the form of The Pacific Telegraph Company Ltd he trounced Audley Coote for the 'fallacious' figures he produced in support of the argument with which he hoped to win British Government support. Australia to San Francisco via Norfolk Island, Fiji, Samoa and Honolulu was the route proposed by the promoters of this company who included the Duke of Manchester, Sir Charles Bright and Latimer Clark. But HMG would not be tempted and the company went into abeyance.

Meanwhile Sandford Fleming never stopped dreaming up alternative routes. What about following the coast of the Aleutian Islands off Alaska and then to Japan and down? he wrote to Sir John Macdonald, Prime Minister of Canada on 20 October 1885. The British Colonial Office showed little interest apart from suggesting to colonial representatives who met in London for the Indian and Colonial Exhibition of 1886 that if they insisted on a Pacific Cable why not start it from somewhere in British Columbia rather than American San Francisco? But it would be *their* project; HMG could play no part in it. They gave no encouragement to the document, if indeed they ever saw it, which Sandford Fleming on a visit to London in July gave to Sir Charles Tupper, Canadian High Commissioner, which was a general assessment of his plan for a Pacific Cable and the principles on which a company should be formed to administer it. John Pender kept a watchful eye over all the schemers, and picked up tit-bits of information from his fellow-delegates at the Congress on Deep Sea Telegraph Cables which he attended in Paris in May 1886.

When he returned to London, John Pender heard that a deputation from the Imperial Federation League had asked Lord Salisbury to call a Colonial Conference in 1887 which was to be Queen Victoria's Golden Jubilee. Occasional meetings of colonial representatives were seen as satisfying the advocates both of Parliamentary Federation and the looser Alliance in achieving the 'concert of governments' which would complement continuing consultation by letter, telegraph and diplomatic agent. This first conference of 1887 was confined to Self-Governing Colonies – only Canada had established a national government at this date. A main object was to co-ordinate local defence and induce colonies to share the expense of the empire as a whole. But in his preliminary circular Sir Henry Holland, the Tory Colonial Secretary, said he hoped they would discuss not only military and naval defence but vital peace interests such as 'the promotion of commercial and social relations by the development of our postal and telegraphic communications'.

John Pender was no longer in the House of Commons in 1887. He had lost his seat for Wick in 1885. Like many back-bench Liberals he objected to the Home Rule for Ireland Bill which his leader, Gladstone, introduced in 1886. He contested Stirling in that year against Henry Campbell-Bannerman, the future Liberal Prime Minister, as a Liberal-Unionist – to demonstrate his view that Ireland should remain united with Britain. But he lost, and remained in the wilderness until 1892 when he was re-elected for Wick.

In the activity from which he derived the most satisfaction, however, prospects had never been better. In its fifteen years of existence the growth of Eastern Telegraph had been phenomenal. In 1872 it owned 8,860 miles of submarine cable; owned or rented 1,200 miles of land line; had twenty-four stations and two repair ships. Its capital was £3,397,000 and gross annual revenue £376,900. But by the end of 1887 it owned 22,400 miles of submarine cable and had sixty-four stations – seven in England, one in Scotland, three in France, three in Spain, five in Portugal, five in Egypt, ten in Greece, thirteen in Turkey, one in India, two in Austria, seven in Cyprus, one in Russia, one in Morocco, one in Barbary, one in Gibraltar, one in Malta, one in Perim Island at the foot of the Red Sea, and one in Aden. The capital was £5,900,000 and gross annual revenue £650,971. The following year they increased the capital by another £200,000 to duplicate the South African system.

It was unfortunate that he was not a Member of Parliament during the conference at which the Pacific Cable put down its roots; but perhaps it left him freer for the skilled extra-Parliamentary agitation at which he was so adept. He certainly saw that members of Lord Salisbury's Tory administration were aware that any upset of the current imperial communications scene would meet with his greatest displeasure. In his very opening speech to the conference on 4 April, Sir Henry Holland referred to the possibility of an Australia-to-Canada cable and how it had always been opposed by the companies which owned the existing telegraph lines. 'A very strong case would have to be made to justify Her Majesty's Government in proposing to Parliament to provide a subsidy for maintaining a cable in competition with a telegraphic system which, at any rate, supplies the actual needs of the imperial government.'

Before the conference began, Pender lobbied not only the Colonial Office to which he had entrée, but wrote to the PMG of every Australian colony offering to reduce the tariff from 9s 4d (to which the last International Conference had reduced it from 10s 8d) to 4s a word, providing they guaranteed to his company its average receipts during the preceding three years. His routes, he pointed out, were the most secure in peace-time and the most easily protected in war. During the conference itself an eye-witness described him as being 'in constant attendance, buttonholing the delegates and exerting his influence both inside and outside of the Conference'. But he relied on the Colonial Office to keep him officially informed and wrote to Sir Henry Holland:

> Our telegraph system is now very much in touch with Her Majesty's Government, and we have letters from the Foreign Office to the effect that whenever discussions take place in regard to submarine telegraphs we shall have full information on the subject, and representation during such discussions. I, therefore, hope that the Colonial Office, looking to the vast interests involved in the submarine telegraphic system, will grant to my companies similar recognition on the present occasion.

On the second day of the conference, Sandford Fleming was invited to state his views. A telegraph map of the world, he said, showed Britain's dependence on foreign powers for the security of her communications with Asia, Australia and Africa – particularly the friendship and protection of Turkey. 'Were a cable laid across the Pacific from one British island to another, not only would there be a communication with Australasia but by the cables of the Eastern Telegraph Company, India and Africa would equally be in touch with the centre of the Empire without dependence on any line passing through a foreign country.'

This was the crucial difference between having telegraph routes determined by profit considerations and by national policy.

SIR JOHN DENISON-PENDER, K.C.M.G.

15 Sir John Pender's son by his second wife Emma (*nee* Denison), Sir John Denison-Pender (*above*) became Managing Director of Eastern Telegraph on the death of Sir James Anderson in 1893; and Chairman in 1917.

16–17 Main cable station building, single staff quarters and superintendent's house at Porthcurno at the end of the last century, with the Eastern Telegraph Company's house flag flying from the pole by the tennis court. (*Below*) a group of probationers (in their worn-out garments?) and staff.

18–19 On the death of Sir John Pender in July 1896, the Marquess of Tweeddale (*above left*) took over as Chairman of the Eastern Telegraph group of companies. As Lord William Hay he had been a director for many years. He was succeeded in 1900 by Sir John Wolfe Barry (*above right*), a distinguished civil engineer.

0 The sandy beach of Porthcurno bay would become a hive of activity as another shore end of a submarine telegraph was floated in on buoys from a ship anchored out at sea. As the U.K. terminal of the Eastern Telegraph Company's imperial cables, Porthcurno became a place of national importance. Here a cable comes ashore from the *Colonia* one hot August afternoon in 1906; from Porthcurno the telegraph was extended to London over a line rented from the Post Office.

21 Teamwork, too, was the key to the successful landing of the shore end at its New Zealand terminal of the submarine telegraph from Sydney, Australia. Here the *Edinburgh* and the *Hibernia* complete the lay between Botany Bay and Nelson in April 1875.

22 On 20 July 1894, two years before his death, Sir John Pender presided over a banquet and fête held at the Imperial Institute in London to celebrate the twenty-fifth anniversary of the Establishment of Submarine Telegraphy with the Far East 1869–94; the cover of the souvenir booklet.

23 The southern end of the Pacific Cable built in 1902 link the west coast of Canada with Australia – the fi serious challenge to the Eastern Telegraph monopoly landed on a beach at Southport, Queensland. It was co nected to terminals inside this hut erected by the Paci Cable Board. 'It is particularly requested,' states notice, 'that it may not be interfered with in any wa Should the action of the waves cause the cable to beco uncovered it is of great importance that it shall not meddled with.'

24 When in June 1900 a Chinese secret society, The Boxers, organised a revolt against dominance by 'foreign devils', they seized all the telegraph lines inside Peking – seen here in a contemporary illustration in *Le Petit Journal*. John Denison-Pender dispatched a ship loaded with cable so that stations could be set up at Cheefoo, Taku and Wei-hai-wei to keep the world informed of what was happening.

25 The electric telegraph brought great benefits to the West Indies and Bermuda. Here is the office at 6 Front Street, Hamilton, Bermuda in 1900. The two companies were later absorbed in the Cable and Wireless Group.

26 A submarine telegraph service which enabled the Colonial Secretary to communicate with the Governor of the British colony of Hong Kong was an early provision of the Eastern Telegraph group. Here a pig-tailed local resident looks down on Victoria, Hong Kong Island (now filled with skyscrapers) and the Eastern Telegraph offices – the building squeezed between two higher edifices beside the waterfront.

27–8 A threat to the Eastern Telegraph monopoly was the new form of telegraphy without wires devised by Guglielmo Marconi (*above*) who came to England in 1896. (*Below*) Marconi sitting in the hut at Signal Hill, Newfoundland, in December 1901 with the instruments used to receive the first wireless signal across the Atlantic from Poldhu in Cornwall.

29 Sir Sandford Fleming, Chief Engineer of the Canadian Pacific Railway, who conceived the idea of a Pacific Cable linking Canada (Vancouver) and Australia, to be planned, built and operated by colonial governments, breaking the ocean telegraphy monopoly of the private enterprise, London-based Eastern Telegraph group. He began his campaign in 1879, but the telegraph service of the Pacific Cable Board did not open until 1902.

Fleming then referred to 'the opposition'. The conference had seen John Pender's statements showing the scheme was impractical on physical grounds; that, if capable of realisation, it would be a financial failure. Eastern Telegraph were ready to charge as low a rate as that promised by the Pacific Cable promoters for a smaller subsidy than they were asking for. Pender had claimed, moreover, that it would be a great injustice to the company, which had been the pioneer in establishing telegraphic connection with the Far East, if the British or Australasian governments were to subsidise a rival line. Instead of a Pacific Cable benefiting the colonies, said Pender

> I believe that the laying of such a line would only benefit its promoters, and would be inimical to the interests of the telegraphing public, as it would inevitably lead to a war of tariffs which would eventually impoverish both the Pacific and existing cables, and result in a starved and inefficient service, the only remedy for which would be higher tariffs or much larger contributions from the colonies.

Fleming maintained that the practical difficulties of laying the cable had been exaggerated, and that the greatest depths would be in the region of 3,000 fathoms, not the 12,000 fathoms or thirteen miles hinted at by Mr Patey of the Post Office who seemed to have been impressed by Mr Pender's arguments. The Eastern Telegraph Company should be treated with consideration, but it had no right to the monopoly of the telegraph business with the East.

> This is not the first time that a company or an individual has been called upon to relinquish a monopoly found to be inimical to the public welfare. Is it for a moment to be thought of that Canada or Australia are never to hold direct telegraphic intercourse because a commercial company stands in the way? Are commercial relations between two of the most important divisions of the British family forever to remain dormant in order that the profits of a company may be maintained . . . Is the peace of the world to be endangered at the bidding of a joint stock company? It is time for the British people scattered round the world to set about putting their house in order.

H C Raikes, the British Postmaster-General, came out in support of Fleming's ideas, but predictably refused to promise the active concurrence of the British Government. 'It would be a matter of extreme difficulty, I think without precedent, for the English Government itself to become interested in such a scheme, in such a way as to constitute itself a competitor with existing commercial enterprise carried on by citizens of the British Empire.' The Prime Minister of Queensland said Eastern Telegraph had done good work but they were now considering the matter from a national and imperial point of view. He had the warmest sympathy with Fleming's plans. The representative of Cape Colony said the matter must be viewed from the point of view of the safety of the Empire in time of war. If war broke out, the Suez Canal would be blocked. In future not Constantinople but Table Bay would be the most critical point for the Empire, which was dependent for its telegraph on the Eastern Telegraph line laid in shallow water and thus very vulnerable, down the east coast of Africa.

John Pender thought it was up to him to counter-attack with a positive and precise suggestion. He handed into the conference a letter on Eastern Extension writing paper in which he said he was willing to fix his company's charges on the existing London-Australia cable at whatever amount might be required, say 4s a word, provided the colonies would guarantee any loss which might result from that reduction. On this the Chief Secretary of Victoria observed that whether the Pacific Cable project was adopted or not, the debate had already conferred a considerable benefit on the Australian colonies by bringing the Eastern Extension company to a much more liberal frame of mind.

At its next session, John Pender was asked to address delegates. He said he had nothing to add to what he had set out in the papers they already had. If they were interested in cheap telegrams they could be had cheaper through his system than any other; his companies, too, answered strategical requirements to a very large extent.

> I hold that we being the pioneers of telegraphy are entitled to full consideration. If other competing companies are to come against us, let them come unsubsidised as we are, and I have no fear of meeting such competition.

On being recalled to make his final observations, Sandford Fleming said he thought it most desirable that *all* cables communicating with Australasia and the telegraphs within Australia should be under one management. The Australian colonies should control not only the proposed cable to Vancouver but that portion of the existing system from Australia to India known as the Eastern Extension. The colonial governments could not expropriate what was private property, but possibly arrangements could be made mutually fair to both public and vested cable interests. He considered £4,000,000 would be sufficient to cover the whole cost of establishing the Pacific line and buying out the Eastern Extension company's property on fair and reasonable terms.

The Colonial Conference of 1887 ended with delegates passing two resolutions. The first was that completion of the Atlantic-Pacific railway telegraph across Canada opened a new and alternative line of imperial communication over the high seas and through British possessions, which promised to be of great value alike in naval, military, commercial and political aspects. The second was that a Pacific Cable was a project of high importance to the Empire, and every doubt as to its practicability should be set at rest by a thorough and exhaustive survey.

No view was expressed who should make the survey, but Sandford Fleming went to the Colonial Office next day with a letter signed by 21 of the delegates asking HMG to have the Admiralty do it. The Admiralty said they were prepared to take soundings over the next few years in the course of routine hydrographic surveys but not to send a vessel for the specific purpose of surveying the Pacific Cable route. When the following year Australia proposed that the United Kingdom, Canada and Australia paid a third each of the cost of the survey, HM Treasury said they first wanted an estimate. None of the Australian colonies were ready to contribute to the cost of a survey, and to John Pender's delight the whole Pacific Cable Project was shelved for nine years.

Apart from being the golden jubilee of Queen Victoria's accession, 1887 was the fiftieth anniversary of The Telegraph, and to mark the occasion the Postmaster-General presided over a banquet attended by all the leading telegraph engineers and electricians of the day. Cooke and Wheatstone, whose patent of 1837 the jubilee celebrated, had died in 1879 and 1875 respectively. John Pender celebrated his thirty years in telegraphy the same year, and his achievements in this field were recognised by the state when, in the Queen's Jubilee Honours, he was made a Knight Grand Cross of the Order of St Michael and St George. His friends and colleagues drank his health at a banquet given in his honour at the Hotel Metropole in London on 25 April 1888. In proposing the toast of the evening Lord Derby, who presided, said Sir John was the man to whom more than any other they owed the present development of the telegraphic system of the world. The commerce of the empire involving a thousand million sterling a year was influenced and controlled daily by cable communication, and yet the development of it was only in its youth. There was now as much cable submerged as would go five times round the world. That had been done almost entirely by private enterprise. The cost had been £36 million to the companies and £4 million to governments.

The only shadow on the knighthood celebrations was The Portrait. A leading painter of the day, Hubert Herkomer, had been commissioned to paint Sir John's likeness in oils. Emma Pender who held 'artists parties' at Arlington Street before each Royal Academy, and was the intimate friend of John Millais,* reckoned she knew a good portrait when she saw one and took an instant dislike to Herkomer's picture of her newly ennobled husband. She considered it 'coarse, careless and bad'. She wrote to Marion:

> I have written Herkomer and told him 'That the pictures represent a much heavier, older and more commonplace man; are wanting in expression of the vigour and strength of purpose which mark Sir John Pender's features so plainly'.†

What could the artist do but bow to this formidable woman's opinion and start all over again? And happily this time his canvas pleased.

In this year of 1888 in which Pender's monopoly reached its peak, his acknowledged mouthpiece Sir James Anderson, deputy chairman and managing director, came out in favour of Government purchase of the cable companies. The occasion was a lecture given by John Henniker Heaton at the Royal Colonial Institute on imperial post and telegraph communication. In the discussion that followed, Sir James attacked Henniker Heaton, and according to the latter, who quoted it in his evidence before the Inter-departmental Committee of 1901, said

> With regard to what Mr Heaton said about Government control, I claim to have been the first to put that into his head. At the Berlin conference I proposed the substance of a plan with which Mr Heaton was very much pleased; and if that plan were adopted the Government might take over the telegraphs paying 4 per cent on the present capital. Keeping up the present tariff they would have a quarter of a million to do what they liked with. They could reduce the tariff one half and still make a profit. Private companies cannot do that.

---

*Sir John Millais (1829–96) painted Emma Pender's portrait and a picture of her two daughters Marion (Lady des Voeux) and Anne both wearing red slippers, sitting beside a golden fish bowl. The present Lord Pender's mother has the portrait of her great granddaughter but sold the other picture to the Walker Gallery in Liverpool in 1975.

† Her full letter to Herkomer dated 26 April 1888 is at Appendix B. He was asked to paint an original and a 'replica' to present to the telegraph company's board of directors; hence 'the pictures'.

His proposition was apparently made at Berlin in private conversation with Henniker Heaton who was also attending the conference; but now he was making it public. He was, however, only echoing the ideal that Sir John himself had expressed on frequent occasions beforehand.

For five years no action had been taken on the recommendation of the Royal Commission of 1884, appointed following the Telegraph Congress held on Barbados two years previously, that there should be a direct cable between the West Indies and Halifax. But in 1889 the British Government decided to go ahead with the northern section of this project and link the important naval base on Nova Scotia with the equally important British naval harbour of Hamilton on the mid-Atlantic island of Bermuda some 800 miles off the coast of Georgia. HMG invited tenders for the cable and for operating a service which would be primarily imperial if not strategic, and only to a less extent commercial. They promised an annual subsidy for twenty years. A commercial group formed The International Cable Company Limited to bid for the concession, and when they won it they assigned it to The Halifax and Bermudas Cable Company Limited. Henleys* made and laid the cable which arrived on board their ship the *Westmeath* at Grassy Bay, Bermuda on 8 June 1890.

On 11 July the first messages sped along the new cable and through the Atlantic Telegraph from Bermuda to England with 'Greetings to our Beloved Queen', the Colonial Secretary and other worthies. The day it was opened for public traffic – 14 July 1890 – saw the band playing on Front Street, a firework display from White's Island and a dinner given by the manager of the Halifax and Bermudas Cable Company at the Royal Bermuda Yacht Club for 'representative gentlemen' on the island and those immediately involved with the telegraph.

The West Indies had to wait for their direct submarine link with Europe, but unlike Bermuda the Caribbean islands, as seen, were already connected to the west by the cable to Florida. Sir John Pender played no part in this further example of collaboration between Government and Private Enterprise, nor was he the commercial partner in the operation being projected on the South African continent by someone who, like Pender, had a foot in both

---

*W T Henley had died in December 1882 aged 68. During his career he made and laid some 14,000 miles of submarine cable. His telegraph works at North Woolwich spread over twelve acres and employed 2,000.

political and business camps, but at a very much higher level – Cecil Rhodes.

Rhodes, a British businessman born in Hertfordshire in 1853, went to the Cape of Good Hope in 1870, and by 1888 had monopolised the Kimberley diamond fields in that colony. At 35 he controlled 90 per cent of the world's diamond production, and his interest in the Transvaal goldfields brought him £400,000 a year. His wealth enabled him to acquire a dominating position in South African politics. 'Philanthropy is all very well in its way,' he once said, 'but philanthropy plus 5 per cent is a great deal better.'

There is no record of his having met Sir John Pender when he came to England at the end of April 1889, but it is likely that he did so – and the two would have got on famously. He came to present to the British Government a scheme for developing the areas to the north of the three Boer republics of Transvaal, Natal and Orange Free State, on the other side of the river Limpopo known as Bechuanaland, Matabeleland and Mashonaland, and to extend the railways and telegraphs to the river Zambesi. His plan was to form a British South Africa Company on the lines of the Imperial East Africa Company, and he was granted a royal charter for this venture in October (1889). A clause prevented him from establishing 'monopolies in trade' but railways and telegraphs were specifically excluded from that phrase.

In November he wrote a 'Most Confidential' letter to the British Foreign Secretary in which he offered to extend the telegraph from Salisbury, capital of Mashonaland, to Uganda which HMG were considering taking under their protection (some 1,200 miles), without asking for any government contribution. He would have to obtain permission from the German Government to take the line through German East Africa, and hoped that HMG would do this for him.

> I may add that my ultimate object is to connect with the Telegraph System now existing in Egypt which I believe extends as far as Wady Halfa, but I am fully aware that under existing circumstances at Khartoum such an undertaking cannot at present be carried out.*

Before Rhodes went back to South Africa in 1890 to become Prime Minister of Cape Colony, he ordered 250 miles of telegraph wire from a London manufacturer for the first part of his transcontinental line from Mafeking in Bechuanaland to Bulawayo in Matabeleland.

---

*Quoted by Sir Lewis Michell, *The Life and Times of the Rt Hon Cecil John Rhodes, 1853–1902* (Edward Arnold, London, 1912).

'The railway is my right hand,' he declared, 'and the telegraph my voice.'

Sir John Pender must have envied Rhodes's breadth of vision and the panache with which he projected it to all and sundry. But Eastern Telegraph's position in Africa, with submarine cables down each coast, was a strong one, and he now made it even stronger. In 1889 he had a second African line laid – 'the largest thing that has yet been done in connection with submarine telegraphy'. No other company could attempt to do anything of the kind, he told his shareholders on 16 January 1890. While other countries were striving hard to get a footing in Africa they were the pioneers of the whole thing. At every point of the African seaboard they were in communication with Britain. Their policy, he reiterated, was to follow the flag of England. Africa was in a state of transition and dominated by two British companies. So Eastern Telegraph had laid a cable from Zanzibar, which became a British Protectorate in 1890, to Mombasa, headquarters of the Imperial East Africa Company. Two years later HMG was giving Eastern Telegraph an annual subsidy of £28,000 for twenty years to lay and operate a cable from Zanzibar to the Seychelles and Mauritius.

Sir John welcomed the subsidies and relished the large profits which followed in their wake. But in thanking shareholders for their condolences on the death of Lady Pender which occurred in 1890 he concentrated on the company's higher role in society – 5 per cent plus philanthropy. Their system, he told them, had become probably the most useful and civilising institution that existed on the face of the earth. The work they had done in promoting political, commercial and social progress throughout the world could not be measured.

And he was very conscious of the extent to which the system depended on the people who operated it, and in 1891, the year Eastern Telegraph came of age, he introduced staff assurance for employees who had served ten years and more. 'Manipulators' could retire at 50, but not ships' captains.

The constant high temperatures of places like Pernambuco and Alexandria will have been more injurious to health than the occasional blizzard of the kind that struck Porthcurno in March 1891, and Sir John and his advisers were well aware of this. Announcing the new scheme the chairman said staff never had to work in very pleasant places since the bulk of them were near to where the cable landed. 'The consequence is that they live a quiet and lonely life, and not a life which tends very much to health or life itself.'

It was quiet except for the ticking of the telegraph instruments which carried the messages without voice or feet; but the stillness was soon to be broken, and to the monotonous keying was about to be added the rising and falling tones of human speech.

In 1889, W H Preece, Assistant Chief Engineer and Electrician of the British Post Office, had designed a twenty-eight mile telephone cable to cross the estuary of the River Plate and connect Buenos Aires and Montevideo. He had then forecast that telephonic communication between London and Paris, a distance of 275 miles, was perfectly practicable. The Scientific Summary of Whitakers Almanack of 1889 quoted him as saying that in telephonic matters mere distance scarcely entered into the matter at all; the difficulties arose from the character of the materials used, and the presence either of underground wires or submarine cables.

The French Minister of Posts and Telegraphs had indeed expressed a wish for telephonic communication between the two countries, but Preece had found it impossible to carry speech over telegraph wires, and a main difficulty was the relatively high

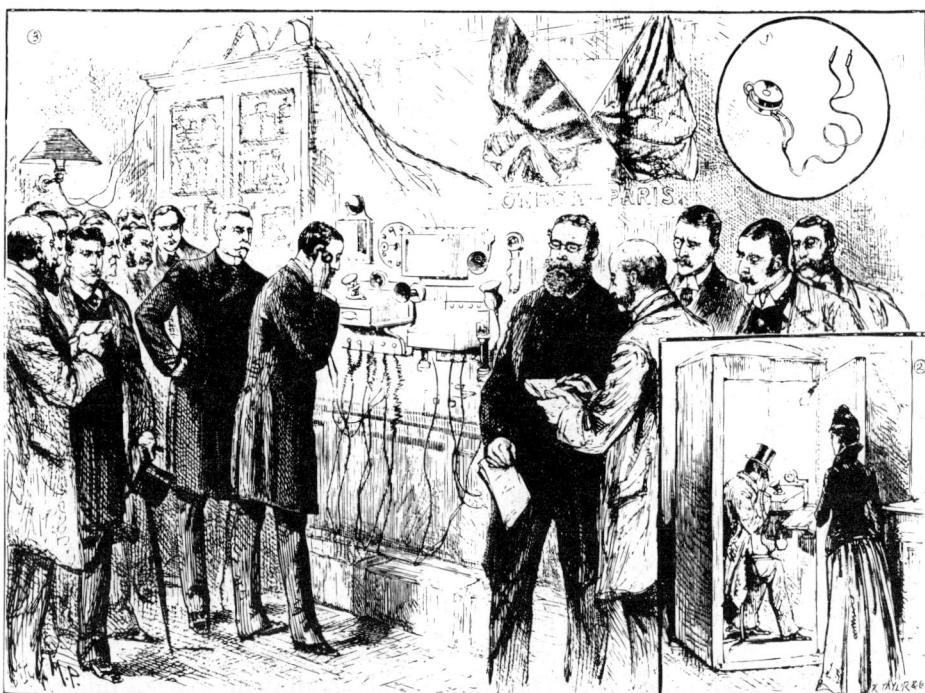

Talking to Paris from the General Post Office in London via the world's first commercial submarine telephone cable in 1891.

capacity of the submarine cable section under the Channel. But a member of Preece's staff, H R Kempe, made experiments, and Siemens Brothers manufactured a telephone cable to his specification at their Woolwich works. The Paris-London telephone service opened on 1 April 1891 with the Prince of Wales speaking to President Carnot. The charge was 8s for three minutes, and most of the traffic, which was large from the outset, was between the London Stock Exchange and the Paris Bourse. Soon there were 250 calls a day. It was the world's first submarine telephone cable.

No research into submarine telephony, as far as can be seen, was made by any of the Cable Companies in Britain. Apart from Sir John Pender's association with the Oriental Telephone Company, Eastern Telegraph chose to ignore the possibility of long distance, international speech communication. In 1891 an American, Silvanus P Thompson, had taken out a patent for ocean telephone cables; but in Britain it was HM Treasury who put up the money for experiments in ocean telephony and the Post Office who carried them out.

4

Bewildering Possibility

1892–1897

It was telegraphy that Sir John Pender had persuaded his share-holders to invest in, and he saw no reason for involving them or their money in any other form of long distance communication. He needed all his powers of concentration to outwit Cecil Rhodes whose grandiose scheme in Africa certainly gave cause for anxiety, though he was well aware that its execution depended on finding the gold reef which was supposed to lie beyond the Limpopo. To his shareholders Sir John could not but praise 'the great ability and adventurous courage' of Rhodes in opening up Africa and in proposing to carry across it up to Cairo a telegraph system which he thought would enable him to transmit telegrams from the Cape to Europe for 2s 6d a word against Eastern Telegraph's 9s. But the proposed line would pass through countries controlled neither by Rhodes nor the British Government. The Egyptian revolutionary leader who called himself the Mahdi had part of it too. By the time Rhodes had got his line to Wady Halfa, Eastern Telegraph would have paid off their debentures and have more capital than the amount Rhodes needed for his African Telegraph. If Rhodes found the gold he was looking for he would be putting messages out through the Eastern Telegraph system for many years to come. So he was not a competitor but a feeder. And when his land line reached Cairo, Eastern Telegraph would be ready to meet him on his own ground. Their submarine line would be able to do the work more accurately, quickly and cheaply than the land line.

Rhodes claimed that his line to Egypt, provided the Mahdi was 'squared', would cost £300,000 to £400,000. Lying to the east and west of the coast were cables which had cost more than £3,000,000 – the Pender system, which charged 9s 6d a word. 'But when we get to Egypt by the telegraph I am referring to' he told shareholders of The Chartered Company, as it was known, in London in 1892, 'we shall be able to do it for 1s 7d from Egypt to England, or in all about 2s 6d against 9s 6d a word.'

Sir James Anderson expressed his unconcern at Rhodes's African Transcontinental Telegraph Company by remarking that it was hardly conceivable that anything worth calling commerce would ever grow beside those 5,000 odd miles of land line, excepting where there was gold, which could hardly be hoped to be found over the whole area.

> We may therefore look forward to continuing to be the carrying line for commercial traffic from the coast to the interior, since the natives who do not wear clothes are not in the habit of sending telegrams.

In the event neither Sir John nor Sir James, who died in 1893 and was succeeded as managing director of Eastern Telegraph by John Denison-Pender, had anything to worry about. The gold reef turned out to be a myth and Rhodes's African Transcontinental Telegraph was only built as far as Tete and Vjiji in Portuguese East Africa.

The line had not been a success financially said Sir Lewis Michell, Rhodes's biographer, but, though only a fraction of the original scheme, had checked the slave trade, promoted civilisation and helped the pioneers replenish the earth and subdue it.

After there had been a twelve-day interruption on the New Zealand to New South Wales cable in 1889, Eastern Telegraph decided to duplicate it. Telcon laid this in 1890, but relations became strained when the New Zealand government intimated she intended laying the next submarine cable herself. Relations with Queensland worsened when the colony refused to join the other colonies in paying Eastern Extension their £1,000 a year subsidy. Queensland's PMG antagonised Pender by cabling him that his government would not join any agreement which assisted a monopoly. But worse, he awarded the contract for a cable to the French island colony of New Caledonia to a French company with a subsidy of £2,000 a year. The New South Wales government contributed £2,000 a year too, and the French Government £8,000. To show there was no ill-feeling the latter gave Sir John the Legion of Honour – rarely presented to foreigners at that time. But disconcerting as it might be as a change of policy for a one-off exercise, had it even more alarming implications as the first section

THE RHODES COLOSSUS

STRIDING FROM CAPE TOWN TO CAIRO.

.·. Mr. Rhodes had announced his intention to continue the telegraph northwards across the Zambesi to Uganda, then, crossing the Soudan, to complete the overland telegraph line from Cape Town to Cairo.

of the Big Project which was the subject of his main bone of contention with Australasia? He soon convinced himself that in agreeing to subsidise the New Caledonia cable the governments of Queensland and New South Wales 'were practically committing the Australasian colonies to the laying of a Pacific cable under French auspices and control'.*

When Victoria protested to the Colonial Secretary, Audley Coote of Tasmania said it was because they were a pawn of Eastern Telegraph.

> The aim has been for the last nineteen years to break a growing and ungenerous monopoly in the telegraph world and, now that we have so far succeeded by giving the contract to the French company, the secret allies of the Eastern Extension Telegraph Company are agitating in the press, and in the minds of public men in some parts of Australia, to try and prevent the Pacific Cable being a success. It is not because the cable touches at New Caledonia, it is because it is opposing the Eastern Extension Telegraph Company.

France of course had a particular interest in Canada which had been wrested from her by the British in the eighteenth century, and still contained in Quebec and elsewhere a large number of French. Coote's letter of 13 May 1893 seemed to confirm Pender's worst fears and its bitterness reflected the degree of opposition that existed to Eastern Telegraph's monopoly. It was proof that there was to be no letting up by the Pacific Cable promoters, in spite of the Eastern Telegraph's delaying tactics which had deceived no one. Along with opposition to the British company came surprising support for the Pacific Cable idea from sources outside those directly interested.

In 1892 the Association of Chambers of Commerce of the United Kingdom meeting in Newport resolved that the extension of direct telegraphic communication between the parts of the British Empire would facilitate defence, promote trade and investments, emphasise community of interests, and generally stimulate the development and consolidation of the Empire. Commercial men recognised that effective defence must operate to their benefit. A telegraph over British territory would not only be serviceable in peace but 'in times of alarm' would reduce the risks to which their property on the high seas would otherwise be exposed, and high charges for insurance.

By now the Admiralty had made a survey of the proposed Pacific Cable route and the results had been satisfactory. The five lines to the East and Australia all passed through possibly hostile countries where they would be useless in time of war, or through shallow seas

---

* His Memorandum to the Colonial Office, of 3 Jan 1894.

where the cables could be easily fished up and destroyed. 'With the establishment of uninterrupted communication across the Dominion [of Canada] an alternative route is now made feasible by the laying of a Pacific cable which would remove this defect.'

The Royal Navy had gone even further. In 1892 HMS *Champion* had seized 'Johnston Island' south-west of her base in Honolulu as a possible landing place for the Pacific Cable. But the British Government felt obliged to restore it to the Government of Hawaii who said it belonged to them, but Britain could land a cable there at any time. There was no question in Sir John Pender's mind that, when it came to the point, HMG would come to him as the leading exponent of ocean telegraphy and ask him to lay the cable – and operate it. On 22 July 1890 he wrote to Sandford Fleming:

> If the various governments interested are determined to have a line across the Pacific, and are prepared to incur the requisite expenditure for the purpose, I am quite ready as I have always told you, to co-operate in carrying out the work on fair and reasonable terms, and in this way the object might be attained more easily and economically than if third parties were employed.

If, however, they decided to establish 'a separate and distinct undertaking from the existing lines' the two cables needed would cost £3,600,000. 'One line could be no more relied upon in the Pacific than in the Java seas where all our cables between Java and Australia were suddenly and simultaneously interrupted by earthquake a few days ago.' In replying to Sir John's comment that this was a very rare occurrence, Fleming asserted that on the contrary the Eastern Extension cables between Java and Australia had broken down thirty-six times in the last eighteen years, and fourteen of these were between Port Darwin and Banjoewangie, one of which caused an interruption of four months and another three and a half months.

In November 1893 Sir John once more assured his proprietors 'when another system is required for connecting the old country with the Australian colonies, you may depend on it that we [the Eastern Extension] shall have a finger in the pie (applause)'.

In the summer of 1893 Hon Mackenzie Bowell, the Canadian Minister of Trade, took Sandford Fleming with him on a mission to Australia, and presented the colonial premiers with a Memorandum setting out a detailed plan (written by Fleming) for a Pacific Cable. If economy, low rates and high efficiency were the aim the telegraph should be undertaken as a public work under Government control.

> Promoters of companies generally desire to make large sums of money. The policy of companies is to obtain from the public as large profits as possible, while that of Governments is to accommodate and benefit the public in every possible manner by reducing the rates to the lowest practicable point, and by giving the most efficient service. The principle of ownership of telegraphs by Government is not new. It has long been adopted in the United Kingdom, in India, in these colonies and elsewhere and in every case I am aware of, where the principle has been tried, the public has derived the greatest advantage.

Various efforts had been tried to have the Pacific Cable established by a subsidised company but none had offered to do it for a smaller subsidy than £75,000 a year for twenty-five years. So he proposed Australia, New Zealand, Fiji, and Canada should be its joint owners and that it should be worked 'as a public undertaking for the public good'. The four Australian colonies of New South Wales, Victoria, South Australia and Western Australia were obliged to pay Eastern Extension an annual subsidy of £32,400 as well as a Cable Guarantee to compensate for the rate being reduced from 9s 4d to 4s 9d – a total of £59,920, which was more than the interest on the capital required to establish the Pacific Cable. This subsidy ended in May 1899, but until then the colonies concerned would not be able to co-operate on equal terms with the others.

Bowell estimated it would take up to £60,000 a year to work the telegraph; the rate from Australia to Vancouver should be lowered to 2s a word immediately the line was opened, so the Australia to England rate would become 3s 3d instead of 4s 9d.

Most of those who attended Bowell's meetings showed neither support nor opposition, though Queensland expressed definite sympathy. Bowell attributed this hesitancy to 'a very proper fear of the Eastern Extension, arising out of a knowledge of the power of that corporation and the possibility of an early increase of telegraph tolls'. The South Australian government, who had built the Adelaide to Darwin line at a cost of £506,000 in 1870, considered an alternative Pacific Cable would injure the value of that public asset.

To what extent it was a deliberate stroke in Sir John's feud with the governments of Canada and Australasia over the Pacific Cable is not clear, but on 28 October 1893, Eastern Extension made an agreement which excluded those countries from making telegraphic communication with Hong Kong. Lord Ripon, the Colonial Secretary, granted Eastern Extension the exclusive privilege of laying a cable to Hong Kong, reserving to the British Government an option to take possession of the cable from Singapore to Labuan and Hong

Kong by giving twelve months notice and paying Eastern Extension
£300,000. It was a considerable coup. Eastern Extension had laid a
cable from Hong Kong to Manila in 1880, to Foochow and Shanghai
in 1888, to Macao in 1884. In 1894 they laid one to Labuan on the
north coast of Borneo. All these cables came in to Telegraph Bay,
Pokfulam on Hong Kong Island and were connected by land line to
the Eastern Extension office at Marine House (on the site occupied
by Old Mercury House in 1979).

Sir John Pender had now been long enough in the business for
some of his first concessions to start running out. In 1892 Brazilian
Submarine's exclusive twenty-year concession of 1872 expired and
that company and Western and Brazilian found themselves for the
first time having to meet fierce competition from French, American
and other British companies. The Director of Brazilian Telegraphs
who succeeded Capanema in the change of government of 1890
conceded the Joint Purse Agreement which Western and Brazilian
and Brazilian Submarine had sought; but that was of little assistance
in combating the rival service of the Central and South American
Telegraph Company which offered a route to Europe via USA on its
cable from Valparaiso. So W and B and Brazilian Submarine formed
The Pacific and European Telegraph Company to link their systems
with that of the West Coast of America company. Sir John Pender
obtained the necessary concessions from the Chile and Argentina
governments and the new company was incorporated on 24 June
1892. In 1893 Sir John was elected chairman of Brazilian Submarine.

The South American Cable Company was formed by Telcon who
laid a cable for Spanish State Telegraphs from Cadiz to Teneriffe and
Senegal, and then obtained a concession from the Brazilian Government to connect Dakar on the West Coast of Africa with Pernambuco touching at the island of Fernando Noronha on the way. The
'Via Noronha' telegraph, as it was known, charged fourpence a word
lower than Brazilian Submarine's rate. They opened the service in
October 1892 and had their office in a building a few yards from
Pender's companies. But in 1898 the Brazilian Government granted
Western and Brazilian exclusive rights for cables to the River Plate
for twenty years, at the same time, however, imposing a tax of ten
gold centimes a word on all traffic passing over Brazilian Submarine's lines – an innovation. But the government agreed not to
authorise any other company to lay cables from Brazil to Uruguay or
Argentina for twenty years. In return, W and B agreed to increase
the capacity of their cables if the government asked them to.

The Portuguese government also prolonged Brazilian Submarine's concession of 1872 for the exclusive right to lay and operate cables between Portugal and Brazil. Having given Eastern Telegraph the concession to lay the cable from Lisbon to the Azores, subject to the approval of the Cortes, the French Government intervened and persuaded the Portuguese that for 'political reasons' the contract should go to a French cable company. But when the latter had not completed the work by the stipulated twelve months the concession was taken away and given once more to Eastern Telegraph (June 1898) who invested £55,660 in The Europe and Azores Telegraph Company formed to operate it. The Lisbon-Azores cable laid by Telcon opened for traffic in August 1898.

Sir John's interests stretched from east to west, but as seen they were confined to long-distance telegraphs whose future he saw as limitless. The wider subject of what became known as 'telecommunications' was never the concern of 'the Cable Companies'. 'Telegraphy must go on; it is impossible to stop it,' he said in 1894. 'I think that the business world would almost come to an end if the submarine telegraph companies which unite the whole of the world were in any way interrupted. . . . Someone said to me the other day that we shall be going to the moon next (laughter). I do not know that we shall do that, but if any part of the globe requires telegraph communication, and can afford to pay for it, you may depend upon it that we shall see that it can be given.'

His vision of one day 'going to the moon' was as an extension of the capability of the present electric telegraph. But he was more stirred by the short term prospect of Eastern Telegraph shares gaining in value. Success was the £100 debentures which earned 4 per cent interest standing at £116. But being an exercise conducted entirely abroad, exchange difficulties – 'the silver question' – resulted in serious losses. To make up he tried where possible to increase the rates to users, a move which gave further ammunition to those who, like the Australian colonies, accused him of being a capitalist first and only indirectly interested in providing a public service. His attitude to dividends, however, much to the disdain of many of his investors, was conservative, preferring to put profits to reserve to pay for new cable against the unforeseen day when it broke, rather than paying high dividends and having to go to the money market for loans. Ocean telegraphs, he shrewdly remarked, were unlike railways in which you could *see* them depreciating; they needed a different financial strategy. A violent volcanic eruption which no one could have predicted had just occurred on the Island of

Porthcurno

Halifax

Fayal Carcavelos Gibraltar

Bermuda Madeira Malta

Cuba Turks Island Alexandria S
Puerto Rico
Jamaica
St. Vincent St. Iago
Colon Barbados Dakar
Trinidad Bathurst
Konakry
Georgetown Sierra Kotonou Lagos
Leone Accra Bonny
São Luiz Principe
Belém São Thomé Gaboon
Fortaleza
Momba
Recife Loanda
Lima Salvador Benguela
Mossamedes Mozamb
Mollendo
Arica
Iquique Delagoa Bay
Antofagasta Santos Dur
Rio de Janeiro
Valparaiso Rio Grande Cape Town
Buenos Montevideo
Aires

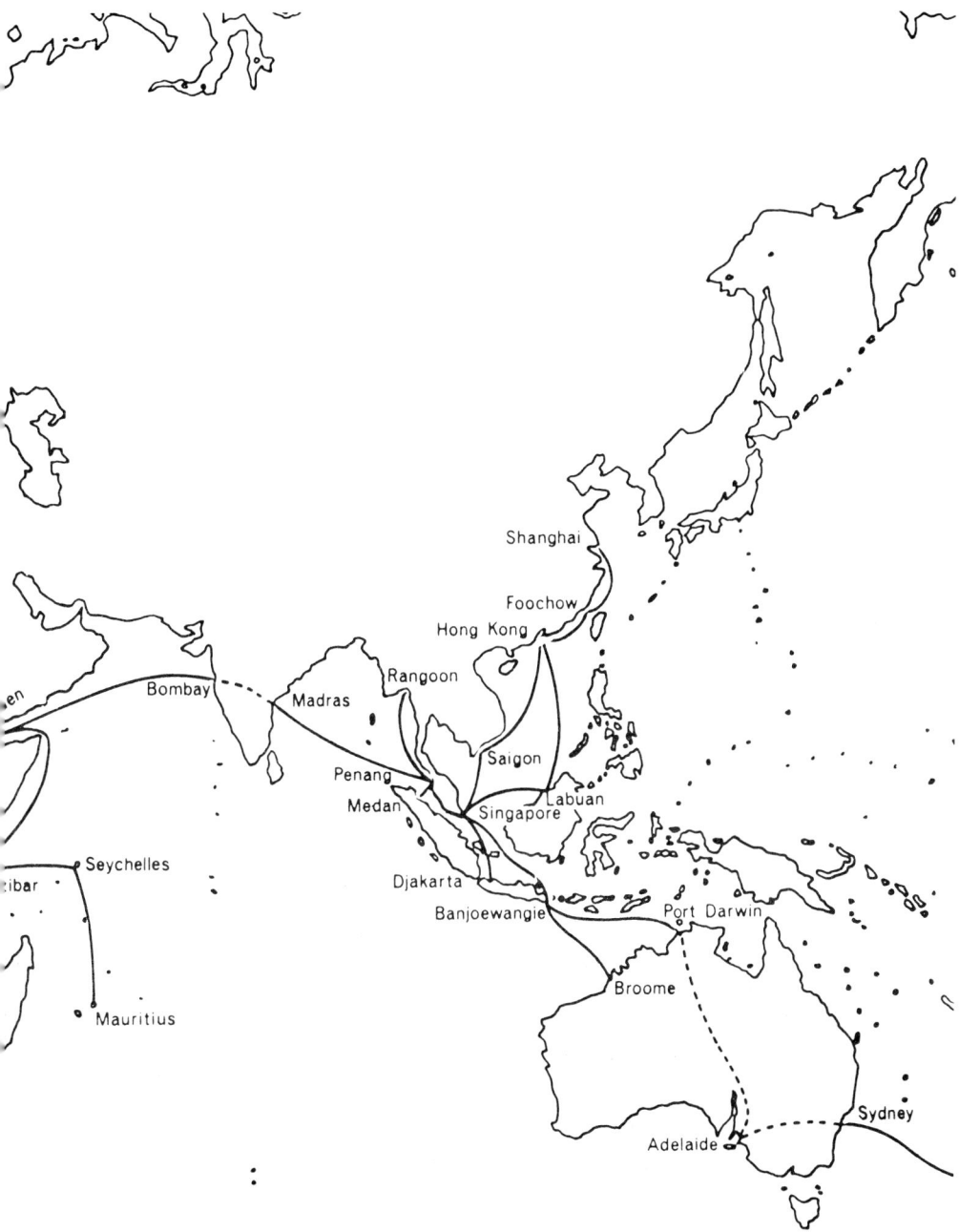

Shanghai

Foochow

Hong Kong

Rangoon

Bombay

Madras

Saigon

Penang

Labuan

Medan

Singapore

Seychelles

ibar

Djakarta

Banjoewangie

Port Darwin

Mauritius

Broome

Sydney

Adelaide

THE EASTERN AND ASSOCIATED TELEGRAPH COMPANIES
CABLE ROUTES - 1898

Zante in the Levant where the Eastern Telegraph had a station. It had not been damaged, but if there was another one it was unlikely to escape again. They had therefore had the cost of removing the station to the Island of Syra.

Within the technological context of his day and his own aspirations John Pender had no equal, and when Eastern Telegraph and Eastern Extension decided to make 1894 the Silver Jubilee of the establishment of Submarine Telegraphy to the Far East (1869–1894), the acclaim was unstinted.

The two companies gave a banquet at the Imperial Institute, London, on 20 July 1894 followed by a reception attended by the Prince of Wales. Some 450 guests sat down to dinner including members of the diplomatic corps, the Lord Chancellor, the Secretary of State for India, leading figures in the world of telegraphy like Lord Kelvin and W H Preece, high-ranking soldiers like General Lord Roberts and Field Marshal Viscount Wolseley, distinguished users of long distance telegraphs like Baron de Reuter, W H Russell and G A Sala, and that pioneer of another new form of communication, the automobile, Sir David Salomons, chairman of the Self-Propelled Traffic Association. In the chair was Sir John Pender GCMG* MP – he had been elected for Wick again in 1892. The Italian Director-General of Telegraphs sent him a congratulatory cable describing him as the King of Cables, and the King of Italy gave him the Grand Cordon with Grand Cross of the Order of the Crown of Italy. The dining hall was illuminated with the new electric light.

Sir John responding to the toast of Submarine Telegraphy in its International Aspect, said there were now eleven cables under the North Atlantic. Thanks to Dr Muirhead, duplex working had doubled their capacity. In the early days submarine telegraphy was regarded as a very risky investment; today it was reckoned one of the soundest in the market. Wherever the British flag was flying and commerce promised a fair prospect of remuneration, cables had been laid. In 1894 there were 152,000 miles of submarine cable in the world, of which 90 per cent had been provided by private enterprise. If these were added to the two million miles of land lines, the combined capital represented £106,000,000.

He referred briefly to the projected Pacific Cable. When it was really required, he said, the companies he represented would not be found backward in meeting public requirements. The Australian colonies had assisted the Eastern Extension with a substantial subsidy over the last twenty-five years to duplicate the cable between

---

* He was elevated to Knight Grand Cross in 1892.

Penang and Australia. The majority of them too had entered into an arrangement under a guarantee of half risk to reduce the rates by 60 per cent.

His companies began business by sending 400,000 messages a year; in 1894 it was two million. International telegraphy had prevented diplomatic ruptures and consequent war, and been instrumental in promoting peace and happiness.

After dinner Sir John welcomed the royal guests who had first been received by the Lord Chancellor – the Prince of Wales, the Duke of Cambridge, Prince and Princess Edward of Saxe-Weimar and Princess Louise. They enjoyed the Evening Fete and toured the exhibition which included models of cable ships (among them the *Great Eastern* by then broken up) and 'sea-snakes' coiled round cables recovered from the depths. The accent, according to the *British Australasian,* was on the maintenance of the Empire – 'the Crimson Thread of Kinship'. Sydney was more in touch with London than Birmingham was fifty years ago. No time was allowed for the growth of bad feeling or the nursing of a grievance. The cable nipped the evil of misunderstanding leading to war in the bud. When famine struck in Russia, the cable brought corn-laden ships across the Atlantic. When there was a flood in China, a fund was at once opened in the Mansion House.

The Prince of Wales sent messages by telegraph to the Governor-General of Canada, the Viceroy of India and governors in all corners of the British Empire and to British ministers in the distant capitals of South America. The 'clearing of the line' for this operation was remembered by Eastern Telegraph operators in South America and elsewhere for years afterwards – 'the tension in the instrument rooms as they waited to flash through the royal message to the west coast of South America and then the royal replies to London which followed'.

The Press responded handsomely, still impressed by the magic of telegraphy. Distance was nothing to it, wrote the aptly named *Daily Telegraph.* If they could only establish a medium of communication with the moon and stars, messages would pass as easily thither as to Paris, New York and Sydney.

> And the mystery of all this is still so utterly unknown that the very masters of science make little or no attempt to understand it. No one can surely tell us what this electric current is which becomes developed when two metals and an acid plate create chemical combination. To style it a vibration of the ether carries us no farther than before. We live amid a universe of miracles and this is one of the subtlest. . . . With a slender endless wire shut away from interruption

by a coat of gum, London converses with all the cities of the earth, and they answer back over mountains, through deserts, under the depths of the sea – a marvel so amazing that we have grown tired of being amazed, a mystery so familiar that we have forgotten how mysterious it really i. (21 July 1894)

It was the Dominion of Canada not the Mother Country who summoned the Colonial Conference at Ottawa in June 1894 and decided it should confine itself to peace issues of which the Pacific Cable was a principal item.

Sandford Fleming came to London in January 1894 to put his case once more to the Colonial Secretary, Lord Ripon. He now advocated the second on his list of routes, landing on the island of Necker, even though it would upset the US government. When HMG again showed little enthusiasm, Fleming took it upon himself 'without counting the cost' to hatch a remarkable cloak-and-dagger plot for taking soundings round the island, finding a suitable landing place and planting the British flag. But when the projected coup d'etat came to the ears of the Prime Minister, Lord Rosebery, he was 'very annoyed' and told Lord Ripon to cable HM consul in Honolulu warning him of what was about to take place. After a drubbing down from the Colonial Secretary, Fleming called his men off. The island was immediately annexed by the Government of Hawaii.

Throughout the months leading to the opening of the Ottawa Conference, Fleming's opponent was also re-stating his case in a series of memorandums to the Colonial Secretary. Since the idea of a French Pacific Cable had been condemned by all but Queensland and New South Wales, said Sir John, an agitation for one entirely under British control had been revived with increased intensity. It was an agitation which he claimed was to a large extent based on sentiment. It was acknowledged that the existing cable service was second to none in the world. With telegrams at 6s a word between North America and Australasia representing only $4\frac{1}{2}$ per cent of all Australasia traffic, there was no demand for it. Build up trade with the new steamship service, and then there might be.

He had no qualms about appealing to sentiment himself. It was difficult to conceive, he stated at the end of his long memorandum of 3 January 1894,

> that either the Home or Colonial Governments would act so unfairly towards the pioneer company, to whom they are so much indebted, as to enter into unnecessary and ruinous competition with it. Not only would it be a complete reversal of the policy which they have hitherto

pursued towards submarine telegraphy, but it might result in so
weakening the company that in times of political trouble it would be
unable to efficiently maintain the service.

Once again he assured the Colonial Office that when the time was
ripe and the necessary subsidies were forthcoming, Eastern Extension would be prepared to co-operate in carrying out the work, and
by making use of its large experience, highly trained staff and cable
ships, the mother country and the colonies would be much better
served than encouraging unnecessary competition.

When in a paper read to the Royal Colonial Institute Sir Charles
Tupper, the Canadian High Commissioner, flayed Sir John's case
against the Pacific Cable as largely based on fallacies, he provoked a
series of letters from Winchester House challenging him point by
point. The existing system was not a monopoly; it had had to rely
upon the businesslike and economical principles on which it had
been established and worked for its freedom of competition (!). He
had nothing to say against any who competed on equal terms, but
when they were supported by Government aid was another matter.

Tupper objected to Pender applying the term 'private enterprise'
to his undertakings when Eastern Extension had received £643,000
in Government subsidies and guarantees, and the African companies
£1,337,000. But these Sir John contended were to duplicate lines not
to 'establish communication'. None of such points had any bearing
on the real question at issue, however. In Sir John's view this was:
could the project be recommended to the investing public? His
figures showed it would result in an annual loss of £48,000, in spite of
Income Tax being only eight (old) pence in the pound.

A Postal and Telegraph Conference at Wellington had just passed
a resolution recommending the governments to offer any company
willing to lay the Pacific Cable a guarantee for fourteen years of 4 per
cent on a capital not exceeding £1,800,000, on condition the rates
were reduced to 3s a word for ordinary telegrams, 2s for Government
and 1s 6d for Press telegrams. An 'all-British-cable', Sir John told an
Eastern Extension meeting of 18 April, had less chance of immunity
from interruption in time of war than the existing cables which
followed the principal trade routes. He doubted if a broken cable
could ever be lifted from the Pacific. He had recently asked an
official at the Foreign Office if he would sign a document getting the
nations of the world to neutralise the company's cables in time of
war. The official had replied that he would, but that the GOC and
naval commander on the spot would break the undertaking when
war broke out. Army and naval commanders had the power to enter

the company's offices and shut them up in wartime, but he hoped they would never destroy their cables.

As soon as the principle of the desirability of a Pacific Cable had been agreed at the Ottawa Colonial Conference which opened on 28 June 1894, delegates did not commit themselves to any details. It was considered useless to attempt an exact estimate of cost until the survey had been made. Those begun by HMS *Egeria* in 1888 had been discontinued in 1890 when the ship working south to north had only reached the Phoenix Islands. Delegates asked for it to be completed and agreed that the cost of £36,000 over three years should be borne by the UK, Canada and the Australasian colonies. The New Zealand delegate observed sarcastically that if conference came to no more definite conclusion than that there would be another seven years' wait. Sandford Fleming saw the conference as another opportunity to attack the Eastern Telegraph companies.

> The progress and well-being of Canada, Australasia and the Empire cannot be retarded in order that the lucrative business of a private company may remain without change. Even if the chairman of the Eastern Extension Company succeeded in converting us to his commercial ethics, that the profits of the monopoly he represents must be maintained inviolate, it does not follow that the project of a Pacific Cable would not be carried out in some form, even if Canada and Australia abandon it.

There were unmistakable signs that a Pacific Cable might shortly be carried out by France and the United States. France had already completed a section of 800 miles at the southern end, and the US had recently spent $25,000 in making an elaborate survey of a third of the distance – from San Francisco to the Hawaiian Islands. With a rival line in foreign hands it was easy to see that Eastern Extension would gain nothing and the Empire lose much.

In his report to Lord Ripon, the British representative at the conference, Lord Jersey, said great stress had been laid on the value of the line for imperial purposes. Weighing the pros and cons of having the cable laid as a national undertaking by the contributing governments as against a private company with a government subsidy, he thought state ownership avoided promotion expenses and the danger of amalgamation with other companies, but a private company could certainly do the work and operate the service more economically. Lord Jersey admired Eastern Extension. It was an enterprise which had conferred great benefits on Australasia but that did not entitle it to the cable monopoly in that part of the world.

After the conference, an advertisement appeared in the Press of

6 August 1894 signed by Mackenzie Bowell, Canadian minister of trade and commerce, on whose authority it is not known, inviting cable manufacturers and contractors on behalf of the Canadian Government to state the terms on which they would lay and maintain a Pacific cable with a capacity of twelve (five letter) words a minute. Only four tenders were received: from Siemens Brothers, Fowler-Waring Cable Company, Henleys, and the India Rubber Company (Telcon ?). Sir John Pender wrote to Mackenzie Bowell on 19 October in reference to the advertisement, and renewed Eastern Extension's previous offer to enter into negotiations when data was available. When all the tenders were in by 1 November they averaged £98,000 less than Fleming's estimate – and £1,244,000 less than the sum estimated by the British Post Office.

But after all this activity, the whole Pacific Cable project once again went into a state of suspended animation, though with the appointment of Joseph Chamberlain as the new, dynamic Colonial Secretary, it was certain it would not stay there for long.

A smaller but less peaceful waterway than the Pacific Ocean was the 4,000 mile long Amazon. Attempts to lay a land telegraph beside this great river of Brazil had been thwarted by the speed and density of the forest growths along its banks. But in 1895 the Brazilian Government conceived the idea of a 'subfluvial' cable. They invited tenders but none came. Eventually Richard Reidy, the representative in Brazil of Western and Brazilian, agreed to make the attempt and in April 1895 his company was granted a concession which gave it the exclusive right to lay a cable in the river from Para (Belem) at the mouth 2,000 miles up to Manaos, the seventeenth century capital of the ancient state of Amazonas, with intermediate stations at Pinheiro and other places. Western and Brazilian undertook to operate the line for thirty years, with a government subsidy of £17,125 a year for the first twenty years. Western and Brazilian formed The Amazon Telegraph Company, incorporated in England, to carry out the concession. The telegraph was opened on 12 February 1896. The company established an office in the Western and Brazilian building at Para and appointed a manager who was Joint Superintendent of both their services with a senior clerk and six operators.

The laying of the Amazon Cable was indeed a triumph, but maintaining it was an even greater task. Doubtless Siemens had seen to it that the outer sheathing made it crocodile-proof, but in the rainy season masses of debris, tree trunks and vegetation were carried down the river by the strong currents, scouring out the bottom and churning up the cable from the bed into which it had been so skilfully buried. Amazon Telegraph bought W and B's ocean cable repair ship *Viking* and converted it to meet the very different conditions of the river Amazon with its varying currents and depths and its shifting bed, which caused faults of a kind never encountered at sea. Several sections of the Amazon cable were interrupted almost continuously, and it dawned on the Brazilian Government that perhaps a subfluvial telegraph was not the answer after all.

It is doubtful if any of this drama came to the ears of Sir John Pender. The year 1894 had been exhausting for the 79-year-old chairman of Eastern Telegraph and its associated companies – a year, incidentally, in which a 15-year-old youth called Edward Wilshaw had joined the company as a junior clerk.

For the first time in his life Sir John was unable to take the chair at the forty-sixth meeting of the company on 20 July 1895, as it was polling day at Wick Burghs, the constituency he was again contesting as Liberal Unionist (he was re-elected). His place was taken by the Marquess of Tweeddale who, as Lord William Hay, had been a director of the company for a great many years. Neither did Sir John attend the next meeting of 23 January (1896), but this time it was because he was confined to his house. His speech had been affected by paralysis, and he was no longer able to attend the House of Commons. A 'farewell' speech was made on his behalf by the son of his first marriage, James Pender MP. He never recovered, and died at Footscray Place, his rented Kent home, on 7 July 1896 aged 80.

He left the Eastern Telegraph and its associated companies not only stable but showing signs of steady growth. Together they owned a third of the total cable mileage of the world, carried two million messages a year and employed 1,800 of whom some 650 crewed their combined fleet of ten cable repair ships. When they were formed, the average time taken by a 'cablegram' was five to ten hours; in 1896 it was thirty to sixty minutes. A message to China which had taken eight hours now took eighty minutes; to India thirty-five minutes instead of five hours; to Argentina one instead of ten hours.

Eastern Telegraph's share capital stood at £6,217,000 and it had £845,460 in reserves. Its 26,350 miles of submarine cable and land line together with its stations were worth £5,155,000; its five cable maintenance ships, *John Pender, Chiltern, Electra, Mirror* and *Amber,* £163,500; spare cable £130,800; land and buildings £125,000; shares in other telegraph companies £850,000. For the six months ending 30 September 1896, Eastern Telegraph spent £135,700, of which 'London Expenses' accounted for £9,600 and overseas stations £83,000, repairs and renewals £22,000 and depreciation of spare

The Eastern Telegraph Company Limited,

Winchester House,

Old Broad Street, E.C.

London   12th February 1894

TRAFFIC DEPARTMENT.

TELEGRAPHIC ADDRESS.
"EASTERN LONDON".

No

Mr E. Wilshaw

Dear Sir,

Referring to your application of today's date, we beg to i
inform you that you are appointed to a clerkship in this Department
from the 15th inst.

Your salary will be £25 per annum for the first 6 months
when it will be increased to £30. At the end of 18 months you
will receive £35, and at the end of 2 years £40, after which time
you will be on the permanent staff and will receive annual
increments of £5 per annum.

It is understood that these increases will only be given
if your work and conduct are found to be satisfactory.

Yours truly

for Managing Director

W. Hibberdine

Traffic Acct:

*The letter engaging 14-year-old Edward Wilshaw as a clerk in the Traffic Accounts Department
of the Eastern Telegraph Company from 15 February 1894, at £25 a year.*

cable £5,000. Just over £2,000 went on directors' remuneration and
£2,500 on the Staff Fund. Six months' income tax amounted to
£5,800. With income from messages and other receipts at £420,000
and dividends on their shares in other companies at £23,000, total
earnings for the half year (after deductions) were £435,000. So there
was a healthy credit balance of nearly £300,000. After paying

debentures and dividends, £7,500 of this was put to General Reserve, £20,000 to a Fire Insurance Fund and £10,000 to the Maintenance Ships Reserve Fund. Gross receipts for Eastern Extension for six months ending 30 June 1896 amounted to £313,900 (£264,100 in 1895). Working expenses were down from £95,250 in 1895 to £87,250. On the credit balance of £226,600 the company made a net profit for the six months of £170,700.

Sir John left personal estate valued at £348,179 – worth more than £5 million in 1979? A £1,000 memorial fund was launched, and an endowment fund established for the maintenance of the Pender Laboratory at University College, London and the Pender Chair of Electrical Engineering. A Pender scholarship and Gold Medal were also instituted at Glasgow University. He was succeeded as chairman of Eastern Telegraph and Eastern Extension by Lord Tweeddale; his son John Denison-Pender, aged 41, was already managing director and was at once made deputy chairman. It is said Sir John was nominated for a barony in the 1897 New Year Honours, but as he died before it was gazetted, James Pender his eldest son and heir was given a baronetcy 'in compensation'.

He was by no means the pioneer of ocean telegraphy; he entered the field, as this story has shown, in the steps of John Brett, Charles Bright and many others whose lead he would be the first to acknowledge. But his energy, business acumen, a highly developed sense of diplomacy in the winning of concessions, his selection of able lieutenants and the backing of a dynamic wife, enabled him to turn the conglomerate which became the Eastern and Associated Companies into the largest and most successful international telegraph business of its day, the envy of all other 'Cable Companies' and of the British Post Office. It was the foundation of the enterprise of the 1970s.

The man of affairs and company chairman are seen in his public speeches and confident harangues to shareholders at general meetings; but of the liberal-minded idealist of Middleton Hall, Linlithgow, Arlington Street and Footscray, who saw beyond the material achievement of his wire laying and key tapping, very little emerges. He appears without much imagination or humour. Did he enjoy his wealth? His unsmiling portrait gives little hint of conviviality – though like most men of his means in late Victorian England he was probably as heavy a drinker as the rest of them. For his role in Society, he would have been guided by Emma. But whether he enjoyed her 'artists receptions' and had time for the social round, for reading, for going to the theatre, to the races, to weekend house parties no one knows. None of his private correspondence, unlike Emma's, has survived. But in thanking shareholders for re-electing him to the Board at the last meeting he chaired (25 January 1894), he confessed 'I do not hunt, I do not shoot, I do a little bit of yachting, but the greatest pleasure of my life is to attend to the interests and duties devolving on me as your chairman. . . . I have made submarine telegraphy to a great extent my hobby'.

His eighty years spanned from Waterloo (he was born in 1815) to wireless. For five weeks before he died, one Guglielmo Marconi applied for his patent.

PART II
1898–1938

For Whose Profit?

It may be safely averred that railways, steamships and telegraphs are combinedly our most powerful weapon in the cause of Inter-Imperial Commerce . . . What has so far been done to foster trade between the scattered units of the Empire by direct, efficient and cheap telegraphic communication has been almost entirely due to private enterprise. Today the service is good but costly and imperially speaking incomplete. The tariffs are high due to the costly character of the cable. Apart from this, private companies naturally fix the charges for messages from a shareholders' point of view rather than from the public standpoint . . . No government should be precluded from interfering with private enterprise where desirable in public interests in contradistinction to the interests of shareholder.

Charles Bright, *Imperial Telegraphic Communication,* 1911

5

Pacific Rival

1889–1902

When the Scottish scientist James Clerk Maxwell died in 1879, he left behind no more than a 'theoretical prediction' that there was a new sort of wave which travelled at the speed of light, could be focused and reflected like light, but would pass through solids, liquids, gases and even a vacuum. But in 1887 Heinrich Hertz demonstrated that 'electro-magnetic' waves did in fact exist. From then on many carried out experiments to establish their nature, including Sir Oliver Lodge, Alexander Muirhead, Captain Henry Jackson RN, Lord Rutherford in New Zealand, Professor Righi in Italy, Braun in Germany, Bose in India and Dolbear in the United States. When Hertz died in 1894, *The Electrician* came out with a leading article reflecting on the future of his discovery.

> The practical man will, we suppose, seize upon Hertzian waves ere long, patent their application to the use and convenience of man, start syndicates, float companies and so on.

But *what* use?

The Electrician suggested detecting thunderstorms. Few were ready to start thinking about ways of applying them before the pure scientists had found out what they were. An exception was young Guglielmo Marconi who had first read of the effects of Hertzian waves in 1894. He conceived the idea of applying them to telegraphy.* He carried out encouraging experiments in the attic and garden of his parents' villa in Bologna. When the Italian P and

---

* Professor Dolbear was working on the same lines in America and when Marconi went to US in 1897 he found an American Wireless Telegraph and Telephone Company already in existence exploiting the patents taken out by Dolbear.

T Department showed no interest, he came to England. He was 22. On 2 June 1896 he applied for the world's first patent for wireless telegraphy. He spent the rest of the year wooing the War Office and the Post Office.*

If it occurred to any of those planning the Pacific Cable that the snail pace of their deliberations had led to their being overtaken by a new method of long distance communication which conceivably might cause them to think again, it is unlikely they would have shared the apprehensions which Charles Bright expressed in the concluding pages of his great *Submarine Telegraphs* published in 1897. 'Such inventions as those respectively of Hertz and Marconi,' he wrote, 'might of course in the end deal a veritable death-blow to submarine telegraphy as at present known. On the other hand (if it ever came to that) they would bring about such an enormous extension of telegraphic work all over the world – constructive, administrative and operative – that they would not necessarily be unwelcome to those who are professionally or industrially engaged on it.' It had not come to that and did not look like doing so for many years yet. The medium of the electric submarine cable had proved its worth, and if only they could stop talking and begin *doing,* there was no reason why Canada and Australia should not be linked in eighteen months.

Action still seemed a remote possibility when there came to the Colonial Office the man of determination who saw in the Pacific Cable the means of fostering the 'new imperialism' with which his name is ever linked – Joseph Chamberlain.

The need for a dependable diplomatic link had never been better demonstrated than in Joseph Chamberlain's ability at once to explain to President Kruger by telegraph that the raid of Cecil Rhodes's viceroy in Zambesia, Dr Starr Jameson, into the Transvaal on 29 December 1895 had been carried out without the authority or knowledge of HMG. The case for 'imperial' lines to outposts of Empire was reinforced by the fact that at this time the Eastern Telegraph cables down east and west coasts of Africa were both interrupted, though luckily not simultaneously. If Chamberlain had been unable to get through, Kruger is likely to have interpreted that

---

*According to R N Vyvyan, Marconi demonstrated to William Preece of the Post Office in 1896 that *short waves could be directed in a beam,* using a parabolic reflector; but no further research was carried out on this until 1916. It did not seem to be a promising field, states Vyvyan, as it was known that the attenuation of short waves was very high and the power that could be used with short waves in 1896 was small. Nothing was known either in 1896 of the effect of the Heaviside Layer in the upper atmosphere, nor of its ability to reflect waves. Thus research on short wave propagation was neglected.

Union Jack at the head of Jameson's horsemen as meaning the raid was 'official' and acted accordingly.

Chamberlain at once took up Canada's idea that HMG should call an Imperial Pacific Cable Committee to examine every aspect of the project, and this met in London on 5 June 1896 with the Earl of Selborne, Colonial Under-Secretary, and George Murray of the Treasury representing the British Government, two representatives of Canada and one each from New South Wales and Victoria. It was the first time HMG officially recognised that there was such a project, and that fact immediately brought it into the realm of practical politics. It seemed that action was not far away. The Committee sat until 12 November and heard twenty-seven witnesses including Sandford Fleming, Alexander Muirhead, William Preece and Lord Tweeddale, the new chairman of Eastern Telegraph, who said though they had duplicated their line between Darwin, Penang and Madras the single line was more than equal to carry the traffic.

Chairman: Should it be decided to lay this cable for imperial reasons, would your company be prepared to work and maintain it for an agreed annual sum?

Tweeddale: Not a line from Vancouver to Fanning Island. The risks are so great that the company could not undertake that on any terms.

They would only land on Fanning Island if the Government told them to. The Pacific was much deeper than the Atlantic. A heavy strong cable would be difficult to raise and repair; a small light core would make working on it too slow.

John Lamb of the British Post Office claimed the Government could establish the system and manage it more cheaply than a private company. The associated governments could not raise the money at a lower rate than 3 per cent; but the imperial Government, which would have no promotion expenses, could raise it at $2\frac{1}{2}$ per cent. In England HMG could not 'tout' for traffic to this route as it had to act in fairness to all overseas telegraph companies operating in the UK.

I think that ownership by the Government might lead to demands for cables in other parts of the world of the same kind, and that this would still further disturb the relations of the Government with existing cable companies; in fact the Government would be interfering with private enterprise in a very special manner, and would be stepping on a slide which would carry it into difficulties. I think it would arouse resentment and contention and claims for compensation.

The Government had not competed with telegraphs inside the UK; it had bought them up entirely. The cables to the continent of Europe had been purchased by the Submarine Telegraph Company. The imperial government had never given a subsidy to procure the laying of a competitive cable or bring about a reduction in rates; merely given them to procure communication where communication did not exist.

The Imperial Pacific Cable Committee issued its report on 5 January 1897. It asserted its belief in the technical practicality of the project; urged the need of a full survey while the cable was being manufactured; recommended that the route be from Vancouver to Fanning Island, Norfolk Island, New Zealand and Queensland. They preferred a light cable to give forty paying* letters a minute, or five words of eight letters a minute.

The committee assumed the cable would carry 750,000 words in the first year. They recommended State ownership with a manager in London under the control of a small board on which those governments associated with the telegraph would be represented.

The Imperial Committee's report, for reasons which caused some controversy, was not published until April 1899. The discussions at the 'closed conference' of premiers of self-governing colonies which Joseph Chamberlain called during the Diamond Jubilee Year of 1897 were never made public. But the Colonial Secretary reported to Parliament that the subject of the proposed Pacific Cable had been raised, and the majority of the provinces had wanted it deferred until they had fully considered the report of the 1896 Committee which they had received in January. He reminded members that the United Kingdom was not taking the initiative but that the British Government would help and work with the colonies if they wanted the cable.

Such an attitude could not but lead to further delay. The Australasian colonies were not only disunited politically, they disagreed on the need for the cable. Western Australia and South Australia opposed it; New Zealand, New South Wales, Queensland and Victoria wanted it. These four colonies, each of whom owned their separate internal telegraph systems, would control the trans-Pacific traffic and if they had decided to go it alone, they could have

---

*Certain letters which had to be transmitted in a telegram were part of a service code used by the telegraph company to denote routeing, priority etc and were not part of the message paid for by the sender.

made the cable profitable. But they were unwilling to make a proposal to HMG which was not supported by the whole of Australasia.

For Sandford Fleming it was all a plot by Eastern Telegraph to cause even further delay, if not finally to kill the scheme. Eastern Extension was a remarkably profitable investment which its directors were unwilling to have upset. Telegraph traffic was increasing and with it their profits. It was easy to understand why they had never viewed with friendly feeling the proposed Pacific Cable. 'They have secured a rich monopoly, and their desire is to make it even more profitable, and to strengthen and perpetuate it.' The object of the Pacific Cable was to promote imperial unity.

> The proposition of the Eastern Extension Company submitted to the Conference of Premiers has no such purpose in view. Its subject is indeed the very opposite . . . The policy which animates the company would cause those communities [in Canada and Australia] to remain severed. Is such a policy to be commended? Does not the Eastern Extension Company, when persistently exercising its manifold and widely ramified influence to keep Canada and Australia disunited, assume an attitude of hostility to both countries and to imperial unity?

All the advantages of the Pacific Cable worked by the State could be attained without cost to the tax-payer. The capital would be obtained at the lowest possible rate of interest. There would be no possibility of enlarging it by adding 'promotion expenses' or by 'watering stock'. No dividend need be declared or bonus paid.

The Imperial Conference had left the Pacific Cable project in mid-air, said Fleming; but with Australasia disunited, Canada as the elder brother of the British Family should take the initiative. He was not a member of the government, however, and his appeals from the side-line could have no immediate effect. Queen Victoria's Diamond Jubilee Year came and went with the Pacific Cable Project still marking time, though faith in the future of Submarine Telegraphy was undiminished.

> When many of the passing crazes of the present time, many of its sensational inventions, have been consigned to limbo (wrote Charles Bright in the conclusion to Part I of *Submarine Telegraphs*), our great cable enterprises and the men who carried them out, will be noted by the historians of posterity as some of the most characteristic features and personalities in the civilisation of the latter half of this century.

Still more sensational developments of material progress in the way of communicating with our fellow men may be in store for humanity. We may learn to fly through the air on wings, or Mr Maxim may construct for us a new aerial leviathan, while Mr Edison or somebody else teaches us, with some new fish-like craft, to cleave swiftly and noiselessly through the still depths of the ocean instead of pitching and tossing in a painful manner on its summit. But looking ahead, say, as far as the rising generation and its immediate successors, it is extremely doubtful if any advance in applied science of such comparative importance to material civilisation will be seen for many a year as was marked by the successful laying of the first great submarine telegraph lines.

London management of the three English telegraph companies in South America was consolidated in the summer of 1898 when shareholders in Western and Brazil and London Platino exchanged their holdings for shares in Brazilian Submarine. Then the following June, Brazilian Submarine changed its name to The Western Telegraph Company, with W S Andrews, chairman of Brazilian Submarine, as chairman, J Axworthy manager, and J Hodson Steer secretary. There was little change, however, in the conditions under which the staff had to operate, and of course no re-juggling with the corporate structure could alter the weather.

The Western Telegraph Company Ltd
Rio de Janeiro Station, 21 Jan 1901

David McNeill Esq., Assistant representative.

Dear Sir,
    Enclosed please find application from Mr Browne asking permission to live out of Quarters. The state of the Quarters since the terrible storm on the night of the 19th inst. has rendered the building in most parts dangerous and uninhabitable. I visited the Quarters about half an hour after the storm had passed over. There were only three bedrooms dry, all the others were inundated; beds and personal effects were all drenched. The majority of the tiles were blown off; these I had repaired yesterday. The interior of the building and some of the furniture is most seriously damaged.
    I may mention that the Staff has been very discontented for some time on account of discomfort at the Quarters, and if urgent steps are

---

*This is one of the letters in the letter books selected for preservation by Dr Platt, and received from The Western Telegraph Company in Brazil in 1965 by the D M S Watson Library of University College, London, by whose kind permission it, and other letters from the deposit, are quoted. Padbury's letter is in the bound volume labelled 'Rio de Janeiro, Company Superintendent, January 1901 to July 1901'.

not taken to give the clerks more comfortable dwellings I am afraid several others will apply for permission to live out.

Percy Geo. Padbury, Superintendent*

There were eighteen young men in the Mess at the Western's Rio branch at this time, and the Mess President, Charles Martin, was allowed fifty dollars a month for breakages, but thirty dollars of this went regularly on replacing 'lamp glasses' – the glass funnels over the wicks of oil lamps which were particularly vulnerable. But Martin had other cause for complaint.

To O Fell, Superintendent   Rio de Janeiro   June 7, 1901

Mess Utensils

I am compelled by the great disadvantage we suffer through want of an adequate equipment of table necessities at the Quarters to apply for permission to purchase a new supply. Our present stock is barely sufficient to allow one knife and fork to each man when we are all assembled at meals. In spite of frequent purchases of table glasses we are still obliged to take the tumblers from the bedrooms upon occasions when we are a few more than usual at the table. Our supply of spoons is practically non-existent and our crockery is old, cracked and of divers patterns, besides being conspicuously deficient in many particulars.

Uncomfortable living conditions will have affected their operating accuracy. Rises in salary depended on their ability to transmit without error. The superintendent made out an Analysis of Errors for each clerk. When David McNeill received an application for a rise from J Evoy he referred to the operator's record for the quarter ending June 1900 and saw he had made sixty errors. Though he noted 'majority trivial/good conduct', McNeill wrote on Evoy's memo, 'Cannot be granted. Error average sixty. Matter referred to Head Office.'

Evoy and other 'clerks' like him who landed up in places far from home like Rio de Janeiro had been instructed initially in mirror reading and key manipulation at the company's London head office, which in 1902 moved from Winchester House, Old Broad Street to Electra House, Moorgate. Here circuits were operated between 'A' and 'B' rooms in the basement. The training equipment consisted of cable keys, vibrator recorders, hand perforators and pendulum-clockwork transmitters. After about twelve months in London, the probationer went for further training of twelve to eighteen months at

Porthcurno where one error in every thousand words was accepted as pass standard, and any greater inaccuracy deferred promotion or brought punishment.

In the mid-1890s there were three main working cables at Porthcurno: 'Por-Car-One' (to Carcavelos, Lisbon), 'Por-Vigo' and 'Por-Car-Two'. By 1898 the pattern of working circuits had changed; hand-translation was in fashion as the forerunner of chain working. Messages on tape were now gummed down instead of being written by hand, which enabled operators at Porthcurno to handle 4,000 messages a day. In 1898 'Por-Gib-Three' was added and in 1901 a cable was laid to Madeira. In that year there were 86 living in the main building and 105 sat down for the midday meal. The Instrument Room had doubled in size, half being the Landline Section and the other half the Submarine Cable Section.

At the turn of the century the name of the village had a 'w' at the end, as Arthur Graves recollected in 1971. The heading on the Mess notepaper was   Exiles Club
Porthcurnow,
Treen R.S.O.,
Cornwall.

Porthcurno, or 'PK' as it was always known (and still is), was then more remote than any outpost in South America. Once or twice a week a horse-drawn bus ran to Penzance, but the people of the valley rarely saw a face from the outside world. On the land which is now the car park grazed the heavy white horse, Simplex, which pulled the cart twice a week to Penzance for stores. A Penzance hairdresser named Coulson, who had a shop in front of the Western Hotel, drove into PK in his pony trap and set up his chair in the theatre in which the young men presented boisterous smoking concerts and plays. On a table he put a display of sweets, ties, collar studs, handkerchiefs, gloves and 'PK Caps'. For chaps whose school traditions had involved winning their colours and wearing house ties, crested blazers and straw hats with striped hatbands, the production of a PK Cap was a masterstoke and they sold like hot cakes. Shut off from domestic comforts they wrote home from the 'Exiles Club' in darkest Cornwall, the name given to it by some unknown wag which attached itself to all Messes patronised by those who had the experience of the original Exiles Club at PK. R.S.O. stood for 'Rural Sorting Office'. It was one of five which composed a twice daily, five mile, postman's round.

In those days PK had its own Rule Book, which was administered, if not written, by William Ash the superintendent on whom head office relied to keep discipline. Ragging and ritual had been the order of the day at the boarding schools these youngsters had just left. What their elders called 'undergraduate humour' predominated – indeed for most of them PK was their university. New arrivals from London in their best clothes would be met at Penzance station by a rowdy reception committee who cut the buttons off their jackets and threw their stylish bowler hats and boaters on trees. They were assigned to one of the four sitting rooms which constituted rival 'clubs' each with their own initiation ceremonies, after the manner of public school Houses.

The Rule Book forbad probationers to visit the Logan Rock Inn at Treen, but that made heroes of those who could boast of having bought a glass of cider from German Annie of the fair countenance and fearsome tongue. All pubs were 'out of bounds', as were all beaches other than Porthcurno for bathing, and even there on Sundays. Strong currents made sea bathing dangerous and there were frequent drownings among summer holidaymakers. Probationer Philip Saw followed in the tragic footsteps of Probationer Baglehole in trying to reach a kestrel's nest on the cliff below where the Minack Theatre now stands. He slipped and fell to his death. But for the most part surplus energy was absorbed by games of rugger, soccer and cricket.

Smoking was forbidden in the Instrument Room until after ten in the evening, and never by anyone under 18. Old Man Ash knew better than to challenge the young man of apparent under-age who he had been warned had had his eighteenth birthday that day. Pipe and tobacco had been bought for ceremonial lighting as ten o'clock struck.

Accused by probationers of lacking a sense of humour – that is, *their* sense of humour – in fact William Ash exercised considerable patience in his reactions to the practical jokes which tended to repeat themselves and become traditional as intake succeeded intake. The story of Cap'n Ash and the Cow Pat became legend. When Dick Peck drove a cow through the front entrance of the Quarters along the passage, and out by the back, Ash eyed the deposit it had made on the carpet. 'Aha, cow been here, eh?' he exclaimed in mock surprise to the little group of disappointed young gents who had waited to see him vent his anger.

The idea of forming an Eastern Telegraph Company Volunteers Corps at PK affiliated to the Electrical Engineers Volunteers during the Boer War was not taken up. And when in 1901 the War Office suggested that the Instrument Room should be housed in a bomb-proof shelter at a cost of £1,200, head office offered to provide the land if the War Office would build the shelter, but Eastern Telegraph were not prepared to allocate money.

In later years a favourite first posting for probationers was Bermuda, the island off the Atlantic coast of the United States which for long had been an important British naval base. A cable had been laid from the other base in Halifax, Nova Scotia as seen, and in 1898 the line was completed through to the West Indies as planned at the Barbados Telegraph Congress in 1862. The first section to Turks Island was opened on 31 January and to Jamaica via Windward Passage early in 1899. Thus the West Indies got their alternative all-British route to Britain and Europe – Halifax of course had direct access to the Atlantic Telegraph.

Most messages handed in at the only 'Cable Office' on Barbados, however – the first floor room ('upstairs') of a building at the corner of McGregor Street and Beckwith Place, Bridgetown – were for shorter distances. Since 1875, the Barbados Legislative Assembly had voted the West India and Panama a £2,500 annual subsidy in five-year grants, but in 1899, just before renewal, they decided to show their displeasure at the delays and interruptions by reducing the next subsidy to £1,500 and having to renew it every year. The trouble had been that on the strength of the company promising to duplicate their West Indies cable, the government had made them a ten-year grant, but the breakdowns continued in spite of duplication. This was taken very badly by the conscientious West India and Panama manager, J R Meade, who came from Montserrat and held the reputation of the company as a personal responsibility. In charge over a long period (he only resigned in 1930) he ruled his staff with a firm hand, but the high ideals of service which he instilled into the many young men he trained as operators had wide effect, and he became a legendary figure in the history of Barbados communications. It was largely due to the high regard in which he was held on the island that the government soon relented and restored the subsidy to £2,500 and five-year grants.

Staff in the West Indies and South America doubtless worked as hard as anyone else, but their brothers in other parts of the world tended to look down on them as an inferior breed. Eastern Extension regarded themselves as the upper crust – the cavalry of the service – superior to Eastern Telegraph, who in their turn considered Western Telegraph, West India and Panama and Halifax and Bermudas as very much the poor relations. A probationer who was posted to an Eastern Extension station thought himself highly favoured. One who was appointed to the staff of the Hong Kong branch in 1898 would have found a brand new Marina House opened that year on the site acquired in 1895. In the following year Hong Kong acquired its first 'overseas' telephone service when a line was opened between the police station on Hong Kong Island and the New Territories by means of a submarine cable under the harbour. The internal telephone system had opened on Hong Kong Island in 1882 with fifteen subscribers in the Victoria district, but it failed to attract any more, and closed in 1885. The service was re-opened two years later by the China and Japan Telephone Company.

Britain's trade with China represented a capital investment of only £30 million, a little more than 1 per cent of her total investment abroad, but it was half the whole China Trade. China stood at the end of the trade routes and telegraph lines. Neither goods nor messages went *through* China. But China was important if only by virtue of its size; the potential was enormous. In 1864 Palmerston had commended the commercial advantage to Britain of having unimpeded and uninterrupted commerce with one-third of the human race. Thirty years later France and Germany had joined her in the race to pour goods through the 'open door' of China which she was quietly 'colonising' in the informal way she had done in South America. China was dying as an integral empire, and Britain, France, Germany and Russia, to say nothing of Japan and the United States – 'the living nations' – were vying with each other to share her partition. In 1898 they competed to lend her money and exercise influence by acquiring concessions to establish ports, railways, telegraphs, banks, mining rights. In 1898 Russia occupied Port Arthur; Germany Kiaochow; Britain Wei-hai-wei and more of Kowloon on the mainland behind Hong Kong – the so-called New Territories. France opened the treaty port of Kwangchowwan west of Canton. China was in danger of disintegrating, and at a time of drought and plague in the north the Chinese decided they had had enough. In June 1900 they mounted an uprising in Peking to oust the 'foreign devils', spear-headed by a secret society the English called The Boxers who attacked foreign legations and killed the Japanese and German ministers. The remainder of the diplomatic corps sought refuge in the British Legation to which the insurgents then laid siege. The Chinese seized all the telegraph lines inside Peking. The nearest submarine cable was the Eastern Extension's at Shanghai some 800 miles to the south.

An international naval force captured the Taku forts on the seafront to the east of Tientsin, and sent a force made up of detachments from the various ships to try and reach Peking. But it was ambushed and had to retreat. The Chinese then disrupted rail and telegraph links to Peking. For weeks the area was cut off from all communication with the outside world. Rumours reached London that the British Legation had been stormed and everyone in it slaughtered. Arrangements were being made for a memorial service in St Paul's Cathedral when a message got through that the report was untrue.

John Denison-Pender, managing director of Eastern Telegraph, at once dispatched a ship loaded with cable from London to the Far East to join the cable ships of the Eastern Extension and Great Northern companies – the two operated a joint purse arrangement – who combined to extend the cable from Shanghai to Cheefoo and from there to the danger point at Taku. Another link was later made to the main British naval base at Wei-hai-wei.

With the establishment of a cable terminal at Taku close to a fort on the south bank of the river Hai-po, shortly after the relief of the legations in Peking, it was possible to keep the outside world posted on how the Boxer Rebellion was progressing. Telegraph stations were set up at Cheefoo, Taku and Wei-hai-wei. The operators maintained the high tradition of Eastern Extension by providing a twenty-four hours service in extremely trying circumstances. In 1932, A C Ussher, Controller of the Eastern Extension branch in Tientsin, wrote an account of these events in which presumably he took part himself.

> Amid the din of war, the stir and bustle of the constant movement of troops and many nations and the commotion from the landing of stores and munitions from a fleet of transports, the pioneer staff at Taku commenced their arduous day and night struggle to cope with the heavy traffic which included long and important government communications, full accounts for the newspapers concerning the operations and the peace negotiations.

The military laid down a field telegraph from Taku to Peking, though according to Ussher, communication with the city was also established 'by a private line erected by Mr Poulson'. But Eastern Extension and Great Northern at once re-erected the land lines of the Imperial Chinese Telegraph Administration which had suffered in the siege of Peking. By December 1900 they had completed the line to Tientsin, and in February 1901 it was open for traffic to Peking. A branch was opened in Tientsin in temporary offices inside the French Concession worked by Chinese operators under Eastern Extension's supervision and control. On this Ussher comments:

> Apparently it was not at first made sufficiently evident that the office and landlines were being worked and controlled by the company. The antipathy to being dependent on the Chinese in so important a matter as telegraphic communication was then particularly strong and those whose interests were opposed to the companies'

caused trouble by taking advantage of this to make people doubt the real control of the companies. The commercial community, however, were not long in perceiving the true position and gave their support more and more to the companies.

It was a stirring chapter in the history of Eastern Extension and added to the image of superiority of that part of the group which in 1901 claimed as a whole to control half of the 200,000 miles of cable then submerged, valued at £41,000,000.

It was the Jubilee of Submarine Telegraphy 1851–1901, for which a committee was formed to organise suitable means of commemoration and also 'to memorialise the leading part taken by Sir John Pender in the commercial development and extension of submarine telegraphy throughout the world'. 'The event of which the jubilee occurs to-day,' stated the leader in *The Times* of 23 August, 'marked a great step in the progress of humanity. Like all such steps it was taken in defiance of the opinion of the world in the face of obstinate lethargy and even active opposition. It succeeded because it was made by Englishmen of energy and willpower believing with all their heart in the possibility of what they meant to do and determined to prove their belief well-founded.'

The country's dependence on the telegraphy system established by Sir John Pender – and its not entirely satisfactory nature as a medium for conveying the subtleties of diplomatic thinking on which lives can depend – was demonstrated by the part it played in the events leading to the outbreak of the South African war. It was the only means by which the Colonial Office in London could guide the negotiations which Sir Alfred Milner, High Commissioner and Governor of the Cape, was conducting with President Kruger over the future status of the Transvaal. Joseph Chamberlain was determined to avoid war; Milner was ready to take the risk. Milner and Kruger met at Bloemfontein on 31 May, but when after six days each rejected the proposals of the other, without reference to London, Milner declared 'this conference is absolutely at an end'. In fact a cablegram was on its way from the Colonial Secretary begging Milner not to break off hastily, but the rupture had already been made. Milner took a week to reply to Chamberlain's message and then admitted he was wrong in breaking off 'quite as quickly'; he would not have done so if he had received the telegram from London in time. He was sorry they had been at cross purposes, but this was inevitable owing to the pressure and the impossibility of explaining everything by cable.

President Kruger issued his ultimatum on 9 October 1899, and the British Government faced a war in South Africa for which militarily they were little prepared. Neither was their only means of instant communication with the new theatre of war, the same as that used by Chamberlain and Milner, quite the strategic one the War Office and Admiralty would have desired – the highly sectionalised telegraph of Eastern Telegraph down the west coast of Africa or the lines from Porthcurno to Gibraltar, Malta, Alexandria, Suez, Aden and the east coast. In March, Cecil Rhodes had been received by the Kaiser in Berlin and come away with an agreement for the British South African Company to carry the lines of the African Trans-continental Telegraph Company through German East Africa; but there was little possibility of his Cape to Cairo land line being ready to help conduct the hostilities which broke out in October.

The fact that Cecil Rhodes was still actively pursuing his scheme and had had this success with the German Emperor, probably contributed to Eastern Telegraph's decision at once to lay a third African submarine cable from Cape Town to St Vincent in the Cape Verde Islands via St Helena and Ascension Island. That apart, the frailty of the two existing routes, and the risks in wartime of their passing through so much foreign territory and shallow coastal waters where anyone ill-disposed to Britain could easily cut them, were only too apparent.

Eastern Telegraph had Telcon dispatch its cable ship *Anglia* to Cape Town with 2,000 miles of cable, and the *Seine* with another 800 to St Helena. By 26 November the *Anglia* had laid the cable from Cape Town to St Helena; and by 15 December the *Seine* had connected St Helena with Ascension Island where a station was opened. The *Anglia* returned to England, collected another 2,000 miles of cable, and between 10 and 21 February 1900 linked Ascension Island to St Vincent, from where a line already ran into Porthcurno.

Ascension Island had been occupied by the British since 1815. Rear Admiral Sir George Cockburn, who landed on St Helena with the ex-emperor of the French in October of that year, sent one of his captains, James White, to take possession of Ascension Island in the name of His Britannic Majesty to prevent the French making it a platform from which to rescue Napoleon. The Royal Marines took over HMS *Ascension* in 1817, and a garrison were still there when a team of engineers from the Eastern Telegraph Company landed with their cable from St Helena at Comfortless Cove in December 1899.

With the 'aid' of the marines, they set about connecting the cable hut on the cove with their office in Georgetown some one and a half miles inland. But they had not reckoned on the officious attitude of

Captain R H Morgan, the senior officer of Marines, who continually felt it necessary to refer to the Admiralty (via the new telegraph) for orders, and took every opportunity to put the civvies in their place. Eastern Telegraph's superintendent, C A Paine, was the soul of tact, his patience with the moustachioed, monocled captain was exceptional; but the rocky terrain which needed blasting, if it was not to be avoided by tortuous routes through volcanic ash beds, and the haphazard methods of the Kroomen (Liberian negroes) who did the dirty work, meant that building the one and a half mile buried land line was not completed till 1904!

Landing rights presented no problem on an island which had been part of Britain's colonial empire by seizure since 1815. But it was not always as easy as that. Applying the most effective procedures for obtaining landing rights from foreign governments, often with the aid of the British Foreign Office, had become part of the admired expertise of the Eastern Telegraph group, but it took time. Granting landing rights for limited periods, and threatening not to renew except with certain stipulations, became a principal means whereby a government, used to operating its own internal telegraph, could control the private enterprise submarine companies. It gave them important bargaining powers. Appreciating the significance of this, in 1899 the British Government, who up to then had had the Post Office handle applications, appointed a Cable Landing Rights Committee to advise the Board of Trade who, as the department responsible for foreshores, regulated the granting of the rights. This was a standing committee of representatives of the Admiralty, the War Office, Foreign Office, Board of Trade, India Office and Colonial Office. It issued reports on each application and up to 1901, at any rate, all were unanimous. But two principles had always been in conflict – the protection of British interests and making England the 'telegraphic centre of the world'. In the early days the policy had inclined towards protection, but by the dawn of the twentieth century it was favouring free trade.

That it took four years to build the land line from Comfortless Cove to Georgetown on Ascension was unfortunate, but it in no way detracted from the triumph of establishing the new Capetown to Porthcurno link within so short a time after the outbreak of the Boer War. King Edward expressed his gratitude by making John Denison-Pender a Knight Commander of the Order of St Michael and St George. And to help them defeat the Boers the British Government enlisted the help of the very latest means of communication which the young Italian inventor had at last convinced the authorities could be relied upon over long distances without excessively high aerials, by sending a number of Marconi installations out to South Africa.

Proving the practicality of Wireless had taken time. During the three years of his stay in Britain, Guglielmo Marconi had resisted 'many tempting offers' for his patent rights from 'commercial gentlemen' which he did not positively seek. In his report to the Secretary of the Post Office in the summer of 1896, William Preece recommended that the Government should acquire them for £10,000, but the memo was pigeon-holed. They were not yet certain that it *worked,* or that Marconi's patents were valid as against those of others like Sir Oliver Lodge. Preece was asked to report on the practicality of Wireless for shore to ship communication – lightships. The company which the commercial gentlemen considered forming would have manufactured apparatus for giving ships warning of rocks, shallows and other ships ahead in foggy weather. A German, Professor A Slaby who went on to develop the rival Telefunken system, suggested to Marconi that he came to an arrangement with Algemeine Elektrizitats Gesellschaft of Berlin. The idea attracted him but he failed to agree terms. Finally, in July 1897 with the aid of his wife's English cousin, Jameson Davis, he decided to exploit his patents himself and formed the Wireless Telegraph and Signal Company; whereupon the Government's use of public money in allowing the Post Office to collaborate with Marconi in his experiments stopped. But a 'friendly relationship' continued, and the War Office carried out experiments on Salisbury Plain. The PMG's report for the year ending 31 March 1898 stated trials had taken place 'with special reference to its adaptability for lighthouse, lightship and other communications' but no practical results had been obtained – Marconi objected that this should have read 'no practical *commercial* results'. Lloyds also expressed interest and became one of his first and most important customers.

But if The Wireless Telegraph and Signal Company wanted to run a telegraph service as well as make and sell wireless apparatus, it could only do so, so far as the British Isles were concerned, under licence from the British Post Office which held the telegraph and telephone* monopoly as a state service. He took the first step in this direction in December 1897 when he applied for a licence to provide

---

*The Post Office had given The National Telephone Company a licence to operate the sole telephone service in the UK until 1901. In that year they renewed the licence for a final ten years and opened the first Post Office telephone exchanges. From 1911 the Post Office had the field to themselves.

Trinity House with a wireless telegraph by which the captain of the East Goodwin lightship could communicate with the keeper of the South Foreland lighthouse, and both of them with the 'stations' which Marconi's company had established on the coast to help him develop the system. The company undertook to hand over any messages received at these shore stations to the Post Office's Inland Telegraph System. The Admiralty, the War Office and the Board of Trade whom the Post Office first consulted had no objection, and the Postmaster General recommended to the Treasury that the company be granted a licence for this particular, limited operation 'subject to a royalty of a very moderate amount'.

When in August 1899 the chairman of The Wireless Telegraph and Signal Company sought clarification from John Lamb of his position vis-à-vis the Post Office, Lamb said they wanted to acquire the right to use the system. So two weeks later the company offered to allow the Post Office to use it within the UK for £30,000 a year. *If the Post Office paid £50,000 a year, that use could extend also from the United Kingdom to British Possessions overseas.* When on being told by the Treasury such charges were excessive, and the Post Office broke off negotiations, Wireless Telegraph and Signal made no approach to the operators of the biggest overseas telegraph service in the world who were based like themselves in Britain; neither did the directors of the Eastern Telegraph group see fit to approach Marconi's directors.

In his report to the Treasury on the validity of Marconi's patent, Preece pointed out that in 1899 Marconi had no commercial circuit at work anywhere. There was no case yet for granting his company a general licence 'to make money'. It was not a matter of filling a public want, for 'public appreciation of Wireless Telegraphy is only sensational, following the wonder of a solution of an apparently impossible problem.'

> A licence granted now would cause the [City] syndicates to play upon this ignorant excitement. A new company would be formed with a large capital, the public would wildly subscribe to an undertaking endorsed by the imprimatur of the Postmaster-General, and the Government would encourage another South Sea Bubble.

The Admiralty still expressed an interest, however, and were able to use Marconi apparatus under powers granted by the 1883 Patents Act. The practicality of Wireless over a distance, and across water, had been demonstrated in the summer of 1899 when Marconi succeeded in sending signals thirty-two miles across the English Channel between North Foreland and Boulogne. And then, in

November, at the request of the War Office, as seen, a number of Marconi installations were sent to South Africa for the Boer War. Owing to the difficulty in procuring suitable masts where the British forces were in action, they were transferred to the Royal Navy squadrons in Delagoa Bay, where they were so successful the Admiralty purchased thirty-two more sets.

In February 1900 the name of The Wireless Telegraph and Signal Company was changed to Marconi's Wireless Telegraph Company. With the British Post Office refusing to give the company a licence under whatever name it operated, the Telegraph Acts of 1868–9 prohibited it from instituting a competitive inland message-carrying service. However, there was nothing to stop it putting on a commercial basis the ship to shore service with which it had been successfully experimenting, so long as the equipment was supplied on a rental basis; and on 25 April 1900, Marconi's International Marine Communication Company Ltd was formed to do just this.

As far as Marconi's apparatus was concerned, there had been no change in its design for four years, but in April 1900 Marconi took out the famous Four Sevens master patent (no. 7777) which solved the problem of 'syntony', the tuned circuit which overcame the disadvantage of spark transmission shock-exciting a broad band of frequencies into being, by which a receiving station within range of several transmitters received a garbled mixture of all their signals. Marconi was the first to find a working solution to this inhibiting lack of selectivity, and from then on adjacent stations were able to transmit wireless signals without interfering with each other. When, as chairman of the Inter-departmental Committee on Cable Communications, Lord Balfour asked Cuthbert Hall, manager of Marconi's company, in August 1901 if wireless would be affected by a gale of wind, Hall replied 'Not at all. Nor by fog.' Only thunderstorms would affect signals, but the new tuned system had almost completely eliminated trouble on that score too. They were prepared to fit up installations transmitting 750 miles over land and 1,000 miles across sea. With cable costing £200 a mile, wireless telegraphy was cheaper than a submarine cable telegraph for any distance over ten miles. Their only permanent installation at that date was in the Sandwich Islands (Hawaii). They charged sixpence a word – half the cable company rate. As they had no licence from the British Post Office they had to keep outside territorial limits.

It was the syntony breakthrough that encouraged Marconi to embark on the 'wildcat venture' of defying all known laws and try to send a signal round the curve of the earth 1,700 miles across the Atlantic Ocean. The story of how Marconi, with G S Kemp and P W

Paget, managed to transmit the Morse signal for the letter 'S' from Poldhu in Cornwall to Signal Hill, Newfoundland, on 12 December 1901 has been told many times. It was, as Sir Oliver Lodge, Marconi's rival, wrote, 'an epoch in human history on its physical side, and was itself an astonishing and remarkable feat'.

But the Cable Companies, who were not interested in developing tele-communications as such, saw it differently. Their aristocratic managements were not trained to detect the significance of technological advance. They saw financial management on behalf of the shareholders of the company as constituted, and a certain paternal obligation to their staff, as the limit of their responsibilities. If they had been engineers, or had had the imagination to see how Marconi's telegraph system based on electro-magnetic waves with no need of wires which had been cradled in Britain, could be weaned and brought to maturity in Britain too, in partnership with the electric cable system, this story would have been very different. But they missed the opportunity which presented itself in 1901. So far from welcoming Marconi and his wireless telegraphy into the fold of those who provided instant communication services from Britain to the rest of the world – a telegraph centre with or without wires – they declared war on him. On 16 December Anglo-American, which Sir John Pender had helped to found and whose transatlantic cables Marconi had used in setting up his Atlantic experiment (which had sent no message for gain), instructed their solicitors to write him a letter pointing out that Anglo-American had a monopoly of the communications throughout Newfoundland, and forbidding him to make any further infringement of their rights under pain of legal action.

It was an action that Marconi never forgot. It now became firmly Cable *versus* Wireless; whereas in 1901 it might well have become Cable *and* Wireless. He at once dismantled his apparatus and took up the offer of Graham Bell, whose company had as much cause to fear competition from wireless as John Denison-Pender's, to continue his experiments at Cape Breton, Nova Scotia. Newfoundland could not disguise its contempt at Anglo-American's action. The Old World, conservative and traditional as ever, led by the Press, at first thought Marconi's claim too astonishing to be true and treated it with some disbelief.

It was the moment when Britain lost the initiative to the United States, as Marconi was to recall at the thirtieth anniversary celebration in London in 1931. 'The announcement of my success was received with scepticism by most scientists – principally in Europe,' he said. 'But not so in America. All the great American scientists believed in me; and the American Institute of Electrical Engineers was the first technical and scientific body to endorse my statements.'

On 1 April 1902, he registered the Marconi Wireless Telegraph Company of America to which he transferred the American rights in his inventions for £50,000.

To shareholders of the Eastern Telegraph Company at their fifty-ninth general meeting on 29 January 1902, Sir John Wolfe Barry, who had succeeded Lord Tweeddale as chairman in 1900, put up a brave show of confidence in the future of cable communication in face of the sensation caused by Marconi's transatlantic feat. He was a distinguished civil engineer, and he assured them that he and his fellow directors did not see wireless telegraphy as being in any way in competition with submarine telegraphy, and gave as the former's disadvantages, lack of secrecy, slow speed, liability of interruption, confusion of messages. But the management were worried nonetheless. What the chairman did not tell shareholders was that the company had contracted with N Holman & Sons of Penzance to erect a 170 ft pitchpine mast and hut on ground rented for twenty-one years at £12 a year above the cliffs near the cricket field at Porthcurno to monitor what Marconi was doing at nearby Mullion on the Lizard. This was the headland Pedu-Meor which became known as Wireless Point. In the small, two-roomed wooden hut undefined equipment, designed to pick up signals from Marconi's wireless transmitter, was installed by professional entertainer and prestidigitator Neville Maskelyne who had spent much time and effort in trying to prove that 'wireless' was a conjuring trick. This experimental listening post was visited from time to time by technical members of Eastern Telegraph staff from London to see, as Dan Cleaver describes in his *History of PK,* 'what could be picked up on the ether on gear they knew little about'. They sometimes heard calls, he said, from Marconi at Poldhu, from stations on the Scilly Isles, and from the Lloyds Signal Station at The Lizard Point. Now and then they picked up wireless telegraphy signals from a ship like the *Minneapolis* and *La Savoie.* There were suggestions for carrying out tests with the cable ship *Mirror* 'but not much came of them'.

This early exercise in industrial espionage was known to 'the enemy' from the start. Marconi's men were well aware of the purpose of the mast on the cliff, and it is on record that the first message the Porthcurno set was able to pick up was one from Poldhu informing an unknown station that the aerial was up but they did not know whether the cable company's instruments were capable of receiving them.

Cleaver quotes William Ash as writing:

> I understand from Mr Maskelyne that he is paying a visit to a neighbouring village for a holiday and will be here next Saturday. This is the day the Prince and Princess of Wales will be at Poldhu. As you will no doubt like to have all the information we can get, and as Trennan is here with Brown's apparatus, will you please tell me your wishes. We have as yet no proof that Brown's apparatus will be reliable. We expect the metal screen tomorrow. In case we get no satisfactory results from the Brown Chemical Radioscope, may I let Mr Maskelyne have a free hand to get a better record of what Poldhu is doing?

Later Ash wrote: 'Could you please ask Mr Maskelyne to send me a few more conjunctures, as I find different conjunctures give different results according to other conditions.' What, asked, Cleaver were 'conjunctures'? Probably part of the same apparatus that enabled Maskelyne to jam the lecture and demonstration which Professor Ambrose Fleming gave the following year at the Royal Institution to which Marconi sent signals from Poldhu. Fleming had no idea where the mysterious messages which came up on his receiver uninvited emanated from, but it is likely they were being transmitted from that hut on the cricket field at Porthcurno.

Spying on what they regarded as the enemy from their vantage point at Porthcurno was merely a diversion from the task of undermining the plans of their main adversary, Sir Sandford Fleming, Engineer of the CPR and leading champion of the projected Pacific Cable. Fleming had been knighted in 1898 and on 28 October of that year he wrote to Joseph Chamberlain from Ottawa taking the idea a step further. 'Is it not desirable that the whole British Empire should have a state controlled Cable System?' he asked. His plan would connect fortified and garrisoned coaling stations at Hong Kong, Singapore, Trincomalee, Colombo, Aden, Capetown, St Helena, Ascension Island, Jamaica, Bermuda, Halifax, Esquimault (Vancouver Island) and Thursday Island; and 'defended ports' like Durban, Karachi, Bombay, Madras, Calcutta, Rangoon, Adelaide, Melbourne, Hobart, Sydney, Brisbane, and Wellington. As far as pan-Britannic Commerce was concerned 'would it not be in the interest of a great commercial people to have all these and all such points in the outer Empire connected by a means of communication so perfect as the electric telegraph?'

> I am satisfied that the Pacific Cable would prove to be the entering wedge to remove forever all monopoly in ocean telegraphy, and free

the public from excessive charges; that it would be the initial link in a chain of state cables encircling the globe with branches ramifying wherever the British Empire extends, and that it would be the means of bringing into momentary electric touch every possession of Her Majesty.

He saw an imperial cable system in three sections – in the Pacific Ocean, the Indian Ocean and the Atlantic Ocean. Some 23,000 miles of new cable would be required, costing between five and six million pounds.

In a formal letter to Lord Salisbury, the Tory Prime Minister, Lord Tweeddale once again said Eastern Telegraph would be willing to establish a Pacific Cable provided it went via Honolulu. The arrangements the Government now contemplated were serious interference with private enterprise and were without precedent. HM Postal and Telegraph Department had hitherto always acted on a principle of alliance not interference with private enterprise.

Salisbury sent the letter to his Colonial Secretary for reply. Joseph Chamberlain gave his considered answer to all the company's points in a twenty-four paragraph letter signed by Lord Selborne, dated 10 July 1899. In this he observed, not without a hint of contempt, that it appeared that Eastern Telegraph were now prepared to do for nothing what two years before they had said required a subsidy of £32,400. He could not admit that there was any rule or formula of universal or permanent application limiting the functions of the State in regard to services of public utility, as the Eastern implied in their accusation that HMG was departing from principle.

> With the progressive development of society (wrote Selborne), the tendency is to enlarge the functions and widen the sphere of action of the central Government, as well as of the local authorities, and to claim for them the more or less exclusive use of powers and the performance of services, where the desired end is difficult to attain through private enterprise or where the result of entrusting such powers or services to private enterprise would be detrimental to the public interest through their being in that event necessarily conducted primarily for the benefit of the undertakers rather than of the public.

The tendency was specially manifested where the public were deprived of the safeguard of unrestricted competition, or where the State considered it inexpedient itself to enter on competition. It was a misapprehension to allege that the action of the Postal Department had been by any general principe. In certain spheres it had been in competition with private enterprise; in the case of inland telegraphs and cable communication with the continent of Europe it had entirely superseded the private companies.

HMG supported the Pacific route because the colonies wanted it; though it was the case that the route via the Cape would offer greater advantages from a purely strategic point of view. It was the duty of HMG to respect colonial feeling, 'the more so as it is the first time in the history of the Empire that two great Colonial groups have approached Her Majesty's Government with a view to a joint undertaking for the furtherance of commercial, political and social relations'. A scheme that diminished the cost of telegraphing and so promoted commercial intercourse deserved consideration, even though for a time it might diminish the receipts of the section of the public with pecuniary interest in existing cable systems. Mr Chamberlain was unable to see on what grounds the company claimed compensation. There was no general principle of non-competition by the State with private enterprise, It would, however, be the duty of HMG to avoid unnecessary injury to the interests of Eastern Telegraph's shareholders, though their primary duty was to the interests of the public. The company could have little cause for complaint if it was now exposed to competition when, by the time existing subsidies expired, Eastern Extension would have received nearly a million pounds from the Australasian governments, sufficient to cover a large part of the capital outlay.

> If the competition arose from another private company there would obviously be no claim for compensation, and Mr Chamberlain is unable to see why, so long as the project is conducted on commercial principles, fair competition by the State should give rise to a claim for compensation, which would not be suggested for a moment if the competitor were a private person or company, however wealthy or influential.

HMG could not admit the case of the acquisition by the Government of the English Channel cables in 1870 and 1889 afforded ground for the contention that the Eastern Telegraph group should receive compensation. Parliament aided railways and steamship lines by grant of special powers, mail subsidies and the rest.

In the memorial Tweeddale sent in advance of a deputation which was received by Chamberlain and the Chancellor of the Exchequer on 29 June, he said Eastern Telegraph had changed their policy to meet the 'sentiment' that had developed for all-British cables. At considerable expense they had given up their land line across France so that British telegrams might be transmitted entirely by cable, 'without the intervention of alien assistance'. The company did not claim favour, but was entitled to 'justice'.

Tweeddale summed up his case in a final letter of 19 July, replying to what had been said at the deputation and the long letter of 10 July. He accused the Colonial Office of shifting their ground during the two and a half years' wait before publishing the Pacific Cable Committee's report.

> Mr Chamberlain, when receiving the Companies' deputation, expressed the opinion that the Companies' fear that the Government was entering into this competition without regard to its own profit or loss was entirely groundless. I respectfully submit that I have shown the contrary. Mr Chamberlain concluded by saying, "if any government were mad enough, or the House of Commons were inclined, to allow any Government to spend the money of the taxpayers in order to enter into a violent competition for the purpose of destroying a private industry, then no doubt your position would be a dangerous one; but that idea appears to me to be wholly chimerical. The Government is not at all likely to do that. If any Chancellor of the Exchequer were found weak enough to allow it, the House of Commons would step in and prevent it".

But by the time that letter was written, Joseph Chamberlain had held a meeting putting the Pacific Cable Scheme into operation (4 July). The management of any telegraph service that might be decided on should be entrusted to a Board of eight members, three representing HMG, three Australasia and two Canada. HMG would guarantee the capital and control the method of raising it, and contribute five-eighteenths of any loss on working and maintenance. The Board should be set up at once and obtain tenders for the cable specified by the 1896 Committee. The Board would report the result of the tendering to the governments of the interested colonies, who would then have to decide unanimously whether to proceed on the lines recommended by the Committee or not, and what the rates would be.

By the end of the summer, Tweeddale realised that the Government had made up its mind, and on 28 September wrote to the Colonial Office admitting there was no purpose in continuing the correspondence. The work assigned to the Pacific Cable Board at the conference of 4 July was delegated to a Pacific Cable Committee of eight constituted in the same way. It met monthly and organised the establishment of the telegraph service which the Board would operate once the cable was laid and functioning. The Committee acted swiftly. They appointed Clark, Forde and Taylor their consulting engineers who had received four tenders by October 1900. The Committee found none acceptable; but Telcon's, whose tender was the lowest for each of the three parts, afforded a basis for negotiation. Their price was £1,795,000. On 26 November the Colonial Office authorised the Committee to give them the contract, and Telcon received it on 31 December. It was for making and laying the entire cable, estimated to be some 8,000 miles long, in five sections: from Doubtless Bay on the northern tip of New Zealand to Norfolk Island 500 miles to the northwest; from Southport, just south of Brisbane, Queensland, also to Norfolk Island; from Norfolk Island to Suva, Fiji (1,980 miles); from Suva to Fanning Island 2,040 miles north just across the Equator; and from Fanning Island to Bamfield on Vancouver Island.

Having settled these details the Government then asked Parliament to vote the money, which when the Pacific Cable Bill was considered in Committee on 30 July 1901, several MPs thought was treating the House with contempt. Austen Chamberlain, the Financial Secretary to the Treasury, sought authorisation to issue £2 million at 3 per cent out of the Consolidated Fund to be lent to the Pacific Cable Board. The money was to be raised by terminable annuities so it could be paid off in a reasonable time. The British Treasury had undertaken to raise the capital but were only responsible for five-eighteenths of it, the colonial governments* accepting liability for the other thirteen-eighteenths. Profits (or losses) were to be distributed in proportion to each government's contribution.

When the Bill came up for second reading on 12 August, John Henniker Heaton the Australian newspaper owner who had had to pay high rates to Eastern Telegraph for his cabled news, was a close friend of Guglielmo Marconi and MP for Canterbury, took the opportunity of making another attack on the Eastern Telegraph group.

> I know no monopoly in the world that is doing more injury to trade than the concentrated companies represented by the Eastern Telegraph Company and its six or seven satellites . . . I once described John Pender and Co as an octopus which, with its tentacles in every direction, is sucking the lifeblood out of the Empire . . . I regard the scheme now before the House as a great step forward towards the

---

*A federal union of the six colonies had come into being on 1 January 1901 as the Commonwealth of Australia – the second 'Dominion'. All the undertakings entered into by the three colonies for the Pacific Cable scheme were taken over by the 'Australian' Government's Post and Telegraph Department.

breaking up of one of the greatest monopolies the world has ever seen, and towards the consolidation of the Empire.

Austen Chamberlain said it was only when competition had become almost a certainty that the Eastern Telegraph Company, with its £1 million subsidies, offered concessions they had never contemplated before, in order to avert that competition. An Irish MP asked why the Government, which in 1899 had said the cable should be laid and paid for by the governments of Canada and Australasia, had changed its mind and were now building it. A clause in the Bill laid the entire responsibility on the UK to run it if it did not make a profit. Annual expenses should come from receipts, but 'so far as those receipts are not sufficient, out of moneys provided by Parliament'.

In the Committee stage Austen Chamberlain conceded that HMG's original proposal had been merely to subsidise the running of the telegraph; not take any part in the construction of the cable or the management of the service, and so not to take any of the profits. But that was not acceptable to the colonies. In deference to them HMG had agreed to be joint owners and be jointly represented on the Board of Control which had been set up in February.

For the Tories who introduced the Bill in this somewhat ham-fisted way their majority made its passing a foregone conclusion, and the Pacific Cable Act received the Royal Assent on 17 August 1901.

Surveying the route for the Pacific Cable began on 30 August. Most of Eastern Telegraph's objections were exploded. The bed of the Pacific Ocean did not consist of hard coral, but 'globigerina' ooze – the best possible resting place for armoured cable. The greatest depth was between Vancouver and Fanning Island – 3,400 fathoms or three and a half miles, not the fourteen miles guessed by the scheme's opponents.

Laying was begun the following year (1902) from south to north. Telcon's cable ship *Anglia* laid the Southport-Norfolk Island length between 8 and 18 March, and the New Zealand line to Norfolk Island the following week. On 25 March, Sir Joseph Ward, the Postmaster General of New Zealand, telegraphed Sir Sandford Fleming to say the first section of the Pacific Cable was completed, linking Australia and New Zealand. He hoped the end of the year would see the line at Vancouver to form 'another important link in the chain which binds together the whole of the British Empire'. In discussing the effect of this on Eastern Extension on 16 April, Sir Joseph urged governments to exercise caution in coming to any arrangement with the 'rival' operators. 'The probability is that before twenty years have passed Australia will be the joint owner with Great Britain, Canada, India and New Zealand of the whole of the cables serving these countries.'

The *Anglia* laid the 900 mile cable from Norfolk Island to Suva between 3 and 10 April. There was then a lull of five months. For the long stretch between Vancouver and Fanning Island Telcon had Wiggin Richardson of Newcastle build them a ship capable of holding the 4,000 miles of cable weighing 8,000 tons needed to do the work in a single voyage – a bigger load than the *Great Eastern*'s. They named it *Colonia*. It set out from Bamfield on 18 September and laid the 3,458 nautical miles to Fanning Island in just under eighteen days at an average speed of eight knots – a record. It was the largest section of cable ever laid in one piece – 285 miles longer than the French North Atlantic line from Brest to Cape Cod. This meant slow working on this section – 100 letters, or eight paying words, a minute, but 170 letters by duplex. The final section of just over 2,000 miles from Fanning Island to Fiji was completed by the *Anglia* in fourteen days. The contract completion date was 31 December but the *Anglia* arrived in Suva on 31 October, two months ahead of schedule. In the event the distance as the crow flies was 7,269 miles, but they sank 7,836 miles of cable. Charles Bright's son, also Charles Bright, thought the Pacific Cable would have been laid even more quickly if the work had been divided up between all the contractors who tendered.

The first message along the completed Pacific Cable was sent by Sir Sandford Fleming from Ottawa down to New Zealand and back to Ottawa on 31 October to the Governor-General of Canada. Joseph Chamberlain cabled the congratulations of members of the imperial government. 'They feel confident that the spirit of co-operation between the Mother Country and the Colonies which initiated the enterprise will gather additional strength from its successful accomplishment.' For that great imperial idealist it brought nearer the consummation of a Dream of Empire expressed by poet laureate Alfred Lord Tennyson which he was so fond of quoting:

> Britain's myriad voices call
> Sons, be welded each and all
> Into one imperial whole –
> One with Britain, heart and soul!
>   One life, one flag, one fleet, one Throne.

The achievement evoked as much emotion as the laying of the

Atlantic Cable forty years earlier. It inspired Rudyard Kipling to write the moving piece of which the following is a small part, but reflects the wonder that still possessed the lay majority in contemplating the miracle of the submarine cable.

> Here in the womb of the world, here on the tie-ribs of earth,
>   Words, and the words of man, flicker and flutter and beat –
> Warning, sorrow and pain, salutation and mirth –
>   For a Power troubles the Still that has neither voice nor feet.

6

War Boosts Traffic

1903–1918

The telegraph service from Canada to Australia and New Zealand run by The Pacific Cable Board opened to the public on 8 December 1902. The 'Board of Control' of which Sir Spencer Walpole was chairman, was accountable to HM Treasury to whom it had to submit an annual report and accounts which would be presented to the House of Commons and printed as a Parliamentary Paper. It was no mere policy-making committee but as much a telegraph operating company as Anglo-American or Eastern Telegraph. It had its headquarters in Queen Anne's Chambers, Westminster, with administrative and technical staff under a Mr Reynolds as manager. It owned fully-manned telegraph stations at Vancouver, Fanning Island, Suva, Norfolk Island, Southport and Doubtless Bay.

The Board fixed the rate for cablegrams from Canada to Australia and New Zealand at 2s 4d a word, as against the old 4s 9d. From Britain the cost was 3s a word; cablegrams could be handed in at the office of any of the Atlantic Cable companies or any post office in Britain.

The Board of course only operated the telegraph between Australia and Vancouver, which was truly all-British. Cablegrams sent all the way from Australia to Britain were re-transmitted eastwards across the 'all-British' territory of Canada, except for 270 miles when the line was diverted through the American state of Maine. But the telegraph companies which handled them were American-owned – the lines of the Inter-Colonial Railway which carried the traffic of the English two transatlantic cable companies Anglo-American and Direct United States. As Charles Bright junior pointed out, in the event of Britain going to war with the US, the Pacific Cable would be of no use to her. However, cabling Australia from Britain 'via Pacific' took only an hour instead of a day, and cost half.

In anticipation of this competition Eastern Telegraph had already reduced its rates, which many opponents of their monopoly considered was itself sufficient justification for laying the Pacific Cable. On 1 May 1900 they lowered the rate for messages between Europe and South Australia, Western Australia and Tasmania from 4s 9d a word to 4s. On 1 January 1901 this was reduced to 3s 6d and in another twelve months to 3s.

Eastern Telegraph agreed to make these reductions as their part of a bargain by which they obtained permission to open telegraph offices in the territory of the three Australian states which had refused to associate themselves with the Pacific Cable. Furthermore they provided them with a line to rival the latter by laying a new cable from Western Australia and Southern Australia to Mauritius which was already linked to Durban and thus Eastern Telegraph's coastal cables to Porthcurno – an Indian Ocean Cable. Australia had sent troops to South Africa for the Boer War and her commercial and social ties were increasing, so the idea of a direct link with the Cape, apart from duplicating the Eastern Extension route from the UK which terminated at Port Darwin and Adelaide, was very attractive. It was going to cost them £1,750,000 and they promised the rate to South Africa would be 2s 6d a word instead of the current 7s 1d.

Impressed by this agreement, the government of New South Wales asked to become a party of it, and they were admitted on 1 February 1901. That New South Wales, leading supporter of the Pacific Cable project, should do this only a month after the contract had been placed with Telcon for the Brisbane-Vancouver line, and without consulting her partners in the Pacific Cable, Queensland and Victoria, shocked the management in Queen Anne's Chambers and Sir Sandford Fleming, who said their action would seriously affect the economy of the Pacific scheme.

The laying of the Australia-Cape cable began in mid-1901 from Mauritius to the nearby island of Rodriguez; then northeast across the Indian Ocean to the Keeling-Cocos Islands, and down to Fremantle, the port of Perth. Messrs. Richmond, Dunbar and Davies were among the group who sailed from Mauritius in sailing boats *Chocolate Girl* and *Ebeneezer* with at least one wife and two small children to build a cable station on Rodriguez from scratch. Their 'quarters' were two-roomed bamboo huts with thatched roofs, no windows and doors at the front and back. Other occupants were rats, cockroaches, spiders, land crabs, mosquitoes, bugs, fleas and centipedes which fell off the roof on to the bed and were capable of delivering an extremely painful sting. The cable house was sent out from England on the cable ship *Anglia* packed into 470 cases.

The company leased forty acres of ground on Direction Island from George Clunies Ross the 'King' or, in legal parlance, the owner in fee simple, of Keeling-Cocos Islands. For £5,000 and an annual sum of £250, Clunies Ross gave Eastern Extension permission to land the cable on the island and establish a telegraph station and living quarters and gardens for the staff. The agreement made it clear that there was only one ruler on Cocos.

> The Company in buying from or dealing with the natives of the said island (stated Clause 5) shall not barter or pay in spirituous or alcoholic liquors or arms, gunpowder, dynamite or munitions of war; and in case any dispute shall from time to time arise between the Company and any native . . . .the Company will cause such dispute to be referred to the said George Clunies Ross, as the final Arbitrator, so that he and he only shall have the exclusive right of determining and settling what is right to be done in reference to such dispute.

All payments to natives had to be in parchment currency obtained from Clunies Ross. The concession was granted for thirteen years terminating 31 December 1919, and was renewable. The cable party from Singapore landed on Direction Island under a Mr Cameron on 6 August in a ship called the *Giang Ann* carrying stores and materials. They were housed in temporary buildings for two years until a ship came out from England with materials for a permanent station. The line from Rodriguez was landed on Cocos on 3 October and opened for traffic the next day. A line was then laid to Perth which opened on 1 November 1901. The following year it was extended from Perth to Adelaide.

The Australia-Cape (Indian Ocean) cable was a useful demonstration of the efficiency and speed of a private enterprise operation at a time when Lord Balfour's Inter-departmental Committee on Cable Communications were taking evidence from MPs like Sir Edward Sassoon who submitted a plan for laying 15,700 miles of new all-British cable costing £6 million. In addition, said Sassoon, the state should purchase all existing cable companies and expropriate their 51,000 miles of line now worth £4,580,000. He estimated the value of the cable when new was £7,575,104.

> Lord Balfour: Your policy is all to lead up to international communication being in the hands of the Government?
>
> Sassoon:　　That is it.

John Lamb of the Post Office said the principle of restraint, as opposed to free competition, was sometimes beneficial to the British Empire. Mere cheapness was not as important as security, rapidity and British control. If private companies failed to get a return on their capital and the Government paid them large subsidies, the foreigner would get the benefit of low rates at the expense of the British taxpayer. Four shillings a word was too high a rate to India. The India Office should suggest to the Persian, Russian and German governments that if they did not help to reduce the rate they would make other arrangements for through traffic when the concessions came up for renewal in 1904 and so deprive them of their tolls.

John Henniker Heaton proposed that the British and American governments should jointly acquire the property and rights of the existing cable companies at a fair valuation, and establish a common state monopoly in cable communication. The committee listened to him with rising exasperation and their report took little heed of his plans for reform. They were strongly opposed to a general purchase of cables by the state. But every important colony and naval base should be connected to the UK by at least one cable touching on British territory or neutral land. There should be a variety of alternative routes to ensure telegraphic communication in war. The normal policy should be to encourage free trade in cables.

The purchase of cables by HMG and the colonial governments was considered by the premiers who came to London for King Edward's coronation – Queen Victoria had died in January 1901. At Joseph Chamberlain's invitation they met in conference throughout the summer of 1902 and, without having the subject on the agenda, had the report of the Inter-departmental Committee 'laid before them' and a Memorandum written for the occasion by the tireless Sir Sandford Fleming 'On the Pacific Cable and the Telegraph Service of the Empire' outlining once again his plan for a state-controlled imperial system. No record was made of discussions at these conferences but the Parliamentary Return of October 1902 stated

> The attention of the members of the Coronation Conference was drawn to the risk, pointed out by the Committee, of hampering and checking the construction of cables by private companies if any general scheme of State construction or purchase were adopted; and the decision arrived at was expressed in the following resolution: 'That it is desirable that in future agreements as to cable communications a clause should, wherever practicable, be inserted reserving to the Government or Governments concerned the right of purchasing, on equitable terms and after due notice, all or any of the cables to which the agreements relate'.

30 In 1902 the Eastern Telegraph companies moved their headquarters to Electra House, 84 Moorgate (*above*) which, apart from being the administrative centre, was 'London Station', extending to Tower Chambers at the back. It was still standing in 1979 – the City of London Polytechnic – surmounted by the huge engirdled globe held aloft by sturdy cherubs.

31 On 1 January 1920 Eastern Telegraph opened a London Training School in this house in Shepherd's Walk, Hampstead, capable of taking 200 operators at a time, a figure which had become 350 by 1922, the year in which wireless training was added.

32 Third from the left in the middle row of this group of students at the Training School in Hampstead of 1925, is H. L. N. Ascough who was Engineer at Hong Kong branch when the Japanese raided and occupied the colony in 1941. He spent the war as an internee and on being freed was the first to give London the news of Hong Kong's liberation.

33 When this photograph was taken on 9 November 1916 these cheerful members of the Eastern Telegraph staff on Cocos Island were the prisoners of that armed German soldier in the topi on the extreme left. Shortly afterwards the German raiders withdrew without putting the station out of action for the rest of the war.

9th November, 1914

German Cruiser "Emden's" Raid on Cocos Cable Station.

COCOS STATION

H.M.A.S. SYDNEY.

EMDEN (WITH DUMMY FUNNEL)

Appreciation

by the Secretary of State for the Colonies, and Lloyd's Committee, of the Staff's prompt action in notifying the "Emden's" arrival, resulting, in great measure, in the Cruiser's destruction as expressed in following letters:-

Colonial Office,
Downing Street,
23rd November, 1914.

Sir,

With reference to the letter from this Department of the 13th of November I am directed by Mr. Secretary Harcourt to inform you that he has noted with much satisfaction the prompt action taken by the Company's Staff at the Cocos Islands in notifying the arrival of the German Cruiser "Emden", to which he feels the successful action fought by His Majesty's Australian Ship "Sydney" was in great measure due.

2. Mr. Harcourt trusts that an expression of his appreciation and his thanks may be conveyed to the Company's Staff.

I am, Sir,
Your obedient servant,
(sgd) Henry Lambert
for the Under Secretary of State

The General Manager,
The Eastern Extension Australasia and China Telegraph Coy. Ld.

Committee Room, Lloyd's,
Royal Exchange, London, E.C.
2nd December, 1914.

Sir,

The Committee of Lloyd's have recently had under their consideration the question of the good services rendered by the Telegraph Operators on Cocos Islands on the occasion of the visit of the German Cruiser "Emden", and the landing of a party of Seamen from that vessel on those Islands.

The Committee are impressed by the foresight and vigilance displayed by the Operators in keeping a look-out for the "Emden", and with the prompt measures they took immediately on sighting her to make her presence and movements known to the outside world by wireless communications and cable messages.

It seems undoubtedly due to this prompt action that His Majesty's Australian Cruiser "Sydney" was able to arrive so quietly on the scene, and to engage and sink the "Emden" before she had time to escape and commit further depredations on British commerce.

The Committee of Lloyd's therefore desire to express their high appreciation of the service rendered by the Operators on Cocos Islands, and they beg that you will be so good as to convey their thanks to the gentlemen in question.

I am,
Sir,
Your obedient servant,
(sgd) H.F. Inglefield, Rear-Admiral,
Secretary of Lloyd's.

The Manager,
Eastern Extension Telegraph Coy's,
Electra House,
Moorgate, E.C.

Cocos Staff.

D.A.G. de H. Farrant, Superintendent.
H.S. Ollerhead, Surgeon.

A.H. Cherry	R.A. Gowen	E.M. Preshaw	E.F. Shields
R.J. Saunders	W.H.C. Rowley	G.H. H. La Nauze	

Supervisors.

R. Cardwell	A.M. Griffin	A.R. Senthall	M.M. Stewart	E.J. Munro
F.A. Redfern	J.S. Trigg	H.M. Campbell	A.P. Peskett	F.P. Poltock
R.H.E. Green	P. Croft	E. Hall	J.S. Radford	F.B. Siddington
A.W.J. Peake	E.J. Beauchamp	E.W. Burnett	J.S. Gollhard	

Cable Operators.

J.W. King, Engineer.

34 The certificate of appreciation from the Colonial Office dated 23 November 1914 for the warning given by Eastern Telegraph staff on Cocos Island that enabled H.M.A.S. *Sydney* to sink the German cruiser *Emden*.

35 Vigo in north-west Spain had been an important Eastern Telegraph station from the early days – here is the staff in 1919.

36 The 'mess' at Suez decorated with hangings and flowers for the New Year's Eve party of 1920.

37–8 With the advances in wireless telegraphy the days of the submarine telegraph cables and cable ships seemed to be numbered, but the latter survived to serve the new co-axial telephone cables and were in greater demand than ever. The 1,984-ton c.s. *Lady Denison-Pender* (*above*) was built in 1920 and had a long and useful life ahead of her. The 1,850-ton c.s. *Mirror* (*below*) was commissioned in 1923 and served as a layer and repair ship for forty-one years. She was sold for scrap in 1964.

39 The seriousness of the competition from telegraphy without wires was brought home to Electra House when on 22 September 1918 a message from William Hughes, Australian Prime Minister, in London after visiting World War I battlefields, was transmitted by the long-wave Marconi Transatlantic Station at Carnarvon to E. T. Fisk, Managing Director of Amalgamated Wireless (Australasia) Ltd, who received it in Sydney 'instantaneously'.

40 The Marconi Company was selling 'the Twentieth Century Route' with some forceful advertising in the 1920s, indirectly knocking the 'old-fashioned' cable. The map shows how already in 1923 there were direct commercial wireless telegraph services via Marconi from London to Moscow, New York, Rio de Janeiro, Buenos Aires, Cairo, Beirut.

THE FIRST DIRECT
WIRELESS MESSAGE
FROM
ENGLAND TO S. AFRICA.

The Wireless Telegraph Company of South Africa Ltd

MARCONIGRAM

| Please quote this number in any enquiry regarding this message | No. | 1 | MILNERTON |
| RECEIVED *Via Marconi* at 12.5 A.M. m. | | | 3.12.24 RADIO |

PLDHU SJB*** NR.1.
LONDON VIA POLDHU
AND
MILNERTON

FROM *   TO*
THE HON. SIR EDGAR WALTON K.C.M.G.   GENERAL HERTZOG
     HIGH COMMISSIONER FOR S.AFRICA   PRIME MINISTER OF
                                                    S.AFRICA

TO GENERAL HERTZOG GREETINGS. WE SPEAK ACROSS
SPACE AND SOME DAY WE SHALL SEE AS WE SPEAK STOP MAY THE
NEW POWER PROMOTE PEACE BETWEEN ALL NATIONS STOP*
                                          WALTON

This Message was Transmitted from
Poldhu (Cornwall) Radio Station on
December 2nd 1924 at 10 p.m. Green-
-wich Mean Time Employing the Beam
System Operating on 92 Metres and
Received at The Milnerton Cape Town
Radio Station of the Wireless Telegraph
Company of South Africa Ltd

41 A forecast of television – 'we speak across space and some day we shall see as we speak' –
was contained in the first direct wireless message transmitted from England (Poldhu) to South
Africa (Cape Town) on 2 December 1924. The message went by the short-wave 'beam'
system, the patent superiority of which made Eastern Telegraph ask the British Government
to rescue them from financial ruin and led to a merger of wireless and cable interests.

42 John Cuthbert Denison-Pender, grandson of Sir John Pender founder of Eastern Telegraph, joined the company in 1900; was a prisoner of war in World War I; returned to Electra House in 1919 and became an MP and a member of the London County Council. He became Chairman of Imperial and International Communications Ltd in 1932 and was created the 1st Baron Pender of Porthcurno on being succeeded by Edward Wilshaw in 1936.

But HMG were deterred from establishing any new cable routes at this time, not for fear of upsetting the private operators but by the belief that the new means of distant and instant communication, wireless, might prove more efficient. The management of Eastern Telegraph had been disturbed by the ominous statement from Austen Chamberlain, now Postmaster-General, that he was 'awaiting events' before making further decisions on cable communication. But the morale of the cable companies was sustained by statements such as that made by Sir Sandford Fleming to the Canadian Press Association in February 1902 when he said while it was greatly to be wished that the highest expectations of wireless would be realised, they must recognise that some doubts had been raised, notably by Professor Oliver Lodge, Sir William Preece, Dr Muirhead and Lord Kelvin.

Asked about the future of wireless at this time, Sir William Preece said

> Causes of disturbance which may or may not be remedied in the future are at present existent, and fatal to the establishment of a practical and reliable system of commercial wireless telegraphy. I am therefore very clearly of the opinion that submarine cable enterprise has nothing to fear in a commercial sense from the competition of etheric telegraphy.

Sir Oliver Lodge confirmed this view:

> To the best of my belief submarine cables will for a long time be pre-eminent for the purpose of long distance telegraphy. It is manifest that wireless or open methods cannot compete in point of secrecy or certainty with closed or cable methods, and can only compete with them in point of speed and accuracy by aid of great improvements and new inventions involving little less than discoveries.*

The cable companies derived comfort from the fact that while the Government were sitting on the fence, which looked like being for a long time, they refused to grant Marconi the commercial facilities to enable him to prove himself. At the end of 1902 he confronted the Government with a request for an exclusive licence to practise wireless telegraphy between England and ships at sea, between England and foreign countries, and between England and the British

---

*One such invention came with the patenting on 16 November 1904 of the thermionic valve by Dr J A Fleming, Professor of Electrical Engineering at University College, London and technical consultant to the Marconi company. This detected the incoming signal more efficiently and sensitively than the 'coherer' and the magnetic detector.

colonies. He had understood, he said, from an interview he had had
with the then PMG, Lord Londonderry, in April 1901 that such a
licence might be forthcoming. But in a letter of 31 December 1902
his successor, Austen Chamberlain, disabused Marconi of any such
idea and told him categorically that it was not in his power to give
him the licence for which he now applied again.

Even if Austen Chamberlain had not let the cat out of the bag by
his statement in the House regarding 'awaiting events' before
ordering new cable routes, the intention of the Government would be
very clear, thought Cuthbert Hall, Marconi's managing director. In
a memorandum to his chairman he said:

> They see that wireless is fraught with all sorts of possibilities and
> that finality has not nearly been reached. They wish, therefore, to
> remain in a position to take the whole thing over without paying
> compensation to us when they can work it successfully themselves, or
> to shut the whole thing up if it suits their convenience for naval
> purposes or because of the cable interests.

To be fair to them, some of the cable interests did see a possibility
in making partners of the wireless interests. Sir John Wolfe Barry
told shareholders of Western Telegraph in London on 2 November
1904 that the company had been approached by different people for
coming to an arrangement by which telegrams might be sent
internally across short land distances by wireless to the station run
by a wireless company, and then re-transmitted by Western
Telegraph by submarine cable. This had been done by *The Times*
correspondents in China at the time of the Boxer Rebellion via the de
Forest wireless system at Cheefoo. 'I see wireless telegraphy,' said
Sir John, 'as a kind of "hand-maid" to the great cable systems of the
world.' That November Sir John Denison-Pender had a talk with
Lee de Forest (1863–1961), the American inventor of the triode-
valve, with a view to Eastern Extension taking a financial interest in
his syndicate, but the Board voted against an association with any of
the wireless companies.

Coping with the political aspect of making his system acceptable
must have been as exacting a process for Marconi as wrestling with
the technical problems. Marconi had his views on the equally critical
matter of commercial competency, but in this he more often than not
had to bow to the rulings of the 'experts' who ran his company.
There had been good publicity from the transmission of the first
transatlantic wireless telegram from King Edward VII to President
Theodore Roosevelt on 19 January 1903, and while in the United
States Marconi had claimed he could send messages across

the Atlantic with ample profit at 1 or 2 cents a word. But his fellow directors on the Board of his company, schooled in the conventional attitude that higher charges made for higher profits, insisted on 10 cents. It was a dramatic reduction from the 25 cents of the cablegram, but Sir Sandford Fleming for one, thought they should be able to manage 5 cents and open telegraphy to the masses.

One way of attracting more people to use the telegraph was to allow them to convey long messages by means of few words. In 1902 'to equalise the burden of telegraphic charges over long distances' Eastern Telegraph appointed Robert T Atkinson Manager, 'Via Eastern Telegraphic Social Code', a publication which he compiled and Eastern Telegraph issued at 5s for the benefit of the 'travelling public'. It contained 'a full selection of everyday sentences' which would afford complete facilities for communicating with their friends readily and cheaply. The rules laid down that code words must not exceed ten letters, must be pronounceable and could be in no language other than English, French, German, Italian, Dutch, Portuguese, Spanish or Latin. A sender was charged for two words whether she sent 'Belgrave Square' or 'Belgravesquare', as the latter was 'contrary to the usage of the language'. The Via Eastern Telegraphic Social Code was composed entirely of Latin words of ten letters and less, and is a tribute to Mr Atkinson's imagination. For instance any victim of a misfortune while holidaying with his fiancée on the Nile, say, had only but to cable his brother in Belgravesquare

> CAMBITAS   COQUARUM   PINNIRAPUS   CALERENT
> GNAPHALIO LEGAVIT LEUCOCOMOS

to convey

> Collar bone put out. Engagement broken off. Afraid it will cause unpleasantness. Please send supply of light clothing. Your name should not appear. Anything now said or done will be decidedly premature. Do not listen to any pretext but act as arranged.

The Electra House Group was now firmly and widely based, but its failure to stop the creation of the Pacific Cable Board in 1902 was the first real check in thirty years of continual expansion. Moreover the 'All-Red Line' was a threat to its monopoly and to private enterprise telegraphy which it considered 'unfair', and had to be fought with every weapon in its formidable armoury – political, economic, technical. For, once the principle of state-owned international communications was admitted, who knows where it would lead?

They reckoned to attract considerable traffic away from the Pacific Cable on to their Indian Ocean Cable, but it was difficult to counter claims that messages which took twenty-five minutes via Pacific took three hours via Eastern. They blamed their limitation to Fremantle and Adelaide as Australian terminals, and at once, as seen, they sought to remedy this with the promise of reduced rates as the bait. With all the Australian colonies federating from 1 January 1901, the new national P and T Department became responsible for their obligations to Eastern Extension and the payment of subsidies. As a result of further negotiations the company found themselves with telegraph offices in Melbourne, Brisbane, Hobart, Perth, Adelaide and Sydney.

Sir Sandford Fleming saw the Australian Government's concessions as threatening to ruin the commercial value of the Pacific Cable just when the Board was finding its feet. He considered Eastern Extension should be given the opportunity of transferring its new Indian Ocean Cable to the state, and the state should contemplate laying one from the Cape to Britain. The safety and well-being of the Empire was a matter of expediency. 'What would prevent a syndicate of German, French and United States stock operators buying up the controlling power of the Eastern Telegraph group of cables? What would prevent the controlling power of the whole telegraph system of the southern hemisphere passing into foreign hands? What would prevent the cables of the Empire being alienated at the most critical moment?'

The efforts of the experienced and powerful commercial group to harm the infant, whose birth they considered did them an injustice, prevented the Pacific Cable Board from growing as quickly as its partners would have liked, but nonetheless its early record was not unimpressive. To meet the interest on the £2 million capital and replace it in fifty years, the Board paid back to HM Treasury some £77,544 a year out of its revenue, and put £35,500 a year to a Cable Renewable Fund. In its first year of operation, 1903–4, gross message revenue amounted to £81,832, but owing to these two payments on top of normal operating expenditure, the accounts showed a deficit of £89,810 which was made up by the contributing governments. As the service got into its stride the gross message revenue went up and the deficit down:

	Gross Message Revenue	Deficit
1904–5	£84,301	£75,849
1905–6	£94,456	£72,556

By 1906, £84,000 had been spent on the Board's telegraph stations, plus another £29,000 on instruments.

Considering Sir Spencer Walpole and his colleagues had had no experience of running an international cable service – though presumably Mr Reynolds their manager had (a renegade from Electra House?) – the Pacific Cable Board were not doing too badly. It was not committed to making quick profits, or indeed any profits, and regarded its commission as essentially long term. They were providing a public service to accelerate inter-national business negotiations and the circulation of commercial intelligence. As Sir Edmund Barton, Prime Minister of Australia, observed in June 1903: 'The Pacific Cable is not for a day, but, it is hoped, for all time. Do not limit the view to momentary advantages but look ahead to a permanent benefit.'

As Sir John Denison-Pender had forecast, the splitting of the comparatively small traffic from Australia between two carriers was proving unhappy for both of them. 'This severe competition,' Sir John Wolfe Barry told his shareholders in May 1904, 'has entailed a considerable increased expenditure at all our Australian stations. The result of the competition is that the company's cable system in Australia is scarcely remunerative, the receipts barely covering working expenses.' He suggested a pooling arrangement but Sir Sandford Fleming thought that would only be the narrow end of the wedge by which Eastern Extension took control of the Pacific Cable. All attempts to come to an amicable understanding between Electra House and Queen Anne's Chambers broke down.

The forces which opposed Eastern Telegraph were formidable, but its second generation management were less capable of meeting them than the pioneers. They lacked the incentive to keep the momentum going in the direction the times demanded. Sir John Denison-Pender was not of the calibre of his father. After thirty years, hidebound by its purely cable terms of reference and actuated by inflexible economic principles, the company had settled down, become cosy. It was at this time that the Eastern Telegraph Company began to acquire a reputation as primarily the source of a fascinating but safe job; a company to which, once admitted, a young man, who had had to find a reputable city magnate to introduce him or be the nephew of the manager in Pernambuco, knew he had a job for life. There would be a place too, he well knew, for his son or brother. In March 1906 the company launched a family magazine, a monthly, price twopence, called *The Zodiac.* In an opening editorial the anonymous editor (J V Burke?) wrote that it would not be supporting any political party. It rested with his readers to say whether they would reach maturity. He invited contributions. It seems to have been a vehicle for staff literary aspirations, for there is no reference in the first issue to cables or telegraphy, though the last page carried photographs of the 'E.T.C.' and 'Coms' (Commercial Cable Co) football teams which had been playing each other.

The Zodiac would have helped to maintain the unity of a far-flung group and the Family Tradition which was giving the Eastern Telegraph Company its conservative image, defending its past and resisting innovation. Electra House was slow to tune in to the new mood, the new Liberalism which had come in with the resignation of Arthur Balfour in 1905 and the ascendancy in the political field of men like Henry Campbell-Bannerman, H H Asquith, David Lloyd George and John Burns, to say nothing of Ramsay MacDonald whose Independent Labour Party won twenty-nine seats in the general election of 1906. The Edwardians of Electra House found it difficult to react with resilience to trends which they resented as an intrusion on the past, which a greater awareness should have told them would never be halted.

Difficult but not impossible. In 1907 in South America, Western Telegraph under their concession from the Portuguese Government, were engaged in an important subsidiary wireless installation 'for moderate distances' in an area where a wire telegraph was difficult. On being asked for his views on wireless telegraphy in the light of press reports that a wireless telegraph service was shortly to open between Ireland and Canada, Sir John Wolfe Barry gave shareholders that piece of information as an example of the company's stance. He did not look upon any system of wireless telegraphy as a competitor because of the fundamental difficulties and imperfections, for instance lack of secrecy. He believed that about 25,000 paying words passed to and from the UK in a busy hour; New York expected to receive and answer stock exchange telegrams with London in four to five minutes. If that number of messages were to be attempted by wireless, there would be considerable confusion in the ether. Wireless was of great value in certain localities and for certain purposes. It was not seen as harmful to cable enterprise, but as of assistance. When a Mr Newby said he thought it might be advisable for Western Telegraph to have both the wireless system and existing cables, Sir John said the company was willing to carry out wireless installations at any station where cables were not suitable.

Keeping wireless at arm's length seems to have been the policy. With the aid of hindsight it is easy to label Sir John Wolfe Barry a King Canute defying the electro-magnetic waves to swamp his carefully structured enterprise, but in the first decade of the twentieth century the balance of probabilities was against wireless ever becoming the reliable method of later years. There was an element of wishful thinking in the company's public pronouncements, as would be expected, but erosion of their cable network by forces other than wireless interests were more pressing. Western Telegraph were fully occupied throughout 1908 in placating the nationalist aspirations of the Brazilian Government and fending off competition from France, Germany, the United States, and a rival British operator.

Siemens Brothers of London applied for permission to lay a cable from North Brazil to Mexico. So Western Telegraph decided to present a formal 'memorial' to the President of Brazil offering to lay the Rio-Ascension cable, to pay the Government's Telegraph Department their total rates on all local traffic to and from the new stations which the company suggested opening in connection with the new cable. In return the Brazilan Government was to ratify the company's 'right to free counter and delivery service and to sub-office in Rio' and extend the 1893 contract to 1933.

It all led, sadly, to litigation, with the Western Telegraph Company suing the Federation of Brazil. The Western's hold on South American communications was then further undermined by the US Government. Robertson, Rio, to Hodson, London, 23 August 1910:

> American Ambassador's representations against any extension of the Western Company's monopoly are said to have been accepted by the Brazilian Government. This would mean virtual annulment of our recent decree, the clauses of which are still under discussion.

London to Rio, 24 August 1910:

> Although we have been unable to learn anything definite at the Foreign Office, there is no doubt that the American Ambassador has made representations against any extension of our monopoly, and that these representations have been accepted in principle.

Relations between Electra House and the White House had been strained ever since Admiral Dewey of the US Navy had peremptorily cut the company's cable at Manila during the Spanish American War of 1898. Eastern Extension had claimed against the US Government, but there was no response until December 1906 when, in a message to Congress, the President had recommended reimbursing them for the expense of repairing the cable. No cash was received in London, however, till very much later.

The new Commonwealth of Australia was also displaying nationalist aspirations. In 1908, Eastern Extension's forty year concession to operate a telegraph between Victoria and Tasmania expired. It involved a £4,200 a year subsidy, and the government declined to renew it, announcing that they intended laying and working a cable of their own to Tasmania. They ordered the company to close its telegraph stations and removed its two cables, whereupon the company sued the government on a technical point regarding payment of a sum they said was due to them in connection with a guarantee. Though the Australian High Court upheld the company's case and ruled that the government's action had been illegal, it was far from good for Eastern Extension's image as a self-confident, financially stable provider of an international public service world-wide.

Much of the increase in revenue in recent years was derived from unplanned circumstances like the outbreak of the Russo-Japanese War* in 1904 (and subsequent falling off when peace was declared) and the earthquake in Valparaiso which had interrupted the rival telegraph in 1907 and diverted their whole traffic to the Western Telegraph line. As had often been said before, the easier part of the exercise was laying a cable and setting up the service – like the 820 mile telegraph from Cocos to Java for which a concession had been obtained from the Dutch Government in 1907, even though Captain Fawcus had been suspended for insobriety during the laying expedition and an outbreak of beri-beri among the Chinese workmen on Cocos had caused seven deaths. More difficult was what they laconically called 'maintaining the cable', that is operating a telegraph service along it and keeping it open for public traffic without interruption, particularly in places like the Indian Ocean where, at the beginning of December 1907, a typhoon caused severe damage to Eastern Extension's station on Direction Island.

But it was not the group's inability to cope with natural disasters that angered those who went to the Mansion House, London, on 11 December 1908 for an 'influentially-attended meeting of City merchants and others' chaired by the Lord Mayor to consider 'cable reform'. It was their unwillingness to lower their charges. Sir Edward Sassoon MP, whose views before the Pacific Cable Committee have already been noted, was in no doubt about the importance of the gathering. One day, he said, it would be looked back on as epoch-making. What they were trying to obtain was easy to realise and immediately practicable – a uniform system of cable rates, low-

---

* In which Marconi had his apparatus in the Russian army and the Japanese navy.

priced and accessible to the masses. He read a letter of support from Lord Curzon, who was on a health cruise, in which the ex-viceroy of India said he held most strongly that cable communications ought to be administered, not in the interests of classes [i.e. those in the upper income brackets able to afford the high rates] and corporations, but of the entire community. 'I have already lived long enough to see many things achieved which, to start with, were denounced as visionary or impracticable or inexpedient; and my own belief is that in a quarter of a century's time, and I daresay long before, our present cable connections and methods will be regarded as hopelessly obsolete and imperfect.' Sassoon pointed out it mattered little to the rich merchant or wealthy manufacturer whether his cable charges amounted to £1,000 a year more or less, but it was for the merchant in humbler walks of life, the artisan whose son or daughter was earning his or her livelihood in our colonies, that they pleaded. 'They can no more think of using the cables, these thousands and tens of thousands among us, than of dining off truffles and champagne.' The system they now had to rely on was obsolete and anomalous. What they wanted could be brought about by state-owned and state-controlled cables.

Rodolphe Lemieux, postmaster-general of Canada, said a logical sequence of the Pacific Cable was a state-owned cable service across the Atlantic. The population of Canada was seven million and emigrants were pouring in, 95 per cent British, at a rate of 300,000 a year. A uniform charge of 10 cents a word between Canada and Britain was needed instead of 25 cents. A state Atlantic cable would be the harbinger of a Pan-Britannic cable and telegraph system encircling the globe. Captain Muirhead Collins, representing the Australian Commonwealth, supported the plan.

A 'Cable Committee' of MPs had been formed, and the Duke of Argyll moved a resolution pledging the meeting's support of its efforts. He advocated the introduction of deferred telegrams at a cheaper rate out of business hours. Lord Milner, in seconding the resolution, said with rates so high compression of messages into shorter forms led to incomprehensibility. Sir Albert Spicer MP, as the representative of the London Chamber of Commerce, realised that existing private rights should not be disregarded, but the time had come when in the truest interests of the whole empire, cables connecting its various states should be owned and controlled by the different governments and not for any private interest. He moved that HMG should call a conference of the empire's P and T departments to concert measures, and Lord Jersey seconded. The meeting ended with agreement to send a deputation to the Prime

Minister asking him to submit the meeting's resolutions for the urgent attention of the government.

In one respect the urgency was greater than they knew, for within two years the sole remaining British transatlantic cable company, Anglo-American/Direct USA, was to pass under American control. It was bought by the American combine Western Union/American Telephone and Telegraph (ATT). It was the kind of event with its political overtones which Sir Sandford Fleming had warned could overtake any commercial telegraph company, even the Electra House group which was no less a 'property' to be bought and sold on the stock exchange than a shipping line or a textile mill. If it had happened with Anglo-American, it could happen with Eastern Extension. Thus in 1910 the England to Australia link via the Pacific Cable was by that amount less all-British.

The London merchants, the colonial governments, now the newspaper proprietors demanded lower rates – at the Imperial Press Conference held in London in June 1909. They too demanded state-owned 'electric' communication across the Atlantic, though whether with or without wires they were not yet prepared to say. The inventor of the wireless version was at the conference to help them decide. Marconi had opened the world's first commercial trans-atlantic wireless link with a limited service for Press and commercial messages between Clifden in Ireland and Glace Bay, Canada, in October 1907. At the Imperial Press Conference he promised them that in August he would have a regular service across the Atlantic at twenty-five words a minute costing half the rate charged on the sixteen transatlantic cables. When the limited service began his rates were a shilling a word for business and private messages and fivepence for Press messages. The new rates would be fivepence for the public and twopence halfpenny for the Press. Spurred by this, the Pacific Cable Board announced their intention too of halving the charge for Press messages.

When on 1 January 1905 the Wireless Telegraphy Act became law and Marconi's company was given licences for all its ship-to-shore stations for eight years, it looked as if he was going to be allowed to develop as a private telegraph operator in Britain after all. But then in 1909 the Post Office offered to buy his stations for cash. Since the Post Office seemed unlikely to renew the licences when they expired in 1913, the company accepted. The nine stations were transferred to the Post Office in September 1909. In the next four years they reorganised the service, re-equipped many of the coastal stations and built others in more suitable sites.

Perhaps Marconi's company was fated to become primarily an

equipment manufacturer selling mainly to governments? In 1909 it received £100,000 worth of orders from the Portuguese and Greek governments. In October 1910 the Governor of the Straits Settlements issued jointly to the Marconi company and Eastern Extension a licence to operate a ship-to-shore wireless service on the Keeling-Cocos Islands. Marconi made no charge for supplying the apparatus, nor Eastern Extension for installing it on their property. Marconi provided the operators, and the two rival companies shared the revenue equally when the station opened for traffic in March 1911.

The enterprising Barbadians made a bid for his services in 1913. After withdrawal of the Anglo-French Company, with whose aid the Barbados Government had planned to build a wireless station on the island in 1912, the Marconi company submitted proposals to erect one at their own expense in Government buildings and collect 37½ per cent of the station's earnings, or alternatively to sell them one outright for £4,256. Neither offer was in fact taken up, but a number of local wireless enthusiasts built a spark transmitter and crystal receiver with an aerial on St Ann's Fort which could work to ships 100 miles off the coast. These pioneers included A de V Chase who had gone to England for the coronation of King George V and obtained an instructor's certificate from the Army School of Signals at Aldershot. As a member of the Barbados Volunteer Force he formed the island's own Corps of Signals whose ranks included Basil Armstrong, later Director of the Marconi Research Unit at Chelmsford, and Harold Rose, for years the Pacific Cable Board's Wireless Engineer, West Indies. In favourable conditions their set could work Trinidad. The station, with call sign VPO, went on the air on 9 August 1914. It soon became part of the island's defence organisation. Harold Rose is thought to have picked up signals from the German raider *Karlsruhe,* and told the Royal Navy who promptly sank her.

However, the full transatlantic commercial wireless service which, in spite of the Glace Bay station being destroyed by fire, began in July 1909, gave Marconi hope of further development as an operator. The Australian Government had set up its own internal wireless system, linking Fremantle and Sydney. HM Treasury had told the Pacific Cable Board it could not consent to a new direct Australian-New Zealand cable as it might interfere with developments in wireless; and in 1909 a resolution was passed at a conference in Melbourne to establish a system of Pacific wireless stations. 'I do not believe,' said Sir Joseph Ward, ex-PMG and now Prime Minister of New Zealand, 'that even the Pacific Cable Board will be so retrogressive as to suggest that we do not establish wireless stations in the Pacific Islands for commercial purposes'. The rates would be the same as the cable rates. To keep those islands outside the area of the commercial world because that might injure the Pacific Cable, he said, was going too far. But in the event it was the Treasury – persuaded by Electra House? – who refused to grant the money unless the governments concerned gave an assurance the stations would not be used for commercial messages – and that was given.

With the thinking in the Empire sympathetic to wireless, Marconi presented the Colonial Office with a plan, conceived in 1910, for linking the entire British Empire with a network of eighteen wireless stations, and asked for licences from the Post Office for the home terminals and the help of HMG in securing licences from the self-governing colonies. His company would build, maintain, own and operate the stations at their expense – that is, without a subsidy. In promoting this idea he had the help of his dynamic new managing director, Godfrey Isaacs, brother of Sir Rufus Isaacs, Attorney General in Mr Asquith's second Liberal Administration of 1910, the first middle-class, 'left wing' modern government which laid the foundations of the welfare state, and whose measures were the subject of bitter opposition from the ousted Tories.

The Colonial Office referred the plan to the Cable Landing Rights Committee who submitted its report in March 1911, though of course it was not published. It rejected the plan, but suggested that the Government took it over. It recommended a state-owned imperial wireless system which the Marconi company would be invited to build.

No mention it seems was made of the Marconi Plan at the Imperial Conference which opened in London in June 1911 at the time of King George V's coronation, but the New Zealand Prime Minister, Sir Joseph Ward, moved a resolution that such an empire wireless system should be established at once. He referred to the resolution made at Melbourne in 1909 but not to any concept formulated in Chelmsford. Supporting the idea, Herbert Samuel the British Postmaster-General, named possible points to form the initial links in the proposed chain – England, Cyprus, Aden, Bombay, Straits Settlements and Western Australia. He suggested 'one of the wireless telegraph companies' be invited to erect the stations. If their terms were not favourable, the Admiralty would do it. But whoever built them, *the Post Office would work them.* An empire chain of wireless stations was desirable both for strategical and commercial reasons. But it should be state-owned. 'If it were in the hands of a company it could not fail to be a monopoly, and in an even higher degree than the cables are a monopoly, because while it is possible to lay various competing cables it is exceedingly difficult to have competing systems of wireless telegraphy along the same route on account of the danger of interference.'

A resolution proposed to the conference read:

> That the great importance of wireless telegraphy for social, commercial and defensive purposes renders it desirable that the scheme of wireless telegraphy approved at the Conference held at Melbourne in December 1909 be extended throughout the Empire with the ultimate object of establishing a chain of British, state-owned wireless stations which in emergency would enable the Empire to be a great extent independent of submarine cables.

It was the burden of the resolution they did in fact pass. It seemed that Marconi the operator was going to have to rely for his main project on the Ireland to Newfoundland service, though the conference laid plans for seeing that here, too, he would be confronted with state competition. Delegates were concerned that the England to Vancouver section of the All Red Line to Australia was far rom all-British. Sir Joseph Ward thought the governments should take over the Atlantic and cross-Canada cables and pay the market value to the owners. It would be one of the finest things fo the empire that had ever been done. It would not injure the shareholders of such companies any more than those of the telephone and telegraph companies had been injured which had been taken over by the state in the UK. The present capital of the cable companies at par was £27,902,000 and their receipts £3,163,000. The capital should be subscribed by the home and colonial governments who would own and administer the cable network jointly. There should be an Imperial Cable Conference in London. John Henniker Heaton had said it was contrary to public policy to leave a monopoly of communication between the several portions of the Empire in the hands of speculators. He (Ward) agreed with that.

On 15 June 1911 the Imperial Conference passed a resolution 'That in the event of considerable reductions in transatlantic cable rates not being effected in the near future, it is desirable that the laying of a state-owned cable between England and Canada be considered by a subsidiary conference'.

The Eastern Telegraph Company were not unduly worried at the prospect of a state-owned transatlantic cable. They could meet any competition, whether from the Post Office or the Marconi company, so long as it was fair and not subsidised by the Government – 'State

Socialism'. Electra House were still refusing to regard the new medium as a real menace. Their Wireless Hut at Porthcurno was wrecked by lightning in 1909 but, as Dan Cleaver remarked in his *History of PK,* 'throughout the years the Radio Watch on the Marconi installation continued to observe that nothing transpired which might lead to serious competition to the now long-established world-wide monopoly of the submarine cable system'.

But whoever was responsible for the 41,000 miles of the latter, state or company, their most valuable asset was still People – the maintenance engineers, telegraph operators and managers who *ran* the system. At the time the Imperial Conference of 1911 was sitting in London there were 150 operators at Porthcurno, of whom a large proportion were the Junior Brigade, lads 'well brought up' and of good education at minor public schools who had been enlisted between the ages of 16 and 17 and sent to Cornwall, as seen, after twelve months in London learning the rudiments of telegraphy.

The day at PK started with the first 'watch' from 8 am till 12 noon; then four hours off; on watch from 4 pm to 8 pm, and the rest of the day free. In the summer evenings there was just time for a game on the six-hole golf course which had been laid out across Trendrennen Fields. The other watch was 12 noon to 4 pm, and 8 pm to midnight. The seniors who turned up for the midnight watch would dress informally, generally a dressing gown and tasselled smoking cap. One of these senior men was in charge of each corridor of the unmarried juniors' quarters. For the juniors lights out was at 10.30; for the seniors, 1 am.

A good night's sleep was essential; tired operators produced the errors which gave the company a bad name. But not all mistakes were as unfortunate as the one which occurred in transmitting the press agency report of the sod-turning ceremony by Sir Arthur Kennedy for a new railway in Australia. The message

GOVERNOR GENERAL TURNS FIRST SOD

was received as

GOVERNOR GENERAL TWINS FIRST SON

From this the sub-editor who received it wrote the news item

> Lady Kennedy, wife of the Governor General of Queensland, gave birth yesterday at Government House to twins, the firstborn being a son.

Unfortunately Sir Arthur was an 80-year-old bachelor.
The chance of making such mistakes was reduced when at this

time a Creed Perforator, Auto-Transmission and a 'Through Numbers' system of handling telegrams were introduced. Instead of each message being re-numbered on each leg of the journey the traffic was grouped and zoned, so the message could pass unaltered to its destination. For instance, London's letter was A, and Bombay's K; so any message pre-fixed AK denoted a telegram from London to Bombay.

In 1912 William Ash retired. He had had forty-one years as superintendent at Porthcurno. The number of young men now spread all over the world who owed their training (and much else) to Old Man Ash could be counted in their hundreds. Stories of his eccentricities and kindnesses were told over and over again in messes from Vigo to Alexandria, Aden to Hong Kong, Singapore to Buenos Aires. The men of Electra House were greatly in his debt. He died in 1918 aged 73.

The new superintendent at PK was C Chevalier. It was difficult to follow a legend, and he only lasted three years. Holiday makers in motor cars started coming to the beach in the summer of 1912, and an added attraction for them was being allowed to see the operators at work in the Instrument Room. But the privilege was withdrawn when the superintendent thought it put the terminal in danger of attack by suffragettes. These tourists became the dupes of an ingenious confidence trickster who sold them fake 'local Cornish butter' which in fact was the cheap variety imported from New Zealand mixed with yellow colouring in his 'factory' in the valley. He sold it at twice the price he paid for it. He employed a carpenter to plane the New Zealand trade marks off the wooden cases it came in, and substitute 'local' labels. All inhabitants of Porthcurno were shareholders in the enterprise and drew large dividends. The man who thought up this scandalous exercise made a fortune.

A more sensational event of 1912 was the sinking of the new passenger liner *Titanic* after she hit an iceberg on her maiden voyage. Among the ships in the area which went to the rescue was the Eastern Telegraph cable ship *Mackey-Bennett* which happened to be at Halifax. After it had picked up a number of bodies from the sea, the cable ship was sent back to the scene of the disaster to locate the wreck and fix its precise position for the underwriters. On reaching the spot, the *Mackey-Bennett* lowered a grapnel which hooked the *Titanic* on the seabed, hove the rope tight and stayed there for twenty-four hours to take careful sights. The rope then had to be cut as it was not possible to detach the grapnel.*

---

*These facts are related by Sir James Bisset in his book *Tramps and Ladies*. Sir James was second officer of *Carpathia* which picked up most of the *Titanic*'s survivors.

Inevitably there was controversy over the part played by wireless in the *Titanic* rescue operation which to many seemed to take an unnecessarily long time to get under way. But in 1912 Marconi was well aware of wireless telegraphy's shortcomings and was enthusiastically attending to the problems they presented. What he, or rather his company, was not prepared for – he was never personally involved – was the sabotage of a non-technical nature which erupted at this critical moment, the 'scandal' which brought delaying action and not only retarded the development of W/T but deprived Britain of a communications aid at a time of her greatest peril.

The Imperial Conference of 1911 which agreed to establish the chain of imperial wireless stations entrusted a committee, of which Herbert Samuel the British PMG was chairman, to negotiate with the Marconi company on behalf of the Post Office for building them – as the Cable Landing Rights Committee, to whom Marconi had given the idea in the first place, had recommended. By March 1912 the two sides had agreed terms in principle and signed a statement (a 'tender') setting out the clauses which would be put into the legal contract for ratification by the House of Commons. The PMG gave no statement on this to Parliament, but the Marconi company announced the terms of the tender, which was signed on 7 March – a step which many, particularly the Tory opposition, thought ill-advised. Placing the order with the Marconi company was by no means a *fait accompli,* but many thought it was. The draft contract was tabled in the House of Commons on 19 July. By it the company was to build the first six stations at £60,000 each, and receive a royalty of 10 per cent of the gross takings at each station for twenty-eight years – so it was to have a continuing stake in the system as an operator as well as a manufacturer, though only a financial one. The respective P and T Departments would own and operate the stations in the distant countries.

The tabling of the draft contract came at the end of the Parliamentary session which left little time for the debate which the Tories were demanding. They considered the terms over-generous to the Marconi company, and in any event why not invite tenders from one of the other systems like Lodge-Muirhead? Could not the Admiralty build the stations? Was it, rumour said, because certain government ministers had a financial interest in the Marconi company? Between August 1911 and April 1912 its shares had risen from £2 10s to £9. The Press took up the story; by the summer of 1912 it had grown into a 'scandal'. Rufus Isaacs they said, had made £160,000 out of it and Herbert Samuel £250,000. Was not Rufus the brother of the Marconi managing director?

Isaacs and Samuel defended themselves against these 'wicked and utterly baseless slanders' in the Commons debate of 11 October on the motion that a Select Committee should be appointed to enquire into the Marconi contract. But Isaacs did not see fit to mention a transaction he had made in the shares of the American Marconi company in April. The Select Committee was not appointed until 25 October and its report repudiated the charges that the ministers had been influenced in their public duties by personal considerations or had used privileged information to gamble in Marconi shares. Rufus Isaacs was censured for his lack of frankness over the American shares in the Commons debate.

Another Select Committee, more important to this story, was appointed under Lord Parker to report on the Marconi System in comparison with others. This did not sit until January 1913. Sir Alexander King, permanent secretary to the Post Office, testified that neither the Telefunken, Lodge-Muirhead nor Poulsen systems had ever worked over the distances required by the proposed imperial chain. Marconi were the only people to have worked over 2,000 miles regularly. The creation of the chain was urgent, and that was why the Post Office had chosen the system that was ready *now*; whether Poulsen might develop a cheaper and better system in five years' time was irrelevant.

The Parker Committee issued an interim report in April 1913 that there was no doubt that 'at the present moment the Marconi Company alone has had practical experience of the sort of long distance work required, including experience in putting down stations, organising the traffic and staff, and in coping with the difficulties that arise in a new industry; and the value of such experience and organisation may well outweigh other considerations if rapid installation and immediate and trustworthy communication is desired'. This technical committee went on deliberating until July; on 4 July Samuel made a statement to the House; on 8 August the contract was once again debated and this time it was ratified with the basic clauses unaltered. Marconi, whose staff had been standing by all these months, were then given the contract. Ignorance and prejudice had nearly prevailed but now it was back to square one.

The democratic process had taken two and a half years. The Liberal Government had a majority in the Commons only by leave of the Irish members. If they had been in a stronger position no doubt they would have been able to ratify the Marconi contract before 1911 was out, and the first six stations of the chain would have been in position and working in time for the outbreak of the Great War on 4 August 1914. As it was, with the twelve months at their disposal, R N Vyvyan, whom the Marconi company put in charge of the contract, was able to select sites in Egypt (Abu Zaabal), India and England (at Leafield and Devizes), erect a few aerial masts but do little else. It was too late. The 'Marconi Scandal' had deprived Britain of a potentially powerful device for helping defeat the Germans, and at the end of 1914 the Post Office had no alternative but to cancel Marconi's contract. The mast system which had been erected at Leafield was used for intercepting enemy signals. Otherwise Marconi's imperial chain of wireless stations, conceived in 1910, was stillborn. His company sued the Post Office for £600,000 damages – and got them.

For the record, in 1915 the Post Office decided to replace the 1913 contract with a new offer which the Marconi company accepted, but no contract ever materialised. However in November 1914 the Admiralty gave the company a contract to install and man thirteen long-range rotary discharger stations, the first on Ascension Island. By June 1915, stations were in operation at Bathurst (Gambia), Ceylon, Durban, Demerara, Seychelles, Singapore, St Johns, Aden, Hong Kong, Mauritius and Port Nolloth. Later a station was built in the Falkland Islands.

Britain managed to have a Marconi long-wave transatlantic transmitter operating at Cefn-Dhu, a few miles east of Carnarvon, just five months before the war began – it was operated at first by land line from the receiving station at Towyn. But Germany had had her central wireless station operating at Nauen, just outside Berlin, since 1906. It was considered the most powerful in the world, and was based, of course, on the Telefunken system. She had another high power station at Hanover.

Outwardly the cable companies viewed this activity in the rival medium with calm detachment. 'As to the competition of wireless telegraphy affecting our revenue,' Sir John Wolfe Barry told the Eastern Extension annual meeting on 12 May 1914, 'I may say that it has not affected this company in the least. What little effect it may have on us may have been in giving us a few messages which came from ships to our station at Cocos. Consequently we have not yet experienced any loss from wireless telegraphy (applause). Personally I do not think we shall, but that must not be taken as a prophecy from your chairman.' The company had just laid another cable from Suez to Hong Kong via Colombo, costing £460,000; their gross receipts in 1913 had been £738,000 and their working expenses only £355,000 – a net profit of £343,000. They paid a 5 per cent dividend, plus 2 per cent bonus, and topped up their reserve fund which stood at £730,000.

With juniors leaving to join Kitchener's Army, it became difficult to find enough young men to maintain the group's vital world-wide cable service, and in November 1914 recruiting offices were instructed not to accept 'cable men'. One who joined the Training School at Electra House the day Britain declared war was Arthur Grass, under age for the armed forces, who in 1979 recollected that to be sponsored by Hon Arthur Brodrick (later Lord Midleton), one of the directors, as 'a nice chap to have in the company', was all that was needed to get him a place. After Porthcurno, where J G Marsden had succeeded Chevalier as superintendent in 1914, he was sent to Madeira for six months to finish his training on key perforators. He was given an inadequate £50 overseas clothing allowance to buy an obligatory list of items which included 'Balbriggan Drawers' (?), cholera belts, a cap, a pith helmet and an umbrella. His first real posting as a trained operator was to Montevideo at £72 a year; and then to Buenos Aires to become a member of the prestigious 'AO Dreadnoughts' operating the lengthy and faint cable to Ascension Island (AO).

As the terminal of the Eastern Telegraph group's network of cables linking Britain with so much of the outside world, Porthcurno was an obvious target for Britain's enemies. On 31 July 1914 a detachment of forty-three soldiers arrived at PK and were billetted in the theatre. They sand-bagged the windows of the office, ringed the buildings with barbed wire, built block houses and dug trenches on the beach.

Under the landing licences by which the cable companies were allowed to operate, the government had the right to requisition the companies' offices and control their traffic. Accordingly on 2 August the British Government took over the Eastern group's offices in London; on 3 August the transmission of all public code and cipher telegrams was suspended, and the public were only allowed to send messages in plain French or English and had to pay full rates. But the company sent any government message regarding British killed and wounded free, and telegrams exchanged between soldiers, sailors or nurses of the Expeditionary Forces and their relatives at quarter rates.

By its prohibition of codes and by insisting that messages were sent in plain language, the Government imposed its censorship on all overseas cable traffic. On 4 August a censor took up his duties at Porthcurno. The same day the Senior Inspector of Telegraph arrived and took control of the Wireless Hut on the cricket field which had been monitoring Marconi's station at Poldhu and been rebuilt after being hit by lightning in 1909. The aerial was taken down, the hut was locked and two soldiers put on guard. Emergency circuits, spare coils of wire and apparatus were buried in hidden caches round the valley. One night three weeks after war began a number of unidentified boats were said to be landing on Porthcurno beach. There was indiscriminate firing out to sea from the soldiers but there was no sign of an enemy landing. The commanding officer reconnoitred the shore, stepped on a pick handle, fell and broke his ankle. Penzance Hospital soon received its first war casualty.

The Cable War began at once. Eastern Telegraph had nine cable ships and the Royal Navy and the Post Office several others. Between them they cut the five cables of the German Atlantic Telegraph Company (DAT) and the German South American Telegraph Company, within hours of the outbreak of war, at the point where they entered the English Channel. The northern ends were buoyed in readiness for possible diversion at a later date. Some German operators at the distant stations continued to communicate with what they thought were their German terminals but turned out to be engineers on board British cable ships. These were Germany's only cable link with the USA. The cables to Britain were also cut. Germany's land lines to the Far East were blocked since all of them had to pass through enemy country – Russia and India. But with their wireless station at Nauen they could call the world.

It was not difficult to predict where the Germans would retaliate. Their main chance of totally disrupting Britain's imperial communications was to create a single, irreparable gap in each of her Pacific and Indian Ocean cables at their weakest and most accessible points, Fanning Island and Direction Island, Cocos. It must have occurred to the British War Cabinet that this would be a likely German tactic, but if it did they did not see fit to give either of these locations naval protection, build defences or provide them with artillery. On Fanning Island it was left to the twenty-six members of the Pacific Cable Board staff (and four white women), who were the link between Vancouver and Fiji, to make what preparations they could; and in the first week of the war they buried spare instruments, ammunition and small arms around the island in places they described in a document which they then locked in the safe together with $3,000. The Royal Navy kept them informed of the movements of the German cruisers in the area, and at the end of August sent a message stating that the *Nürnberg* and *Leipzig* were due off Fanning Island any day. The operator who received this pinned it up over his desk. At the insistence of the womenfolk a man kept watch at night, and early on the morning of 7 September he reported sighting a ship flying the French flag accompanied by a collier heading for Fanning.

The Pacific Cable Board superintendent, A Smith, assembled his staff and ordered a boat to go out and greet the French commander. When they saw two boats lowered from the French ship which then made for shore, the welcoming party stayed where they were. Before the boats grounded their occupants jumped into the water waist deep, and brandishing fixed bayonets and revolvers called on the group on the shore to surrender. The German marines from the *Nürnberg* which by now had hauled down its French flag and was flying the imperial eagle, soon had a Maxim machine-gun trained on the cable station, while a detachment surrounded it and an armed posse led by officers pushed through the door of the operating room. Smith and his men had been thoroughly deceived and Fanning Island had no option but to 'surrender'.

'Take your hands off the keys!' shouted the German captain in English. 'All of you!' he snapped as a clerk with a headphone over his ears continued to tap away unaware of what was happening. He lined the operators up against the wall and his men smashed the instruments with pick axes. The captain saw the pinned up message regarding the *Nürnberg*'s presence in the area, and took it as a souvenir. 'Rather interesting, don't you think?' he said with a wry smile. Failing to find the shore end of either the Vancouver or Fiji cable, marines blasted what was visible with dynamite, while others blew up the engine rooms, boiler rooms, refrigerating plant and dynamo rooms with gun cotton. They borrowed a saw from one of the staff and cut through the giant flagpole with the Union Jack at the top. The pole was cut into sections which, together with the saw and the flag, were taken aboard the *Nürnberg* as souvenirs. Out at sea the crew of the disguised collier grappled for the cable to wreak further destruction.

On looking through papers they had taken with them to the *Nürnberg* the Germans learnt of the hidden instruments and arms, and of the $3,000 in the office safe. So they returned to the island for a second visit, uncovered the buried apparatus which they then blew up, and took away the rifles and revolvers. While dynamiting the safe and taking the money, the German captain apologised to Superintendent Smith. It was the first time in his life, he said, he had ever acted the part of a burglar. He also expressed surprise that they had been offered no armed resistance. They thought that Britain would have taken every precaution to defend so important a cable station. In their twelve hours on Fanning the Germans acted with the utmost courtesy. The private quarters of the staff were untouched, and no one was hurt. The *Nürnberg* steamed away towards the Marshall Islands that evening with plenty of bric-à-brac to adorn their mantelpieces, leaving behind $150000 worth of damage which her captain ensured had not been repaired by making a surprise second visit two days later. The German cruiser was later sunk off the Falkland Islands by HMS *Kent*.

When the Pacific Cable went dead that morning of 7 September the Pacific Cable Board office in Vancouver took immediate action, and on the 17th, Captain E L Tindall was despatched with the steamer *Kestrel* with a cargo of instruments and cable to restore through working once again. They reached Fanning on 25 September. The Germans had towed the ends of the Fiji cable far apart, but one of the engineers, Hugh Greig, managed to buoy them and make a workable connection with ordinary covered wire within two days. The Vancouver end was lifted after three weeks. The German party's destruction of the operating room had not been as thorough as it might have been, presumably through their ignorance of submarine telegraphy. So the undefended Pacific Cable, which might easily have been put out of action for the rest of the war, was only interrupted for a month, and the cost of repairs amounted to a mere £5,000.

In the same period the German Imperial Chain had been effectively neutralised by the Royal Australian Navy. On 29 August they captured the German wireless station at Samoa; then their station on Nauru in the Marshall Islands and on Herbertshihe on New Pommern Island. The Duala station in the Cameroons was seized on 27 September and on 9 November the Japanese, who were on the side of the Allies, occupied Kiauchau and its station. The last powerful installation of the German chain fell in May 1915 when a force sent by General Botha from South Africa captured Windhoek in German West Africa. It was in Windhoek Gaol that the four operators of Britain's only cable station in enemy territory were imprisoned on the outbreak of war. They were removed to an internment camp 50 miles out and then marched 120 miles to Otjiwarango. After eleven months in prison they were released by General Botha's troops who repaired the cable.

When the gunners on board the German cruiser *Emden* shelled Madras on 22 September 1914 they were probably hoping to destroy the cable station and hole the link between Penang and Australia. If that was the case they failed. Though he had captured eighteen British ships, Captain von Muller realised that further activity in the Indian Ocean depended on him destroying the key cable and wireless station on Keeling-Cocos Islands.

At 6 am on 9 November a member of the twenty-nine man Eastern Extension staff on Cocos met a Chinese who told him a ship was lying off the entrance to the lagoon. Climbing on to the roof of the office, the operator saw a four-funnelled warship which he took to be the British cruiser *Minotaur* with which they had been in wireless contact for some days. But the station doctor who joined him was certain that one of the funnels was a dummy, made of canvas, and apparently the ship had no flag. They immediately flashed a message over the wireless, 'SOS Strange ship in entrance'. A few minutes later they learnt that ship's true identity. The message was changed to 'SOS *Emden* here'. The duty operator in the cable room saw a launch leaving the *Emden* with Maxim guns fore and aft, and towing two cutters. On reaching the jetty the boats disgorged a landing party of forty men with three officers and four machine guns. The cable operator who had been 'chatting' with Singapore tapped out '*Emden* at Cocos landing an armed party' and within minutes the news had reached Electra House, London, but more importantly the captains of the Australian warships which were always within steaming distance.

The German landing party split into three sections, each under an officer, which rushed the station, turned out the operators and posted guards. The element of surprise had been lost, and the man at the wireless transmitter continued to send out his SOS calls until the second party of Germans discovered him. The twenty-nine telegraph men, armed with a few twelve-bore shot guns and other small arms, were in no position to resist the forty-three Germans from the *Emden* with their machine-guns and rifles. They were herded into the Mess while the landing party dynamited the wireless mast, blew up the engine room, shattered the dynamos and switchboards and started to cut the cables with a handsaw and then, when this had little effect, with axes and files. They wasted valuable time hauling in and chopping up a spare cable that led nowhere, and only managed complete severance of the cable to Australia, leaving the Java and Rodriguez lines apparently untouched. While they were engaged in this orgy of destruction the siren of the *Emden* blew three times to summon their boats to return. The officers rallied their men, who staggered down the jetty with their machine-guns and spoils, followed by the unarmed British cable staff who took photographs of the departing raiders for the record. The Germans piled into the launches and headed for the *Emden* which was lying only some 500 yards from the beach. But before they reached her Captain von Muller sighted the Australian light cruiser *Sydney* bearing down on her from NNE. It had been alerted by the wireless SOS and cable calls from Cocos. Von Muller decided to weigh anchor before it was too late and manoeuvre into a more favourable position for engaging the enemy. To their horror the landing party saw the *Emden* putting out to sea with every intention of leaving them to their fate. While the guns of the Australian and German cruisers let loose at each other – the *Sydney* finally reduced the *Emden* to a burning wreck without actually sinking her – the landing party put back to Direction Island.

They planted the German flag in front of the Mess, and D A G de H Farrant, the Eastern Extension superintendent, and his men were once again made prisoners of war and put under martial law. Four of the staff managed to evade being rounded up and saw the final stage of the battle between the cruisers. The senior German officer told Farrant that in the event of neither *Emden* nor *Sydney* returning before night fall he intended to commandeer the three-masted topsail, 60-ton schooner *Ayesha* lying in the lagoon, belonging to Sidney Clunies Ross, the governor of the islands, and to try to make their way in it to German territory. They at once went on board and arrested the *Ayesha*'s master Mr Partridge and the two Clunies Ross sons, Edmund and Cosmo, who had left their Island to its own resources, whereupon the 700 native population had panicked and fled to South Island in boats, baths and bailers. The German lieutenant took half of the cable station's four months stores, and promised to cable the Eastern Extension office in Singapore as soon as he could tell them of the station's plight. He said should the *Sydney* return before they got away they would stay and put up a fight, in which case the cable men could make themselves scarce. One of the German sailors thought he had little hope of survival and asked for pen and paper to write a letter to his mother. The cable men were then 'freed', and there was general fraternisation. The stores were loaded on the *Ayesha* together with 150 gallons of water and old clothes like ducks and singlets. That evening the German ensign was broken on the *Ayesha*'s aftermast and she sailed out of the lagoon with a launch and two cutters in tow.

The men on Cocos at once sent a description of the day's happenings by cable to Singapore which was still through – the Rodriguez and Perth lines were tried without success. The *Sydney* returned the next morning to find the German landing party had escaped. The *Ayesha* had both her pumps out of order and took in four feet of water a day, but managed to sail across the Indian Ocean to Bombay where the Germans seized a British steamer and reached land somewhere south of Aden. The party made their way overland to Hodeidah in the Yemen and then marched north 400 miles through the desert to El'Lith where they planned to strike the Hedjaz railway. But they were attacked by Arabs, bribed to do so by the British. After three days' fighting and heavy casualties the survivors reached the railway and ultimately made Damascus and Constantinople, reaching Germany in June 1915.*

On 10 November F E Hesse, the general manager of Eastern Extension, telegraphed their office in Singapore:

> Convey to Cocos following resolution passed at meeting today: 'Board highly appreciate resource displayed and measures taken by superintendent and staff both before and after *Emden*'s visit to station and congratulate them.'

The Colonial Secretary sent a cablegram of thanks on 24 November. The *Sydney* picked up as many survivors of the *Emden* as she could, including the captain and Prince Franz Josef of Hohenzollern, and landed them on North Keeling Island, where the German cruiser had finally run ashore. *Sydney* later left with the *Emden*'s wounded. New telegraph instruments were sent up from Singapore, and Cocos station was in action again within twenty-four hours.

There had been no casualties among the Eastern Extension staff but three operators and the wife of one of them were killed when half of the 5th Light Infantry Regiment of India mutinied at Singapore on 15 February 1915 – Mr and Mrs Woollcombe, L P C Smith and G Wald. The soldiers were about to embark for Hong Kong and, incited by traitors in the pay of the Germans, suddenly turned on their officers, ran amok in the streets, firing as they went, released German prisoners from their camp and held Singapore at their mercy for several hours. The massacre of the cable staff took place in an attack on the Eastern Extension office and on their cable ship *Recorder* whose engineer A S Gardner and first officer T A Flett were both wounded. But the mutineers failed to cut the cable which ran through Singapore.

The enemy's poor firing left the cable station at Madeira undamaged when a German submarine penetrated the harbour, and the vigilance of the Royal Navy saved the important cable junction of St Vincent on Cape Verde Islands when Admiral Sturdee's squadron forestalled an attack by the returning German Pacific Fleet. During the ill-fated Dardanelles operation the company's cable ship *Levant II* succeeded in laying a line from Imbros to Suvla against great odds on the night of the Gallipoli landing for which its captain was decorated. In May 1915 there was an unsuccessful raid on the Pacific Cable terminal at Bamfield, Vancouver Island.

---

* The commander of this party, H von Mugge, wrote the story of this epic escape in *The Ayesha: A Great Adventure,* published by Philip Allan in 1930.

The Pacific Cable and Eastern Telegraph cables survived to maintain their service for the commercial and diplomatic community for the rest of the war in spite of the raids on Fanning, Cocos and Singapore (all of which, with a little more thoroughness, could have been totally wrecked), but within a couple of months of the outbreak there were complaints that the cable companies were profiteering. They had suspended the half-rate deferred telegrams in plain language. Since codes had been abolished all telegrams now had to go in plain language, but Electra House insisted on charging the full amount. A cablegram from England to Australia which would have cost £5 in code now cost £20. Eastern Telegraph had made concessions for cablegrams to the colonies and to Egypt by counting the number of letters and dividing by ten, and regarding each ten letters as a 'word'; but short coded addresses were not allowed and expressions like 'f.o.b.' were counted as three words. Resentment at the 'commercial' operation of essential communications was expressed in letters to *The Times* – 'How long will the Government connive at the accumulation of an illegitimate fortune by the cable companies? . . . the greed of the cable companies that persist in charging full rates is without a shadow of excuse'. But the Postmaster-General insisted that he could not interfere with how the companies charged except by a new Act of Parliament. However, when he renewed the many British cable landing licences which expired in 1915, he added a clause that he could call on them to reduce their rates. But notwithstanding increases in operators' wages and in the cost of cable and ship maintenance, Eastern Telegraph reduced their rate to South Africa from 2s 6d to 2s a word at the end of 1919, and two years later the rate to Canada from 1s to 9d. Weekend messages could be sent to Canada for 2½d a word, to Australia for 7½d and New Zealand 7d. A day letter service was introduced in September 1923 to Australia, South Africa and India at a quarter the full rate.

War, as always, stimulated the use of the telegraph, and Eastern Extension's gross receipts increased by £131,000 in 1915 to £950,000, while working expenses were only £386,000, making a new profit of £563,000 which, with the £31,000 brought forward from the previous year, made a total of £594,000. They paid an 8 per cent dividend in 1915 free of income tax, and decided that year that the directors' fees should also be free of income tax. In 1916 the receipts went up to £1,221,000 and the profit to £791,000, but £246,000 had to be paid in Excess Profits Tax, and Income Tax (6s in the pound). They put a quarter of a million to reserve, and spent £30 18s 3d on wireless telegraphy. In the last full year of the war the company took £1,508,000 gross. Eastern Telegraph's earnings followed a similar pattern – £1,462,800 in 1913, £2,901,500 in 1918, £3,365,000 in 1919.

While the Admiralty presented inscribed gold watches to Superintendent Farrant who sent out the wireless signals from Cocos picked up by the *Minotaur* which alerted the *Sydney,* and to Dr Ollerhead the company's doctor who had treated the *Emden*'s wounded, it seems to have occurred to no one in Electra House to use the inflated wartime profits to raise the staff's general rates of pay. In 1915 The Pacific Cable Board paid their telegraph operators £80 a year on joining and rising by £10 annual increments to a maximum of £220. They paid 'locality allowances' to supplement these basic salaries – for instance, £102 to operators on Fanning Island, £72 in Fiji, £36 in Sydney. Station electricians started at £250, Clerks in Charge at £320, and Superintendents at £500 with £150 allowance for Fanning Island.

The Pacific Cable Board asked their staff to sign on for a minimum of five years, and to retire on reaching the age of 55. They gave a month's holiday on full pay every year, 'subject to the requirements of the service', which could be accumulated up to six months. An 'officer' – all staff were officers – who married before he was earning £150 a year had to resign unless he could assure the management that he had private means to raise his income to that amount, but he would still have to live in unmarried quarters. Officers had to pay 5 per cent of their gross pay to a Provident Fund to which the Board made equal contributions.

Roderick Mann, on the other hand, was paid £5 a month (£60 a year) on joining the Eastern Telegraph Company as a qualified operator of 18 in 1918, plus £2 a month allowance to pay for an *appartement* above a café in Marseilles, to which he had been posted, where the station was run by middle-aged Frenchmen and had no Mess. He, too, was obliged to sign on for five years, but so far from getting a month's annual holiday, he was not allowed home at all in those first five years. In 1920, however, the company introduced 'local leave'. In his third year his salary went up to £6 a month and in the fourth to £7.

But the spartan life at Porthcurno had led him to expect little by way of luxury at the far-away station to which he knew he was fated to be posted. In 1915 the unorganised 'training', which relied on picking up the techniques as an assistant to the qualified operators running the cable station, became more formal. For in that year a definite school was instituted for the first time, giving an advanced course under an instructional staff. Those dozen or so who were selected for this each month from an Electra House Training School class of thirty considered themselves the cream of the probationers. They became Eastern Telegraph employees. Those not sent on to PK were posted direct to a Western Telegraph station in a remote part of South America, and their former classmates rarely saw them again.

But the Chosen Few had to show they had it in them to take more than instruction. The guard on the train from Paddington could always detect the monthly party of Eastern Telegraph boys smartly dressed in their well-cut suits, sporting the latest fashion in curly bowler hats, but never let on what lay in store for them on arrival at Penzance. He well knew that *they* would be the reception committee on the platform who descended on *next* month's party, seized their hats and handed them to the fireman on the locomotive at the front of the train who promptly confined them to the glowing firebox. Hats were not allowed at PK, as already seen. During the long ride in the horse-bus to the lonely valley they learnt that the school uniform was WOGS – Worn Out Garments – and that meant jackets without buttons, shoes without laces, shirts without ties. Wire and string were the fasteners of WOGS, and it was no use protesting, for all that first week the New Boys were 'sent to Coventry' and none of the Old Boys would talk to them. On the first Sunday there was a sing-song at which each entrant had to stand on a table and perform. The ordeal over, the New were accepted by the Old, and they were all Exiles together.

In wartime there was no visit from Mr Coulson the hairdresser. Those who had no bicycle walked, and the others cycled into Penzance making sure to bait the village policeman at St Buryan on the way by shouting and yelling and extinguishing their oil head-lamps. Before making for home they waited for the staff of the American cable company in Penzance to leave their office, then crept in and stole their electric light bulbs. They had to have candles in their bedrooms at PK; the only electric light was in the dining hall, sitting rooms, library and theatre. There was no hot water either, except after those rough rugger matches against the Camborne School of Mines and other local teams. Roderick Mann relieved the routine by accompanying Jackie Marsden, the superintendent, on his archaeological walks looking for flint tools. After four months he took his turn on the twenty-four hour Relay Watch.

Before Roderick Mann left Porthcurno, he shared in the excitement at the appearance one evening in 1917 of a cable ship off Mousehole, a fishing village near Penzance, paying out a 'secret' cable, which many of the PK staff went to help land. The German

Emden-to-Fayal (Azores) cable, the first section of its line to the USA, had been cut in 1914, and now the north end had been picked up in the English Channel and was being diverted to Britain. The link from Fayal to New York had been kept dead since August 1914, since it terminated in a neutral country. But on 6 April 1917 the United States joined the Allies, and on the 24th the cable ship *Colonia,* escorted by a Royal Navy cruiser, cut the cable 400 miles off New York and pulled the Azores end into Halifax, Nova Scotia. Here it was connected to the office of the Pacific Cable Board. The whole telegraph from Halifax to Mousehole, and thence by land line to the Central Telegraph Office of the Post Office in London, was opened for traffic on 18 July 1917 in strict secrecy. As a prize of war the cable belonged to the British Government and the service was operated by the British Post Office, who in this way first assumed the guise of competitor to the commercial cable companies – and without their having the opportunity to object.

Thus war solved the problem of providing Britain with the long desired all-British cable across the Atlantic to link with the Pacific Cable Board's line to Australia and New Zealand. The British end was later diverted from Mousehole to Penzance (and in 1929 to Porthcurno). Their Emden-Azores-North American cable was not returned to the defeated Germany under the peace treaty, much to the United States' disapproval who wanted a link to Berlin and complained at the Washington Communications Conference of 1920. It was retained by the British Post Office who, two years later, acquired a second transatlantic cable, this time by purchase – the Direct US cable which had been laid to New York. But the Post Office had other plans for it and diverted it to Halifax, with a spur to Harbour Grace, Newfoundland. This nationalised transatlantic telegraph service with two cables to Halifax was named the 'Imperial Cable' and the route designated 'via Imperial'. Whitehall sent all its official cablegrams to Canada, Australia, and New Zealand 'via Imperial' and hoped other imperial governments would do the same, which was not always the case. The average transmission time via Imperial between London and Montreal was three-quarters of an hour for full-rate telegrams.

No one, least of all Electra House, could grudge more cables becoming available to help cope with the great upsurge of traffic, reckoned at 150 per cent higher than in 1913, which the cable companies were hard pressed to carry. Before the war Britain had had nine lines of communication with India and the Far East. Six of these were the submarine cables of Eastern Telegraph. But 40 per cent of the traffic had been handled by the two mainly overland

Indo-European lines, and 30 per cent by that of the Danish-owned Great Northern company. These three lines passed through Germany and Russia, and service over them became impossible immediately hostilities began. So the whole of the India and Far East traffic came to Eastern Telegraph, though their burden was eased by Indo-European establishing a wireless link between Bahrain on the Arabian Gulf and the international cable system at Bushire in 1916 – the first messages were transmitted on 7 June. But while this gave good communication to India, the service to the Far East was restricted when the Russian Revolution of 1917 disrupted the routes through Odessa and the Black Sea, which prevented the Great Northern from carrying their share of the China and Japan traffic. Aggravating the situation was the general tendency to send a telegram on occasions when, in peacetime, a letter would have been mailed, in view of the postal delays caused by the withdrawal of shipping for war purposes.

Ill-health caused Sir John Wolfe Barry to resign as chairman of the Eastern Telegraph companies at the end of 1917, and on 22 January 1918 he died. It was his successor, Sir John Denison-Pender, who had to bear the brunt of the criticism at the delays which grew even heavier after the armistice had been signed in November. Already a KCMG, in recognition of the contribution that the group's cables had made to victory in the Boer War, Sir John had just been made a Knight Grand Cross of the Order of the British Empire for similar services in the war just finished. The son of the founder, the first Sir John Pender, he had joined the group in 1881, been managing director of Eastern Telegraph since 1893 and later vice-chairman. To take over as managing director he appointed Captain H W Grant, a serving officer in the Royal Navy who at the outbreak of war had been in command of HMS *Hampshire* in eastern waters and later joined the Admiralty staff in Whitehall where Sir John had dealt with him for some three years. The Admiralty agreed to release him, and he was appointed managing director of the Eastern Telegraph Company on 13 March 1918.

At the first Board meeting taken by Sir John Denison-Pender (in October 1918) it was the 38-year-old Edward Wilshaw who conveyed to the new chairman 'of the largest cable company in the world' the congratulations of the foreign 'Marine Staff' and the hope that he would administer its affairs for many years 'with that unfailing courtesy, kindliness and impartiality so well known throughout the service'.

The excessive cabling delays which the commercial community protested were hampering international trade were caused mainly by the fact that priority had still to be given to Government messages which took up a quarter of all traffic carried by the Eastern Telegraph group in 1918. It was not only the number of messages sent but their unabbreviated wordiness. In 1913 Government cablegrams averaged 3,000 words a day; in 1918 30,000. With the trans-India and India-Far East traffic increasing 327 per cent during 1918 over 1913, there was little wonder that a cablegram from Calcutta to London sometimes took seventeen days. In 1913 the average commercial cablegram contained twelve words; in 1918 twenty-one. With the same network to carry them, the total number of paid words transmitted by the Eastern Telegraph group rose from 64 million in 1913 to 180 million in 1918. News was becoming stale in transit and the Press missed no opportunity to print complaints from those who also suffered from delays – the cablegram which took ten days to reach Hong Kong, the Manchester firm who on 6 March received a cablegram despatched from Egypt on 25 February, only two days faster than the post.

Censorship and the ban on private codes continued until 15 July, but their removal produced more words not less. At the end of 1919 the company was able to charter from the Ministry of Shipping a captured German cable ship, the *Stephan,* one of the largest in the world, and lay a second 1,200 mile fast cable from Porthcurno to Gibraltar, designed to be the first section of a new 7,000 mile Far East submarine cable through the Mediterranean via Bombay and Madras to Singapore. In the other hemisphere, Western Union agreed to bring a cable down from Miami, Florida to Barbados to join the cable Western Telegraph were laying to the island from San Luiz de Maranhao in Brazil. A relay station was built on Barbados for this North-South America cable at Dover Beach, Christchurch. The joint Western Telegraph-Western Union service was opened from Barbados to South America in 1920, and to North America the following year. This was the first appearance of Western Telegraph in Barbados, and it was their staff, not West India and Panama men, who operated the station. The first superintendent was K C Wood and from the outset he employed locals to assist expatriate staff. One of the latter, whose posting to the Dover Beach relay station was popular with the island's tennis community, was A Vernon Hayne who had been Open Singles Champion of Rio de Janeiro, and captained the Savannah Club against Tranquillity of Trinidad.

When questions were raised in the British House of Commons about the cable delays in the summer of 1919, they were answered by John Cuthbert Denison-Pender MP, Sir John's 38-year-old son, who joined Eastern Telegraph when he left Eton in 1900. In 1906 he married Irene de la Rue, and in 1913 was elected Tory MP for Newmarket. He joined the army on the outbreak of war, served in France and Belgium, and was made a prisoner of war in July 1917. In 1918 he rejoined the House of Commons as member for the Tooting division of Wandsworth, and in the light of his known connection with Eastern Telegraph he was able to speak more authoritatively than the Assistant Postmaster-General. Apart from delays, excessive errors came in for criticism. When war broke out, explained John Cuthbert, there were a great many Eastern Telegraph men serving in the tropics due for leave. But all of them had stayed at their posts. The usual time to keep an operator in a tropical station was three years.

> These men have been out there from the very beginning, and all through the war, and as well they have been doing about three times as much work as they did before the war. I think hon. members will see that you cannot have the same excellence of work from men who are over-tired and over-strained, as these men have been for the last number of years.

Government priority was only suspended on 5 June 1919. With every cable filled to capacity seven days a week there was no spare capacity nor the slack Sundays to take the traffic from routes which became interrupted. Owing to the vigilance of the Royal Navy, none of Eastern Telegraph's cable repair ships had been lost in the war, but the location and repair of faults took time. No new cables or improved apparatus could be introduced during the war, even if there had been sufficient trained staff to operate them (which there was not).

A certain easement had come from the first employment of women telegraphists by Eastern Telegraph in June 1918, but there was little which could be done to bring staff up to strength until hostilities ended. Manual operating ceased at Porthcurno in March 1918, never to return, and the place became a relay station. This meant there was no opportunity for juniors to gain operating experience, and in September 1919 it was decided to close the school (Dan Cleaver said the reason given was 'unruly behaviour'). Then early in 1919 Electra House instituted a series of short technical courses at PK on 'Recent Methods of Working' to bring senior operators returning from the war up-to-date on Cable Code, the change from Sectional to Through working, and other innovations. By the time they arrived the troops had left, the defences had been removed and the old bachelors' quarters were derelict. These courses were merely a stop-gap, and they ceased when, on 1 January 1920, the new

43 In 1929, forty-seven-year-old Sir Basil Blackett became Chairman of Imperial and International Communications Ltd. A distinguished economist, he entered the Treasury in 1904 and represented it in the United States at the end of World War I. From 1919 to 1922 he was Controller of Finance at the Treasury and, until 1928, Finance Member of the Executive Council of the Governor-General of India. He resigned as Chairman of I. & I.C. in April 1932, and was succeeded by J. C. Denison-Pender.

44 It was Lord Inverforth, Chairman of the Marconi Company, with whom Sir John Denison-Pender, Chairman of Eastern Telegraph, had dinner in December 1927 during which the idea of a fusion of cable and wireless interests was first mooted.

45 On 16 January 1928 delegates to the Imperial Wireless and Cable Conference (*above*) began the first of their thirty-four meetings in London under the chairmanship of Sir John Gilmour, Bt., MP (*centre, front row*) who also represented Great Britain. On his right is Sir Campbell Stuart who represented Canada and became Chairman of the important Advisory Committee. The report of the conference recommended the formation of a holding company and operating company which combined wireless and cable interests. The latter began as Imperial and International Communications Ltd but changed its name to Cable and Wireless Ltd in 1934.

46 The Cable and Wireless branch at Hong Kong had become very important by the time this photograph was taken in 1938, three years before operations were suspended 'for the duration' under Japanese occupation, and the British staff and their wives were interned at Stanley. In the front row sit the uniformed messengers who delivered telegrams direct to the addressee.

47 'Via Imperial' messenger boys in London wore peaked caps and black boots in the 1930s. Recruited at fourteen (school-leaving age), they worked a forty-eight-hour week for a wage of 14s.

49–50 Two of the most important stations in the 1930s were at Gibraltar and Aden. (*Above*) the staff quarters at Gibraltar on the left and right of the road. (*Below*) the staff at Aden pose for their photograph in 1939. The branch was only closed in 1978.

48 Three years after the merger of wireless and cable interests in 1929, Guglielmo Marconi introduced the first regular microwave radio-telephone service in the world. His customer was His Holiness Pope Pius XI, seen here with Marconi (in top-hat), demonstrating the ultra short-wave apparatus which linked the Vatican to the Papal summer residence at Castelgandolfo.

51–2 An important link in the Far East chain was Colombo, the capital of what is now Sri Lanka. Cable and Wireless operators are seen (*above*) at the Poththode receiving station. But the company's apparatus often had to be housed in unusual quarters such as this 'office' in the hill village of Kakopetria in Cyprus (*below*).

53–4 The Cable and Wireless building at St Georges, Malta (*above*) was a solid structure, complete with essential tennis courts, which was to survive the terrible strafing from German and Italian bombers in World War II. For staff in East Africa and most of those who posed for their photograph at Nairobi in 1946 (*below*), wartime was less hectic.

London Training School was opened in Shepherds Walk, Hampstead. With an initial capacity for training 200 operators at a time, this was soon restoring the depleted ranks of the group's F.1 (foreign service) staff. In 1922 the intake was increased to twenty a month, and some 350 were under instruction. Significantly wireless training was introduced in 1922 for 'promising students' as an extra course lasting six months.

Youngsters of Eastern Telegraph and Eastern Extension may have been paid less than their counterparts in the organisation which ran the Pacific Cable, but they had a much greater variety of locations in which to work – and play – and greater in the 1920s than ever before. Most transfers to other stations were determined by managers' urgent requests for a replacement centre half or spin bowler, a pianist for the concert party, a bass to sing Pooh-Bah or a comic to play Mr Cattermole in *The Private Secretary.* At Gibraltar there was the fine sailing and rowing, and the hectic games of rugger with visiting naval teams and the RAF on the North Front ground which had once been a graveyard. Everyone had to wear protectors to keep the grit from their knees. The day Pip Parsons scored the only try of the match which gave Gib Exiles the victory over the Royal Navy Combined Atlantic and Mediterranean Fleets in 1924 is still remembered fifty years later.

The Eastern Telegraph Mess on the Fortress just after the Great War had 140 bachelors and another 30 Senior Bachelors, with formal Guest Nights two or three times a month in dinner jackets, and all the regimental ritual. On other days there were two dinners, an informal one for the staff going on duty and a second service at 8 at which collar and tie were obligatory. If a member of the Mess came in late, he had to walk up to the Mess President and apologise. There was no starting with the soup; he had to begin at whatever course was being served. Each member had to volunteer at some time for one of the many Mess jobs which were re-distributed at the Mess Annual Meeting – Mess Caterer, Mess Secretary, Wine Caterer, Billiard Captain, Librarian. The company launched the library with a set of Dickens, Thackeray and the *Encyclopedia Britannica,* but from then on books had to be accumulated by local members contributing to a Library Fund. There was also a Silver Fund with which silver was bought for the Mess dining table, generally through Mr Kimber of Mappin & Webb in Queen Victoria Street, London – rose bowls, candelabra, centrepieces.

A model silver ship was purchased by members of the Mess in Suez for a very special purpose – to stop members boring the dinner table with information about the latest ship through the canal. If anyone started to mention a ship, the senior member present had a servant fetch the model ship from the sideboard and place it in front of the offender who then had to stand drinks all round. But it was only the silver, the choice of books and the faces which distinguished one Eastern Telegraph mess from another; for the furniture and fittings, carpets and curtains, cutlery, china and glass were all of the same bulk order purchased in London. If an operator had gone to sleep in quarters in Gibraltar and woken up in Aden he would never have known the difference.

But that would have been much too quick a transfer for comfort. One of the attractions of life in an Electra House company was the long sea voyage from one station to another. It was as well to make the most of it. When Arthur Grass arrived by ship at Sao Vicente one Sunday morning in 1920 he found himself listed for a night shift that very evening. Along with most who came to the Cape Verde Islands for the first time, he caught 'SV Fever' within a couple of weeks and was whisked away to a convalescent home/holiday chalet in more salubrious mountain air to recuperate.

Many lone round-the-world voyagers and explorers have recorded the succour – both physical and mental – they have received on reaching distant lands where the only inhabitants have been the staffs of wireless and cable stations of Eastern Telegraph companies and their successors. While Arthur Grass was on Sao Vicente, they were visited by Sir Ernest Shackleton, the English explorer, who was the first to attempt to cross Antarctica. The Portuguese aviator Sacaduna Cabral put down on the island in the course of a bid to be the first to fly the Atlantic by single-engine Fairey seaplane.

Having to feed an extra mouth was always a problem, as food was very short. Fresh vegetables were non-existent and there was little meat. In the mess the saying went 'If mint sauce was served, it was mutton; if there was no mint sauce, it was beef; but in any case it was goat'. But then after four years of Saint Vincent diet, Grass was able to build up his strength on a slow boat to Buenos Aires which was his next posting.

The war had brought funds for the reconstruction and expansion of the Eastern Telegraph group's cable system, but it took time to marshal the equipment and the trained men to make it *work*. However, was there not another medium whose help maybe could be enlisted to relieve the strain on the company's wires? In 1918 an Eastern Extension shareholder could still only view wireless tele-graphy as a *threat* to his dividend, though he was probably no longer typical. Sir John Denison-Pender's answer to his question at the annual meeting was doubtless not the one he expected. What effect would W/T have on the company's receipts in the post-war period?

> I can say if the wireless system were in a position to relieve our cables at the present time of some of the congested traffic to which I have already referred, it would be a great boon to the public both as regards social and commercial telegrams.

If the price of Eastern Telegraph shares was a measure of the investing public's concern that Ocean Telegraphy was in danger of succumbing to the competition of Wireless, the healthy 160 at which they stood in 1918 compared with 133 in 1916 seemed to indicate that submarine cables still had an independent role to play. But for how long?

7

Regeneration

1919–1927

If in 1916 Guglielmo Marconi had not re-investigated the possibility, which had intrigued him in 1902, of short wave propagation, instant distance communication by submarine cable would have satisfied the requirements of telegraph users for a very much longer period than came to be the case. The unreliability which characterised long wave transmission and gave the cable companies reason to believe they would never be supplanted, would have remained the factor that kept wireless telegraphy down.

However, the war brought the inventor back to his drawing board – to seek a way of remedying the congestion on the normal wave-bands which was hampering communication between warships in the Mediterranean. His experiments with beams of reflected waves were only successful for line-of-sight ranges but gave a certain security, and they reduced interference. On a wavelength of three metres he got good results up to six miles. C S Franklin, the senior engineer of his English company, whom he had asked to study the problem, worked twenty miles from Carnarvon in 1917. Two years later using thermionic valves he transmitted *speech* the ninety-seven miles from Hendon to Birmingham.

There was to be no quick solution of short wave problems and in the meantime the world's wireless needs had to be based on long wave propagation. The Marconi company were encouraged to build and operate wireless telegraphs to France, Spain and Switzerland, and the Post Office willingly granted them licences in 1919 for the British terminal for this network which they sited at Ongar, Essex. Two years later the company built a central telegraph office in Wilson Street, London – 'Radio House'. Was the moment now ripe to make another approach to the Government on the sensitive matter of a private enterprise imperial wireless system? The Liberal Government of 1911 had refused to consider it, but maybe the Liberal-Conservative Coalition of 1919 would think otherwise?

The plan the company submitted was on the grand scale. By it Britain could telegraph every country in the empire, and they could communicate with all properly equipped ships between 60 N and 50 S. It proposed six main trunk routes to Egypt, India, Singapore, Hong Kong, Australia, New Zealand, West Africa, South Africa, the West Indies, and Canada. It involved building five trunk stations in Britain and twenty-one overseas; fifty main stations and 100 local feeder stations. It required a total staff of 17,170 including 9,000 operators.

The company would construct, build, maintain and operate the system at their own cost, not only without subsidy but paying into local Treasuries 25 per cent of each station's net profit. The entire network would become the property of the governments concerned after 30 years, or earlier by paying Marconi a lump sum and then 10 per cent of the gross receipts for the remainder of the thirty years.

The Colonial Office had referred the Marconi Imperial Scheme of 1910 to the Government's standing Cable Landing Rights Committee whose report rejecting it was considered by the Imperial Conference of 1911. But in 1919 Viscount Milner, Colonial Secretary, sent it to the Imperial Wireless Telegraphy Committee which, with Cabinet approval, he had appointed on 24 November 1919 – prompted by having received the Marconi Plan? Its brief was not to report on that plan but to prepare 'a complete scheme of imperial wireless communications in the light of modern wireless science and imperial needs'. The committee contained no representatives of imperial countries and was chaired by Sir Henry Norman MP who, in the Commons debate of 11 October 1912 on the motion to appoint a Select Committee of Enquiry into the Marconi Contract, had been the main Liberal critic of the Government's handling of the situation, declaring the contract to be a bad bargain and one tainted by corruption. The chance of such a committee reporting in favour of placing a new contract with the Marconi company was slim, and in their report the plan was dismissed in three short paragraphs.

> We have examined, as requested, the proposals submitted to the Government by the Marconi Company, (stated the Norman Committee Report of 28 May 1920). We find these too vague to admit of detailed comment, but so far as we can judge the scheme, it appears to be of a scope and magnitude so great, and involving such heavy capital and annual expense, that even if it carried the whole of the traffic handled today by all the cable companies serving the same

regions, it could only be remunerative, if at all, by duplicating the Postmaster-General's system of inland and continental telegraphy, and by competing with the State telegraph systems of the various overseas Governments. We are further of opinion that if fully carried out it would be prejudicial to the interests of free wireless research and independent development.

They did not accept the statement that an imperial service could be satisfactorily carried out only by the use of a particular Marconi patent. They were of the opinion, however,

> that if Imperial wireless traffic were carried by the State, long-distance wireless traffic with foreign countries might properly be left, under suitable conditions, to the commercial companies. Both services would profit by this healthy competition.

The Norman Committee were not prepared to accept the Marconi Plan as a blue print for an imperial system which might be built by the company but operated by Post Offices, as the Cable Landing Rights Committee had done in 1910. They produced their own scheme. The remnant of the chain which Marconi started building for the Post Office and was cancelled in 1914, as seen, were the stations at Leafield and Cairo, and the Post Office were on the point of instituting a service on this circuit with Poulsen arcs which the Norman Committee recommended should therefore be the first link in a post-war chain, which would extend to Nairobi and the ex-German station at Windhoek in South Africa, such stations using thermionic valves. They further recommended another valve circuit linking England, Egypt (Cairo), India (Poona), Singapore, Hong Kong and Australia (Port Darwin or Perth). The structure of the chain was determined by each station having to be 2,000 miles from the next.

Construction would be, not by the Marconi company, but by the Engineering Department of the British Post Office and P & T departments of the various governments concerned, under the direction of a four-man Wireless Commission which would be dissolved once the stations were operating. There would be no overall body administering the telegraph; each government would take the profits of the stations on their territory.* These governments

---

* In 1919 the British Government set up an Inter-Departmental Board of Control which at intervals convened a meeting of representatives of all government departments concerned in communications to discuss any matter that might arise and rule, when required, on situations which demanded a choice between communication by wireless and communication by cable.

were likely to have to carry an annual loss of £100,000 for the first ten years, but after that they should all make a profit. The report made no reference to the relation of the wireless chain to the existing cable network.

Unlike the Pacific Cable scheme of 1902 which was conceived by the interested colonies in partnership, the Imperial Wireless Scheme of 1919 was hatched in Whitehall by the Assistant Secretary of the British Post Office, the Third Sea Lord, the Director of the National Physical Laboratory, a Cambridge lecturer, a London professor and the Chief Electricity Commissioner. There was no one to acquaint them with 'imperial needs' as seen from the outposts of empire as opposed to the centre. On being presented with the Norman Report the Australian Government at once took umbrage at being placed at the end of a chain of stations, the failure of any one of which would cut communication with Britain. In any event it had already established Amalgamated Wireless (Australasia) Limited, and con-tracted with the Marconi company for a high-power, long-wave station for direct contact with Britain, and a series of feeder stations covering the whole Australian continent. It was evident that in the Dominions the thinking tended towards employing or licensing a private undertaking at a time of financial stringency, and so avoid the capital expenditure entailed in the erection of state-owned stations.

The Norman Report was presented to the British Parliament in June 1920, and in August the Cabinet accepted it. When it was debated in the Commons it was poorly received, though the principle of having the wireless service state-owned was generally approved. Britain's imperial partners discussed the detailed scheme for the first time at the Imperial Conference of 1921, and they referred it to the Imperial Communications Committee which on 3 June 1921 passed the resolution:

> That it would be undesirable under existing conditions to modify the decision of the Cabinet by which the recommendations of the Imperial Wireless Telegraphy Committee [the Norman Committee] were accepted, and that the scheme recommended by that Committee be adhered to.

The principle subsisted, but the scheme was forgotten. The Press continually opposed any sort of state wireless scheme. As W J Baker wrote in his *Marconi Company* the Norman Plan 'was put out of its misery shortly afterwards by Mr Winston Churchill (the Colonial Secretary) who as chairman of the Imperial Communications

Committee refused to have anything to do with its outmoded proposals'.

But to be fair to Sir Henry and his colleagues their proposals were not outmoded at the time they were written. It is true that research which H J Round and others were conducting on long wave propagation with thermionic valves was to eliminate the need for the 2,000 mile hops on which the committee's plan was based, but it had not done so in May 1920. Winston Churchill did not become Colonial Secretary until February 1921. It was not until November 1921, eighteen months after the report came out, that Round designed the transmitter with fifty-four valves for the Carnarvon station which upset all previous concepts by working to Australia direct. It was only on 4 December 1921 that the *Daily Mail* sent its first press messages direct to Sydney.

Electra House saw to it that they were well informed of the progress, or lack of progress, on the wireless front, and the intelligence they were able to assemble gave them hope that it would be many years yet before they need have serious cause for concern. They read with some cynicism of the Carnarvon-Australia link and of Professor Fleming's claim to telephone six thousand miles with the aid of his thermionic valves; and they duly noted the effect of sending wireless waves in 'beams' in eliminating atmospherics and interference. But sufficient unto the day. Sir John Denison-Pender told his shareholders in November 1921 that it did not look as if the days of cable were over. 'I will not prophesy, but if wireless communication should prove to be better than cable communication it will cut us out. But it has some work to do before it is in that position, whatever may happen in the future.' And to show his confidence, he increased the capital of Eastern Telegraph by a million pounds to £4 million.

While HMG were making up their mind about a Post Office operated Imperial Chain, in 1922 Godfrey Isaacs sent a cablegram to General Smuts offering to erect and run at the Marconi company's own expense a high powered, long wave wireless station in South Africa within eighteen months, capable of working Britain direct. The Government of the Union welcomed the idea and at once negotiated an agreement which gave them control of the station's profits and rates. Marconi agreed that if and when the British Government decided it was desirable to have an overall Empire Wireless System, that station could be part of it. The company began work at Klipheuval near Cape Town early in 1923; and at the same time started erecting the 1,000 kW station, with its twenty steel masts each 800 ft high, for Amalgamated Wireless (Australasia) Ltd in the Antipodes. They obtained a licence from the Canadian Government to build a station at Montreal.

In a long piece in December 1922 *The Times* made a brave attempt to apologise for a state of affairs which had put Britain, who had weaned the infant Wireless, at the bottom of the communications league.

> Although there has been much official delay in the development of wireless communications within the Empire, the position today is generally much exaggerated by the pessimists. Changes of policy, the war, divergent views and the general financial stringency have all been obstacles in the way of progress. And although it is true that in some cases other countries are now better equipped with wireless, it will be found in the end that Great Britain is able to hold her own, and that while during the immediate past little tangible progress may have been made, invaluable lessons have been learned from the experiments, the progress and the failures, of other nations. It is in the United States that the most rapid progress in the development and use of wireless equipment has been made. There are there ten high-powered stations and six other stations of various grades in private hands.

The British Post Office claimed that their station at Bourne would be the most powerful and up-to-date wireless station in the world. They had put out the specification and called for tenders. It would cover 400 acres and be ready in eighteen months. While the Imperial Wireless Chain as originally planned was not likely to develop, stated *The Times,* the Empire was not to be without wireless equipment.

> No doubt, in general development, there has been serious delay, but part of it has arisen from the desire of too many Government departments to have a share in wireless matters, and part from the overloading of the Post Office. . . . Not the least benefit that will arise from an adequate wireless service will be a reduction in cable rates, and it has been computed that a decrease of one-third may be expected when the system is functioning.

Speculation on precisely when the Government would have that system functioning was doubtless a popular topic of conversation among the staff of the Eastern Associated Companies who gathered at Orleans Park on 8 July 1922 to celebrate the fiftieth anniversary of the formation of the Eastern Telegraph Company in 1872.

The group had bought Meadowbank, a fine house set in the sixteen acres of Orleans Park beside the Thames at Twickenham at the end of 1919, and converted it into a clubhouse and recreation grounds for overseas and UK staff. Opened in 1920, Orleans Park was named The Exiles Club, a reincarnation of the Probationers Mess at Porthcurno now no more.

Staff subscribed to buy replicas in platinum and brilliants of Sir John Denison-Pender's two orders. A R Hardie, secretary of Eastern Telegraph since 1903, presented the gift to the chairman who was the guest of honour at the Exiles Club jubilee garden party. He was also given an illuminated address on vellum containing 8,000 signatures representing fifty-three nationalities. The blue morocco binding bore the seals of the twelve principal telegraph companies in the group. Hardie reminded the gathering that a similar presentation had been made to the chairman's father, Sir John Pender, on 16 November 1894 to mark the twenty-fifth anniversary of Submarine Telegraphy to the Far East. In accepting it as a personal gift, Sir John Denison-Pender said he was aware that it was a tribute from the staff to the whole of the directorate and management at Electra House. Strong as the companies now were, they would develop further, and in the interests of the trade of the world continue to lay and open up fresh communications.

> Thus, gentlemen, we shall ensure the prosperity of the future. We shall ensure the 10 per cent dividend that we are at present paying to our shareholders, and we shall ensure the security of tenure of the various positions held by all of us, to be continued after Time has put us aside for others who must take our places.

A major change was in fact just about to take place, following the death on 22 August of F E Hesse who had been general manager and secretary of Eastern Extension since 1902. He was succeeded by R T Wolfe.

All members of the Pender family were present at the jubilee party including Sir John's sister Lady des Voeux (née Marion Pender) the recipient of the voluminous correspondence from her mother Emma. She had outlived her elder step-rother Sir James Pender, Bart who had died aged 80 in May 1921. Sir John Pender's son by his first wife, James, chose the life of an army officer and country gentleman with a large house in Wiltshire and hunting and yachting as his favourite sports. He was Tory MP for mid-Northamptonshire from 1895 to 1900. Latterly he took up certain City interests, being chairman of Kodak Limited from its formation in 1898 to 1913. But he took no part in the cable business, though he was a director of the Globe Trust, Telcon and the Direct US company.

Sir John Denison-Pender's wife Beatrice had died in November 1920, and the surplus of the money collected for the gift went to endowing a 'Lady Denison-Pender Bed' at St Bartholomew's hospital in her memory. It was unveiled by the Lady Mayoress on 20 July.

The main jubilee celebration took place four days later. It was a banquet and fête in the Royal Botanical Gardens, Regents Park, attended by 700 guests at which HRH the Duke of York (later King George VI) was guest of honour. There were performances in the open air theatre (still there), dancing in the ballroom, sideshows, brass bands, tennis, fireworks and a cinema tent.

The flicks, as they were called – black and white and silent – were no longer a novelty, but five months earlier an unusual announcement had appeared in *The Times* (11 February 1922):

> The first of a series of regular wireless telephone transmissions for the benefit of English wireless amateurs will take place on Tuesday evening. The Marconi Scientific Instrument Company has prepared a fifteen minute programme for the occasion, transmitted through the medium of the 'Cliftophone'. The first telephonic item will be radiated at 7.35 pm on a wavelength of 700 metres from the Marconi station at Writtle, Essex – 2MT.

This story is not concerned with the development of Wireless as Sound Broadcasting. It was an aspect of his invention in which Marconi never took any great personal interest. But his companies could not fail to do so. In 1920 the vice-chairman and managing director of Marconi's Wireless Telegraph Company was the 50-year-old Liberal MP for Bedford, Frederick Kellaway. He had to resign the following year on being appointed Postmaster-General in the last days of Lloyd George's Coalition Government which was defeated by the Conservatives under Bonar Law in the general election of 17 November 1922. But in his brief tenure of the post there took place (on 18 May 1922) the Conference on Wireless Telephony Broadcasting called by the manufacturers of receiving sets and components which led, following negotiations for which Fred Kellaway was largely responsible, to the formation of the British Broadcasting Company. This earned him the sobriquet 'Father of Broadcasting', but it is not as such that he has his role here. When he lost his seat in the 1922 general election he retired from politics altogether, and two years later was back with Marconi as chairman and managing director of his International Marine company.

The formation of the BBC was of course no threat to companies who transmitted point-to-point, rather than broadcast, signals. But the term 'wireless telephony' indicated that the ether had now been made to carry speech, where the submarine cables in which so much money had been invested were firmly tied to telegraphy. But even wireless telegraphy, it seemed, was beginning to erode Electra House earnings, healthy though they were. For the first time since 1914 Eastern Extension gross receipts fell from £2,514,800 in 1920 to £2,394,300 in 1921. It was due mainly to a slackening off of the excessively high traffic at wartime rates which had led to the 'profiteering' accusations, and was not in itself very serious – management were able to afford a 'jubilee gift' of two months' pay to all employees on the books in 1922. Great Northern traffic, diverted on to Eastern Telegraph routes, now returned with the re-opening of their line to China and Japan via Siberia.* Eastern Telegraph traffic was further reduced when Indo-European re-opened their line to India via Germany, Russia and Persia. The success of the Pacific Cable route to Australia, now fed by the British Government's Imperial Cables, was also taking its toll.

The Pacific Cable Board no longer found 'touting' distasteful and had the Post Office frank postage stamps with an invitation to cable 'via Imperial', which Sir John Denison-Pender thought was 'not playing the game'. The campaign was withdrawn. No one expected governments whose field was social management to excel at running a business, but traffic on the Pacific Cable had risen from the three million words a year before the war to ten million in 1921–23. However, agreeing on broad policy matters was still an agonising process, as Sir Campbell Stuart recollected thirty years later.

In 1923 this 38-year-old Canadian had just resigned as managing editor and managing director in London of *The Times* newspaper and been appointed the representative of the Government of Canada on the Pacific Cable Board.

It was my first experience in a Government organisation of this kind embracing different parts of the Empire, (he recalled in his memoirs, *Opportunity Knocks Once,* London, 1952), and certainly we were not a harmonious body. In the particular case of the two Australian representatives there was very plain speaking. There was disagreement between the Canadian representatives and the Board

* Eastern Extension's relations with China had worsened over the last decade. China's monetary obligations to the company had fallen into arrears on payments for the Shanghai-Chefoo-Taku cable laid for the Chinese Government of 1911, and on the repayment of a £500,000 loan given in the same year. China was in chaos in 1923, with several militarist parties all striving to dominate the country.

over the desire of the other representative to duplicate the cable in the Pacific, when the Board owned no wireless which was then looming large as a competitor. The matter was made all the more difficult by the fact that the Government of Great Britain, which was a large shareholder in the Board, also owned and operated through the Post Office the competing wireless service between England and Australia.

Ten million words a year was all the Pacific Cable could take. The Board agreed to relieve the pressure between Australasia and Fiji by laying a second cable from Auckland to Suva which was completed in November 1923. There was less unanimity, however, on the best means of doubling the capacity of the Fiji-Fanning-Vancouver section. Wireless or cable? Those who advocated the former persuaded their colleagues at least to give wireless a trial.

A combined wireless/cable exercise was the basis of the second telegraph which the Pacific Cable Board were now asked to build and manage.

For two years, from 1920 to 1922, the communications situation in the Caribbean had been under scrutiny by the West Indies Cables Sub-Committee of the Imperial Communications Committee who, presumably because of the strategic importance of the area, regarded this as a legitimate area of government responsibility. West India and Panama's concession was going to lapse in 1924 and they were going to use their semi-monopoly to extort a very much higher subsidy on its renewal. But their cables were very old; delays and interruptions had become intolerable. Their own resources would not allow them to renew them. The company was wholly subservient to Cuba Submarine whose system they were obliged to use whenever possible, which meant there was no opportunity to improve or expand their own. Whoever bought West India and Panama would also be bound by this agreement. So the Committee recommended that HMG, in consultation with the Canadian Government and the governments of the various West Indian islands, should lay a completely new cable in the West Indies.

Sir Laming Worthington-Evans, British Postmaster-General, outlined the scheme to delegates to the Imperial Economic Conference which met in London in October 1923. The Pacific Cable Board would operate the telegraph as the agent of the owner-governments. It was never contemplated that this exercise might be the subject of a concession granted to one of the commercial cable companies. Government competition had come to stay, he seemed to be saying, and the sooner they became accustomed to it the better. The tribute he paid to Electra House in the course of his long speech at the conference covering the whole field of imperial cable and wireless communications looked as if it was intended to soften the blow.

> It is difficult to overestimate the debt which the Empire owes to
> this great association of companies which were the pioneers of British
> cable enterprise (he said). The Eastern Company celebrated its
> jubilee last year, and it is a significant example of the enterprise which
> the company have always shown that they should have been able and
> ready to spend very large sums, running into several millions,
> involved in the provision of two new cables to almost the farthest ends
> of the Empire.

Confident in their successful management of the Pacific Cable,
Government could afford to by-pass the cable pioneers, but had
come to the conclusion they could not as easily dispense with the
services of those who had developed Wireless. HMG's wireless
policy which Sir Laming announced to conference delegates was
now two-pronged.

> As regards Great Britain the Government have decided, in the first
> place to provide themselves a station which will be owned and
> operated by the Government; and secondly, to license private
> companies to conduct services subject to an agreement with the
> Government as to the division of traffic between the companies and
> the Government stations.

He was bringing delegates up to date on the new policy which the
Conservative Prime Minister Bonar Law had announced in the
House of Commons on 5 March 1923 in these terms:

> The policy to be adopted with regard to Imperial wireless
> communications has recently been under review by the Imperial
> Communications Committee under the chairmanship of the First
> Lord of the Admiralty, and the recommendations of that committee
> have now been approved by the Government. In view of the
> developments in the science of wireless telegraphy and other cir-
> cumstances which have arisen since the late Government decided
> upon the policy of a State-operated wireless chain, it is not considered
> necessary any longer to exclude private enterprise from participation
> in wireless telegraphy within the Empire. The Government has
> therefore decided to issue licences for the erection of wireless stations
> in this country for communication with the Dominions, Colonies and
> foreign countries, subject to the conditions necessary to secure British
> control and suitable arrangements for the working of the traffic.

This change of policy was a surprise to many, for in a Commons
debate earlier in the year, on a vote for the PMG's salary, it was

confirmed it was in the interests of the state that the wireless service would be totally controlled and owned by the Government. Sir Henry Norman protested that the change had been made without consulting Parliament and was contrary to resolutions passed at imperial conferences and committees.

The Government was not to be deterred. They opened discussions with the Marconi company on how Government and Private Enterprise would work in partnership. But the two found it difficult to agree on how to divide the services between them. The Marconi company had originally expressed a strong preference for a general pooling arrangement. In spite of seeing considerable disadvantages, the Government were ready to start negotiations on that basis and had nearly reached agreement in July when the company changed their minds. The Government had always favoured dividing the traffic on a regional basis – say, Canada and South Africa by HMG, India and Australia by Marconi. The PMG was able to tell conference delegates that at that date (16 October 1923) talks were proceeding on those lines though, as he was at pains to point out, the Government's invitation to the Marconi company had been outside its legal obligations, and the company should consider themselves privileged to be talked to at all.

> The Government station is to be erected partly for commercial and partly for strategic reasons. It is essential that in parting with the monopoly which the Government possesses by statute, we should secure that a sufficient amount of commercial traffic is reserved for the station to operate. Provided this is secured I shall be ready to meet, as far as I possibly can, any views which the Marconi Company may submit to me. The services to be given by the Government station will not be less efficient than the Marconi Company's, and the Government station is likely to be open for traffic before the company's.

Eastern Telegraph were now suffering wireless competition on their cable to Egypt from the British Post Office circuit between Leafield and Cairo; and from wireless links between French Indo-China and Madagascar, and between Holland and the Dutch East Indies. So in 1923 they took advantage of the British Government's change of heart and applied to them for a wireless operating licence; and to the Indian Government too. Eastern Extension began negotiations with the Marconi company over possible co-operation in China, which was a major scheme and in no way comparable with the limited operation on Cocos where the wireless station closed in November 1926. But Eastern Telegraph were to be disappointed if they thought they were about to enter a new era of cable and wireless partnership. Sir John told Eastern Extension shareholders the following year, 'the wireless company took the line that if the cable companies joined in the combination, they were to give an undertaking to maintain higher rates for cables than for wireless. Eastern Extension would not do that unless it got adequate safeguards, as it would tie their hands'. The negotiations with Marconi fell through, and Eastern Telegraph were told by the British Post Office that Government policy regarding granting wireless licences to private companies had been finally determined – another change of mind apparently.

The gesture had come too late. But it can be imagined that failure to establish a bridgehead in Marconi territory will have caused little disturbance to businessmen whose commitment in terms of money, equipment and people to Submarine Cables was total, and whose confidence in their future uncompromising. Since 1913 they had spent £2,158,300 (all out of their own reserves) on 7,200 miles of new cable, and £528,000 on new staff quarters in various parts of the globe. They had just opened new offices in Adelaide, Melbourne and Sydney. They were building a new repair ship to replace the *Recorder* which had seen forty years' service. They guaranteed the British Empire Exhibition at Wembley to the extent of £5,000 – King George V sent a message round the world which was tactfully transmitted outward via Imperial and via Pacific and homeward via Eastern Telegraph and Eastern Extension – 31,500 miles in eighty seconds.

There was no doubt in the mind of Edward Wilshaw, now Secretary of Eastern Telegraph, that the day of the cable was far from over. Electricity had not supplanted gas, nor cable the mail service as many had predicted, he told a meeting in Manchester on 22 May 1924.

> Though wireless has demonstrated its great practical value in very many different ways it still remains true that submarine cable with its world-wide organisation, privacy of transmission and comparative freedom from atmospheric disturbances, is the most reliable, swift and secret method of international communication.

Eight days after Edward Wilshaw spoke to the Chartered Secretaries of Manchester in that fashion, Guglielmo Marconi *telephoned* Australia from England on a 97 metre waveband and his voice was heard loud and clear.

At the beginning of 1923 Marconi had set out in *Elettra* to collect data on long distance reception of short wave signals. He found the further he sailed from a short wave transmitter the weaker the signal became – and then got stronger. In daytime he received signals 1,250 miles from Poldhu short wave station; at night at least 2,230 miles away (the limit of his voyage). Further tests on his return to England showed New York were receiving Poldhu's signals at night, and Sydney were doing so between 5 pm and 9 pm and 6 am and 8.30 am. Short wave transmissions of 200 metres or less had been considered as R N Vyvyan put it, 'too variable and freaky', too unreliable for commercial work. Marconi's experiments on *Elettra* showed this to be untrue. And most important of all, the good reception came quixotically not from more electrical power being pushed out of the transmitter, but *less*. Even when Poldhu's power was reduced to one kilowatt its signals were stronger than those from the Post Office's long wave stations at Leafield and Carnarvon.

There was nothing new in sending wireless signals across the Atlantic; it had been done twenty years before. The novelty was that by the short wave 'beam' system – so called because of the directional aerials – which Marconi and his assistant C S Franklin perfected, they could be sent at three times the speed, using a fiftieth of the power and at twentieth of the cost. It was not just an 'improvement' on the lines of H J Round's discoveries with very high powered long waves using valves which made the earlier long wave system 'outmoded' in eighteen months, but a revolution after which no one could dismiss wireless, as Wilshaw had done the week before, as unreliable and slow. The secret lay in concentrating the signals instead of scattering them in all directions – a rifle instead of a shot gun – so that the 'ionosphere', postulated by Oliver Heaviside and A E Kennelly in 1902 and later verified by Sir Edward Appleton, reflected them forward and downward on to the target, an operation requiring a fraction of the power and maing the signals considerably less prone to atmospheric interference.

To all intents and purposes Wireless had become something else. It began a new life in 1924. In retrospect, forty years on, it certainly appeared a turning point.

> The beam system put radio on a par with cable communication for the first time, for not only were its capital costs and power requirement relatively low, but it was suitable for the transmission of either telephony or high speed telegraphy; the original beams were tested at 300 words a minute and often worked at 200 words a minute, which was a higher speed than was possible on any cable in the Commonwealth system at that time. *(Britain and Commonwealth Telecommunications,* British Information Services, 1963)

The Post Office was deeply committed to its long wave programme but, thanks to democratic delays, not irrevocably. The Government – now Britain's first Labour administration formed by Ramsay MacDonald on 23 February 1924 – felt obliged to ask Marconi for a demonstration, and the Labour PMG informed the House of this decision on 23 July. In the course of a few days, he said he would ask for the Commons' approval of an agreement which the Government proposed making with the Marconi company to erect, as contractors, a short wave beam station in the UK for communication with Canada. The transmitter would have a power of 20 Kw and would be capable of remote control from the Post Office Central Telegraph Office in London. Communication could only take place at night and during one or two hours before and after twilight. Thus it would only be suitable for 'deferred' traffic. The Post Office long wave station, which was now going to be at Rugby instead of Bourne, would provide long distance communications at all hours. If the Marconi company could show that the new beam system was all they claimed it to be in the station designed to work to Canada, the agreement provided for the construction of further UK terminals capable of working to South Africa, India and Australia. The Governments of India and New Zealand, however, had intimated that a beam station would not suit their requirements. The former were proposing forming an Indian company to erect high power long wave stations in that country, though the Governments of Australia and South Africa were disposed to give the system a trial. The Marconi company, said the PMG, had agreed to co-operate whether the decision of the Dominions was in favour of short wave or long wave stations.

Imperial unanimity was still as remote as ever, but one thing was perfectly clear. Sir Laming Worthington-Evans' policy of allowing private wireless companies to own and control stations had been abandoned. The new Labour Government had referred the matter to yet another committee, this time headed by that great champion of private enterprise Sir Robert Donald, chairman of the Empire Press Union. But when, for the first time, he came to weigh *all* the facts, he concluded that the answer was precisely the opposite of what he had always advocated. His committee recommended that a British Post Office *reorganised on business lines* should own and operate all the UK terminals of the latest version of the Imperial Wireless Chain whether it was long or short wave. The Donald Report was published and accepted within a month of the Labour Government taking office.

The Press watched this vacillation with mounting cynicism.

Having reached this decision, the only thing that mattered now, stated *The Times* in its leader of 24 July, was for everyone to work together in implementing it.

> Wireless telegraphy, as the Donald Committee remarked in their report, is the victory of science over space, because it annihilates distance. Hitherto, not because British scientists are inferior to those of other nations, but because of the differences and disputes between the politicians and others on whom it devolved to make use of their knowledge, this country has been left far behind in the wireless race. More than twelve years ago the British Government laid down the principle that Empire wireless should be owned and operated by the State. When Mr Bonar Law was Prime Minister that principle was scrapped in favour of the participation of private enterprise. Under this new policy it was found impossible to combine the element of competition in commercial wireless with pooling arrangements and unified management. It is earnestly to be hoped that there will be no more of the deplorable and destructive delay to which this want of consistency and agreement has given rise. The Government have at least produced a scheme which promises a cessation of the paralysis caused by conflicting counsels and purposes, and the interests of the whole Empire demand that it should be carried through with all possible speed.

It only needed five days after the Government's first announcement in the House for the contract for the short wave scheme to be signed with the Marconi company (28 July). It was a happy change from the three year delay between 1910 and 1913 before his long wave scheme materialised, and then too late to benefit the Allies in their war against the Kaiser. The opposition came this time from the diehards at the British Post Office, but the Labour ministers would not listen to them. The sceptics of St Martins-le-Grand saw to it that the whole project was undertaken entirely at the Marconi company's risk and imposed exacting conditions, enough to make Marconi, Vyvyan and Franklin think very seriously before accepting the challenge.

The site for the test transmitting station to Canada was chosen as Bodmin in Cornwall, with the receiver at Bridgwater. These, the Post Office insisted, would have to be completed within twenty-six weeks of the sites becoming available – though the speed of building can hardly have been a factor in whether the system worked or not. They set a maximum expenditure of £35,120. The contract was for cost plus 10 per cent, but the Post Office refused to make any payment until the station was ready and until they were satisfied by a seven day demonstration that all their guarantees had been fulfilled. They would then pay half the cost. Post Office staff would operate the station for six months, and if all went well they would pay another 25 per cent; the final quarter would be handed over after another six months of satisfactory operation. Considering the system was still in the development stage, with several major technical problems yet to be solved, the company's acceptance of these terms showed considerable courage. But the gamble came off. The Bodmin station survived the ordeal and the go-ahead was given for the whole scheme. By that time Stanley Baldwin and the Conservatives were back in power and enacted the Wireless Telegraph and Signalling Act (February 1925) to supersede the Wireless Telegraph Acts of 1904 and 1905. The new legislation had the effect of strengthening the authority of the Post Office over the whole field of wireless. Its long wave station at Rugby went into commission in 1926.

This was of course basically a wireless telegraphy station, but on 7 February 1926 it established the first radio-telephonic link across the Atlantic. After a year of testing, the speech service was opened commercially on 7 January 1927. Though Electra House had no transatlantic telegraph, here was a long distance medium with the potential to attract yet more customers from the cables whose diminishing traffic they were trying to assuage by further rate reductions (from 2s 6d to 2s 3d a word to Australia from 1 February 1927) and devices like Weekend Telegrams (7½d a word to Australia). As F J Brown pointed out in *The Cable and Wireless Communications of the World* published in 1927, in that year no practical method had yet been devised for adapting long-distance submarine cables to telephony. 'Invention may – and probably will – eventually overcome the difficulties, but for the present the only method of transmitting the human voice for long distances across the ocean (as distinct from long distances across the land) is by wireless.' The fact was submarine cable lacked the capability for telephony. It had to charge and discharge slowly; if it tried to do so quickly the current passed out through its sheath into the sea. Pulses put through at the high frequency required for speech would thus be lost. The cables were only just coping with high speed automatic transmission of telegraphic signals.

With the short wave radio beam telegraph proven, the Dominions hastened to provide terminals to which the English stations could work. Whereas in Britain the stations were owned and operated by the Government, the Dominions and India preferred to have them in the hands of private companies associated with the Marconi

company, with the rates subject to public control.* The latter at
once suspended work on the long wave stations they were building
for the Wireless Telegraph Company of South Africa Ltd, and for
Amalgamated Wireless (Australasia) Ltd (half owned by the
Australian Government) – lucrative contracts they were loth to
cancel – and began afresh with new contracts for cheaper (because
smaller and low-powered) short wave beam stations on the same
sites at Klipheuval and Ballan. The Canadian Marconi Company
began to erect a short wave station at Drummondville; and the
Indian Radio Telegraph Company Ltd was formed to contract with
the Marconi company to build one at Kirkee. The whole of the short
wave Imperial Chain came into operation between October 1926
and September 1927. The rate to Australia was 1s 8d a word.
Eastern Telegraph at once cut 3d off their rate and made it 2s.

Here was no piecemeal threat to the submarine cable services, but
a confrontation across the whole area regarded by John Pender as
his special preserve. 'The effect of wireless competition upon ocean
cable traffic has been closely observed for several years' wrote
Newcomb Charlton, president of Western Union in 1926. 'It is our
experience that where cables can be assured efficient land line
connections, as for example in Great Britain, France, Italy, Germany, Netherlands and Belgium, the cables will hold the business as
against wireless competition.' It was typical of the wishful thinking
of the time.

Wireless telegraphy, radio telephony – and now picture transmission. In December 1924 the Carnarvon transmitter took part in the
first experimental radio transmission of still pictures between
England and the United States, devised not by a British scientist,
however, but by R H Ranger of the Radio Corporation of America.
His method of mounting a print on a rotating cylinder scanned by a
light beam was too coarse and slow for press work – true facsimile
did not come until 1929 with the Marconi-Wright system – but it
was a beginning, and it had not been made by Electra House.
Wireless was now two steps ahead of long distance submarine cable
communication – in telephony and picture transmission – though
the first combined transmitting-receiving teleprinter introduced in
1927 would only work by wire. However, the extent to which the one
had the advantage over the other was now assuming less importance
than the part both contributed to a single operation known as

---

*As in France, Germany and Italy; though elsewhere in Europe they were owned
and worked by Governments. In the US stations were owned and operated by the
private Radio Corporation of America.

'telecommunications'. At last the rival mediums were being seen as complementary.

In July 1926 the Greek Government gave Eastern Telegraph the concession for all their cable and wireless communications with foreign countries for fifty years – the company's first for a combined service.

The system which the Imperial Communications Committee had devised for the British West Indies also involved both cable and wireless as seen, and this was put in hand at once.

When the cable was laid from Bermuda to Turks Island, then a dependency of Jamaica, in 1898 and then continued through the Windward Passage to Jamaica, this halfway point was merely a landing place. But in the Pacific Cable Board telegraph designed to link London with Barbados via Halifax and Bermuda, Turks Island was made a relay station. From there to Barbados the cable ship *Faraday* laid a new duplex cable through the Mona Passage in the spring of 1924, followed by a second and third line to Jamaica and Georgetown, British Guiana, which were completed in June. A fourth cable from Barbados to Trinidad was laid by 10 July.

All these cables met at Prospect, Barbados, near the West India and Panama cable hut, and were connected by underground line via Black Rock to the new telegraph station at what was known as The Reef. Pneumatic drills were seen – and heard – on the island for the first time penetrating the hard rock to create these trenches. The site had been the cricket pitch of the Carlisle Cricket Club opposite Pelican Island.

The wireless part of the scheme was all-embracing – a network, centred on Barbados, linking every British West Indian island.

The ship-to-shore wireless station, VPO, at The Garrison which had served the island so well in the Great War, was closed and the Pacific Cable Board replaced it with their own more powerful installation, also on The Reef site, with 200 ft lattice tower aerials. It was built by RCA under British Post Office supervision. W E Rockingham of the Pacific Cable Board remained in Barbados as the Regional General Superintendent of the Pacific Cable Board's West Indian System, as it was called.

Rockingham opened the new wireless net on 1 December 1924 with the first four stations ready – Barbados, St Lucia, Dominica and Antigua – as well as taking over the ship-to-shore service which had been operated by VPO (it took over its call sign too). Other stations joined later: St Kitts on 10 January 1925; Grenada on

1 February; St Vincent on 1 March; Montserrat on 25 May. The Pacific Cable Board had no radio operators in Barbados and the station was put on the air in the first place by staff taken over from the original VPO, Harold Rose, Carl Fenty, Fred Roett and Percy Croney. But operators were being recruited in England and these arrived some weeks later. Fenty then left to take charge of the Dominica station; Croney became officer-in-charge of Grenada and an ex-Petty Officer Telegraphist from the Royal Navy called West at St Vincent. Shortly afterwards the St Vincent station was burnt to the ground and had to be rebuilt. The officer in charge at Antigua was A H Palmer whose station was offering a ship-to-shore service in November 1924.

It was uncompromising, direct Government-sponsored and Barbados Government-subsidised competition to West India and Panama whose cable system by this time, however, was well-nigh redundant as the British PMG had intimated. The company had suffered innumerable setbacks including the strain on its resources of attempting to make the cable ship *Henry Holmes* seaworthy. It had been lying at the docks of Port-of-Spain, Trinidad for more than a year and piling up charges*.

'The Government competitive system of cable and wireless . . . was opened on 1st December 1924' stated the West India and Panama directors' report of 31 December. 'Our annual subsidies of £26,300 ceased on 30 September 1924, but by request of the Pacific Cable Board the company continued its service between those two dates receiving remuneration for doing so, consequently the new competition is not reflected in the new results.' In November, Cuba Submarine made a bid for West India and Panama's shares, and when the offer was accepted by 80 per cent of the shareholders all the directors resigned (31 December 1924) and their places were taken by directors of the new owners. The last report of the old West India and Panama board was dated 5 October 1925 but its service continued under the new management. Several West India and Panama operators were taken over by the Pacific Cable Board, and one of them, W R Allen, was made Supervisor at The Reef. He was succeeded by Austin Belmar.

The Pacific Cable Board built two tennis courts at The Reef and

---

*This was the second cable ship from which West India and Panama suffered a loss. Its cs *Grappler* had the misfortune to be on duty off Martinique when Mont Pelée erupted in 1902 destroying the town of St Pierre and killing 40,000 of its inhabitants. The resulting tidal wave swamped the *Grappler* which turned turtle and sank with all hands. Lucky crewman was Engineer Cruikshank who had left for England on leave only the day before.

started a club with part of the engine room as the pavilion. Two of the leading players on Barbados ran the tennis club, Gerald and Fred Howard, both above-average cricketers too.

Although the London headquarters of the Pacific Cable Board liked to think they now operated another imperial telegraph from Halifax to Georgetown similar to their Vancouver-Australasia route, in fact, as the map on their letterhead showed, they only ran a 'West Indian System' which began at Turks Island. And unlike the direct Pacific Cable – which by November 1926 had been duplicated between Bamfield, Fanning and Fiji at 1,000 letters a minute – the Board had anything but a monopoly in the Caribbean. As the last in the field they could only supplement services already well established. The Western Telegraph/Western Union operation was half Electra House, half American, financially sound and firmly based. The other competing route provided by Cuba Submarine/West India and Panama, and Halifax and Bermudas was at least all-British, though control of the route lay with the American-owned Commercial Cable.

When Fred Kellaway was asked to become managing director of the Pacific Cable Board's competitor, the Marconi company, when Godfrey Isaacs retired from ill health in 1925, he accepted at once and began to reorganise a company which, under Isaacs, unlike the cable companies which kept to submarine telegraphy, had over-diversified its interests. Kellaway brought in a firm of accountants to investigate the firm's finances which were in a confused state, in an attempt to have the company concentrate on proven profit-earning projects such as the revolutionary Short Wave Beam Radio.

In 1925 Eastern Telegraph had their revolution too – and as a result of an idea submitted through a Staff Suggestions Scheme which was then developed by Telcon.

Electrical pulses get weaker and weaker the longer the wire which

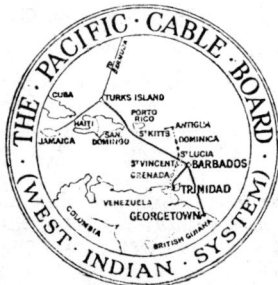

carries them. After a certain distance they grow so weak they can just operate the receiving instrument but flow no further. The exercise must start again with fresh pulses being re-generated by a second human operator and going to a second receiver the operable distance away – and so on and so on in a series of links manned by operators at either end forming a complete, lengthy chain. If only at the point where they grew weak the pulses could be self-regenerating, or by the nature of the carrier wire never weaken!

Regenerator principles were not new but their application to the company's system in 1925 was, and without doubt it was stimulated by the need to improve the cable service in the face of short wave wireless competition – the Cape Town Beam almost immediately took away more than half the South African social cablegram traffic, and the Bombay Beam very nearly as much. Manual operation at transit points on a chain of cable sections was replaced by automatic regeneration and onward transmission, thus reducing the time it took for a message to reach its terminal cable station (On the Cape Town route less than a minute after the tape had been punched at Electra House 1,850 miles away.) Quick clearance of the line was vital when between 5 pm and 11 pm 100,000 words were sent out of Electra House every day.

In addition cables were 'loaded' electrically so that the transmission characteristics were modified in a manner which allowed the speed of operation to be increased. When applied to the London-Porthcurno-Carcavelos-Gibraltar cable in October 1925 it improved the speed by 35 per cent. When the Gibraltar-Malta-Alexandria cable was loaded, its carrying capacity was improved by 27 per cent. The new 1,300 miles loaded cable from Cocos to Fremantle laid in March 1926 had a capacity of 2,100 letters a minute compared with the old cable's 385. The duplicate Pacific Cable from Bamfield to Fiji, which was allowed to adopt the system, could carry four times as many as the first cable, and compared reasonably favourably with beam wireless signals. Loading did, however, require the use of one-way (simplex) working whereas with unloaded cables two-way (duplex) working was the norm.*

By May 1927, 31 Eastern Telegraph cables were operated with the

---

*The hand perforator for punching the holes in the tape which operated the telegraph levers had been replaced by an automatic transmitter driven by clockwork, invented simultaneously in the 1890s by F G Creed and the German/American Kleinschmidt who founded the Teletype Corporation. The more complicated receiving apparatus did not go automatic until Creed invented his printing receiver in 1913 which reproduced perforated tape as sent, which in its turn printed letters on gummed tape pulled along by compressed air.

regenerator, and it was about to be installed between London and India. As a result, the need for hundreds of skilled manipulative operators at branches like Gibraltar, Malta and Suez disappeared almost overnight. This was where the revolution lay so far as Electra House was concerned. As Dan Cleaver observed, 'a mere handful of locally trained operators could readily cope with what was left for them to do. Now the need was for a comparatively small mobile staff of skilled maintenance engineers to keep in order the delicate apparatus which had displaced the human element.' As a result the company was able to reduce expenditure, swiftly rising in all other directions, and halt dwindling receipts.

The Regenerator Revolution gave Ocean Telegraphy a timely boost. But it was not enough.

8

Cable Joins Wireless
1928–1930

Within six months the Post Office beam stations captured 65 per cent of all Eastern Telegraph and Eastern Extension traffic, and more than half of that of the PO's fellow state-owned, state-controlled Pacific Cable – a situation which Sir Hamar Greenwood MP described as Gilbertian.

The Post Office Imperial cables and the Pacific Cable Board's telegraph to Australia and New Zealand, together with its new West Indies System, presented competition which Electra House felt confident could be contained. The assurances of Sir John Denison-Pender, Admiral Grant and Edward Wilshaw that there was no need for panic would appear to have been justified. But the beam service was a much more serious threat. Cablegrams cost 6d the 'ordinary' word rate and 2s the 'full' rate; by beam radio the full rate was 4d and the deferred rate 3d. The short time within which the public switched from cable to wireless to make these savings demanded immediate action. It was a crisis which hit Electra House where it hurt most. As *The Times* commented, 'The inroads made upon the cable companies' profits by beam wireless have driven some of the directors from a position of assumed indifference to new inventions'. Accordingly, at the end of 1927, Sir John Denison-Pender appealed to the Conservative Government, the owner and operator of the service which threatened to put them out of business, to save them from total collapse. He let it be known that if a rescue operation was not forthcoming he would not be dismayed at the prospect of being forced to cease trading and having to pay his shareholders out, which, with the group's £18 million capital and £11 million cash reserves, he could do very generously.

The Government thereupon invited Eastern Telegraph to come

and talk their problems over, and a series of discussions ensued, which were of course private and confidential. Sir John made the company's plight the subject of public debate by issuing an official statement to the Press at the beginning of December 1927 which was followed by a number of articles setting out the cable companies' case.

Their financial position was so strong, boasted Electra House, that they could undercut rates 'under any fair system of competition and could almost ruin the wireless companies.' They emphasised the strategic importance of keeping the cable systems of the Empire. They insisted that 'fading' prevented continuous wireless communication and that supporting cables were essential. Where cable and wireless rates were comparable, as across the Atlantic, the Marconi company had been unable to obtain a reasonable proportion of the traffic.

The Marconi company who had resented not being asked to take part in the discussions with the Government, countered with a statement refuting the Cable arguments, which appeared in the papers of 5 December. They assumed that no decision affecting their interests would be taken without their being given the same opportunity to state their views to the Government. They had never asked the Government for a subsidy and it should never be necessary for the cable companies to do so. Both cable and wireless companies should be allowed to carry telegraph traffic at low rates if they could afford to. It was unfortunate that the cable companies had seen fit to appeal to the Government, as that must involve some form of Government control, which, they agreed with Sir Geoffrey Clarke, was the one thing to be avoided. The relations between the two systems of communication should be discussed between the two companies themselves. The Marconi company had no desire other than to be left free to develop its traffic services under the licences granted to it by the Government (for transmission to countries outside the Empire, such as the United States), currently by long wave but soon to go over to beam short wave too. Most of the claims of the cable companies were inaccurate. A rate war might for a time injure the wireless companies, but in the long run it would certainly be fatal to the cable companies, since the capital cost of beam stations was a tenth of that of cables and the operating costs lower too. The technological advances currently being developed, such as facsimile transmission, were going to enable the company to reduce rates even further. The problem of fading would soon be solved. They called on the Government to prepare a comprehensive scheme of wireless communication for the whole Empire. They were prepared, moreover, to discuss with the cable companies any means of co-operation, provided always that nothing was done to restrict in any way the free development of wireless.

This was in fact the next step. Sir John Denison-Pender and Lord Inverforth, the shipowner Andrew Weir, now chairman of the Marconi company, met at dinner in the house of a mutual friend on 28 December. Here they agreed to negotiate fusion, but only if they could persuade the Government to hand over to the combined companies the operation of the wireless beam stations which the Marconi company had built for the Post Office. They were not prepared personally to hammer out the delicate matter of the financial basis of fusion, but left that to their respective auditors, Sir William Plender (for Electra House) and Sir Gilbert Garnsey (for the Marconi company). The peacemaking dinner party had been prompted by the importance of producing a plan, voluntarily negotiated, which could anticipate any which HMG and the Dominion governments might find it necessary to *impose*. For on 19 December the Postmaster-General had announced the Government's intention of calling an Imperial Wireless and Cable Conference in London on 16 January whose recommendations, in the absence of any other, would settle their differences for them.

Once the decision had been made to co-operate instead of fight, action was swift. On 10 January 1928 the two companies issued a joint statement announcing their intention to make 'a joint report for submission to their respective boards as to a possible arrangeent in the joint interests of their respective companies'. Six days later the Imperial Wireless and Cable Conference held the first of its thirty-four meetings in London under the chairmanship of the Secretary of State for Scotland, Sir John Gilmour Bt MP, who also represented Great Britain. Other delegates represented Canada, Australia, New Zealand, the Union of South Africa, the Irish Free State, India; and the Permanent Under-Secretary of State for the Colonies represented all British colonial territories and protectorates. The secretary of the conference was Sir Norman Leslie who was secretary of the Committee of Imperial Defence. The representative of Canada was Sir Campbell Stuart. Their brief was

> To examine the situation which has arisen as a result of the competition of the Beam Wireless with the Cable Services, to report thereon and to make recommendations with a view to a common policy being adopted by the various Governments concerned.

Their meetings were in private and their proceedings confidential.

They had been sitting for two months, when on 14 March they received a letter signed by Sir John Denison-Pender and Lord Inverforth informing the conference that the Boards of the two groups had reached an agreement providing for a 'fusion' of the interests of the companies through the medium of a proposed holding company, subject to satisfactory arrangements being made with the British Government and the governments of the Dominions and India, and acceptance by the shareholders of the cable companies and the Marconi company. They appended a detailed proposal for the holding company's £54 million capital.* It would acquire the whole of the Marconi company's ordinary, preference and debenture capital, and the ordinary share capital of the cable companies, leaving the preference and debenture issues undisturbed. Cable shareholders would hold 56¼ per cent of the new company's voting power, Marconi shareholders 43¾ per cent. The twenty member Board would be in the proportion twelve Cable, eight Marconi.

According to his biographer, W P Jolly, Marconi could not forgive the Board of his company, reconstituted in 1927 with himself no longer as chairman, from being out-manœuvred by the Cable interests in this way. When he had first mentioned publicly the idea of a merger, it had been from a position of strength, with the cable companies in a panic about the new beam wireless service. But the negotiations resulted in a combined company dominated by the cable interests both in voting power and on the Board.

Details were given to the Press, and *The Times* commented 'the merger is one of the most important industrial fusions that have been arranged in recent years'. It obviously changed the situation which confronted the imperial conference. 'The old system of unrestricted monopoly enjoyed by the cable companies will have to be replaced by a system which recognises that telegraphic communication is a public utility like gas, water and power. The invention of the wireless beam system has revolutionised overseas communication and made possible the use of telegraphic communication as an everyday necessity rather than a luxury.'

On 21 May Walter Baker, Conservative MP for Bristol East, raised the whole matter in an adjournment debate and, from information which had leaked from the 'secret conference', told of his fears of the 'very dangerous situation' which had arisen for the nation and Empire. The state-owned wireless and cable interests, he told the Commons, were about to be sold to private interests; to the

---

*See Appendix C.

Marconi company with its record of 'scandalous management' in which the predominant partner was an American trust. Immense sums of money had been made on the British and American stock exchanges during the sittings of the imperial conference. Eastern Telegraph shares had risen from 136 in January to 242 on 9 May, which showed the very great value of inside information. Early in 1928 Marconi shares stood at 39s 6d; on 14 March they were 67s 6d. The chief result of the conference had been to enrich the share manipulators. The Postmaster-General and his department regarded the conference as 'a scandalous ramp'.

A fellow Conservative, Mr Ammon, referred to Eastern Telegraph complacency. For a number of years after the invention of wireless they had done nothing except make a few technical improvements. Only when the Government began to compete with them did they invent the loaded cable. 'This indicates all too clearly that if the cable companies have a distinct and clear monopoly they are not prepared to give any better service to the country than they are compelled to do.'

Representatives of Eastern Telegraph, the Marconi company and Indo-European were invited to give their views to the conference, which received deputations from the Empire Press Union, the Federation of British Industries and the London Chamber of Commerce. They issued their report on 6 July 1928. The problem before them was complicated, but they managed to dispose of it in twenty-one clearly written pages, which exposed much of the humbug of the often emotional claims made by both sides in their public sparring.

The beam services to Australia, South Africa and India yielded a very high profit, stated the report. They could always afford to undercut the cable rates, 'and if competition were unrestricted could render the cable systems unremunerative'. Yet the cables had a strategic value. Wireless offered a cheap service but was not all-sufficing. The majority of the cables should be kept. If Eastern Telegraph and its associates went into voluntary liquidation and disposed of their assets, the opportunity presented to foreign interests to strengthen their position (by purchasing them) would be considerable. A joint purse pooling system might secure the cable companies reasonable but not luxurious return on their capital, but would not realise large economies. They reckoned that, provided public and Government interests could be safeguarded, the best solution of the problem was 'fusion', not only of Eastern Telegraph and Marconi, but of all cable and wireless interests conducting communications within the Empire (only). Negotiations for this had already been instituted voluntarily. They realised this meant transferring the ownership of the Post Office beam wireless and cable assets of HMG in Britain, and that this 'would involve for His Majesty's Government a departure from the policy hitherto adopted in regard to the conduct of wireless services'.

It was back to Sir Laming, and it was fear of this which had alarmed Walter Baker. For the intended 'fusion' was not to be in a Government, Post Office-like, corporation, but in a private *company*.

The conference recommended the immediate formation of two companies. They must remain under British control.* A Merger company would acquire the ordinary shares of Eastern Telegraph, Eastern Extension and Western Telegraph, and the ordinary and preference shares of the Marconi company. Thus this Merger company would have the investments of the cable companies in enterprises not directly concerned with communications, and the Marconi interests in 'non-traffic' matters, such as the manufacture of radio apparatus and the exercise of wireless patent rights – its activities as a manufacturer rather than as an operator. A Communications company, organised on public utility lines with a capital of £30 million, would purchase the 'communication assets' (the radio stations, cable stations, buildings, cables, cable ships, research laboratories etc) of the two groups, and give them not cash but shares in the Communications company (which in their turn would be held by the Merger Company). The Communications company would buy the Pacific Cable (which had an outstanding debt of £1,233,000 and was one-third owned by HMG and two-thirds by the partner governments) for £517,000; the Pacific Cable Board's jointly owned Caribbean cable and wireless system, which had always worked at a loss and owed £379,000, for £300,000, though they would not have to take over the debt; the Post Office Imperial cables across the Atlantic, wholly owned by HMG and always making a small profit, for £450,000; and *lease* the Post Office beam service for £250,000 a year for twenty-five years (till 1953), on agreeing to make a cash payment of £60,000. The Communications company would manage the entire merged 'imperial' wireless and cable service, and the non-imperial 'international' long wave and short wave radio services of the Marconi company such as its Ongar station working to New York and the circuit to Japan which was soon to come into operation on short wave.

---

*A condition of the merger was that 75 per cent of the share capital of the Marconi company should be British.

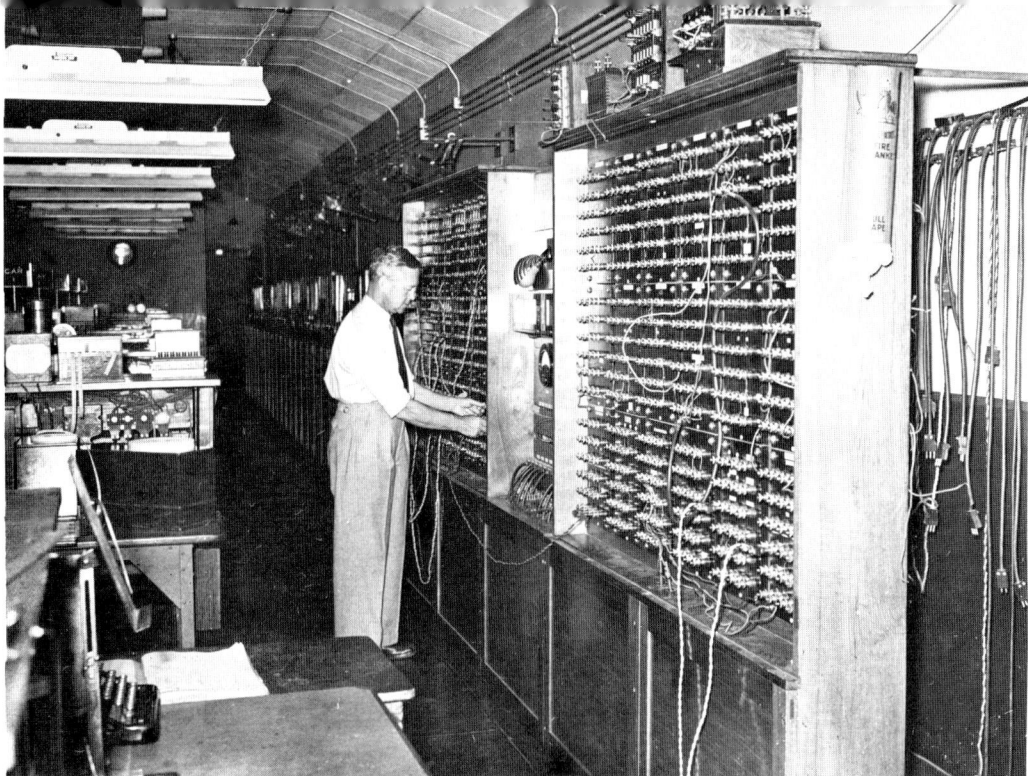

55-7 At work inside The Tunnel, the wartime emergency station opened by Lady Wilshaw in May 1941 at Porthcurno to which all circuits were moved for safety in World War II, is C. A. Harper, Assistant Engineer (*above*). Two hundred Cornish miners excavated 15,000 tons of rock and topped the granite roof of the entrance (*below left*) with reinforced concrete to protect the overseas cable terminal from possible sabotage and enemy action. But no attempt was made to destroy it, unlike Electra House, Victoria Embankment, London, seen (*below right*) after a bomb had hit it in July 1944, damaging the chapel and the chairman's office.

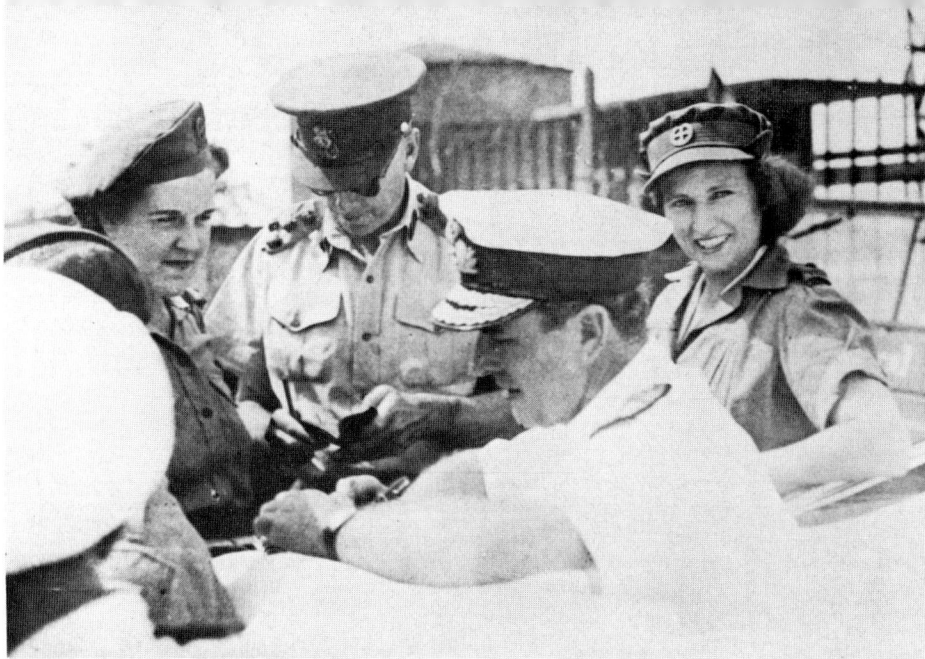

58 Admiral Lord Louis Mountbatten, Supreme Commander South-East Asia Command, signing autographs when he visited the Cable and Wireless 'Press Ship' at Singapore which had transmitted the news of the liberation of Malaya to the world at the end of World War II. The ship was also the means by which commercial services with Malaya were re-established within fourteen days of the entry of the first Allied occupation forces.

59 When Lord and Lady Mountbatten came to see the London end of the Cable and Wireless network, they posed outside Electra House with Sir Edward Wilshaw, chairman, and Lady Wilshaw.

60 When petrol rationing curtailed the use of motorcars in World War II, Sir Edward Wilshaw (above) hired this hansom cab to drive him to his business appointments.

61–2 A major task has always been training the operators and engineers to man the company's vast overseas telecommunications networks. (*Above*) Principal Smith – known to the many hundreds of young men he trained as 'The Prawn' – explains the curriculum to the King of Norway who visited the London Training School in May 1944. (*Below*) London Training School in action in its last days at Electra House before moving to Porthcurno in 1950.

63–4 Once trained, operators had six months 'consolidation' and would then be posted to a station anywhere in the world. However, living quarters and offices all tended to look alike. (*Above*) the operating side of the Instrument Room at Gibraltar; (*below*) inside the transmitting station at Malta.

65 To give staff operating in World War II battle areas the status of war correspondents *vis-à-vis* the Armed Forces, they were formed into 'Telcom' units in battledress. Here W. H. Chilvers operates a mobile battery receiver as part of the Telcom Light Mobile Press Unit in Batavia in 1945. No other accommodation being available, the men lived in a former Japanese gaol. The condemned cell was commandeered by six juniors from New Zealand.

66 Inside the operating coach of one of the many Cable and Wireless 'Blue Trains' – the staff of MWA II at Jerusalem in 1946. In view of the heavy volume of telegraphic traffic arising in Palestine, the company transferred a mobile wireless unit from Italy to Jerusalem to supplement the cable circuit from Haifa to London, capable of handling 40,000 words a day and sending radio pictures.

67–8 After Electra House, Moorgate, was put out of action by the German Luftwaffe in 1941, 'London Station' was moved to Electra House, Victoria Embankment – The Fortress. (*Above*) a general view of the Instrument Room taken in January 1944; (*below*) the Press 'Local' room in November 1945 where press messages were received for transmission to the principal newspapers and agencies of the world.

69 Lord Reith (*left* with Leslie Nicholls and H. H. Eggers) was invited by HMG in January 1945 to make a quick tour of Dominion governments to obtain their views on his plan for reorganising imperial communications. Based on the so-called Canberra Proposals, his report advocated setting up an incorporated central body with local national communications corporations. The British Government called a conference to consider the report's recommendations which Sir Edward Wilshaw's Cable and Wireless opposed.

70 Under the Cable and Wireless Act which took effect from 1 January 1947 *all* the directors of the company were to be appointed by HMG. The chairman was to be part-time and non-executive, and they chose sixty-four-year-old Sir Stanley Angwin who had been Engineering Chief at the Post Office since 1939 (*right*). As Chief Executive and Managing Director they chose John Innes, also a Post Office engineer (*far right*).

71 The centenary of Submarine Telegraphy was in 1950; and an exhibition at the Science Museum in London celebrated the laying of the first Dover–Calais telegraph cable. (*Above*) Lord Pender speaking at the opening of the exhibition on 28 August 1950 at which a message was cabled round the world in 53.6 seconds.

72 Laying the Dover–Calais cable from *Blazer*, 1851.

The Board of Directors of the Merger and Communications companies would be identical, the chairman and one other being approved by HMG on the suggestion of the *cable* companies. Any revenue made by the Communications company over £1,865,000 (6 per cent of the capital), designated the Standard Net Revenue, would be paid half to the company and half to a fund to help reduce rates. This was to prevent users being exploited for the benefit of shareholders. After three years the company would pay the Post Office 12 per cent of any earnings – 'excess profits' – above this standard revenue. Another check was that the Communications company was required to consult on questions of policy, including any alterations in rates, a new body to be called the Imperial Communications Advisory Committee consisting of representatives of the governments of Britain, India, the Dominions, the Irish Free State (which had come into being in 1922). and the Colonies, which would have access to all information in the hands of the company.

The company would also be obliged to consult this committee on such matters as the institution of new services, the discontinuance of any services which became commercially unprofitable, and submit their accounts to it for examination. The Post Office reserved the right to conduct the external *telephone* services of Great Britain, but would agree with the Communications company the terms on which it would have the right to use its wireless stations for telephone purposes. The whole object was to provide cheap and efficient communications services.

The report was placed before the House of Commons in the form of a White Paper. On 2 August 1928 the Government announced it accepted all the recommendations and introduced an Imperial Telegraphs Bill to authorise the sale of the Pacific Cable Board's telegraphs and Post Office cables. It had its second reading on 21 November. 'Believe me,' said Ramsay MacDonald for the Labour opposition, 'this is not a little pettifogging, superficial thing; it is one of the biggest pieces of national policy, of national business that has ever been brought before this House, at any rate in my time.' He opposed the second reading on the grounds that it sacrificed public utility to private gain 'by disposing of valuable State undertakings to private interests.'

One member thought it was as absurd to hand the Government system over to private enterprise as it would be to hand over the Navy, but it was welcomed by *The Times*. 'Left to themselves,' ran its leader of 22 November the day after the Bill's second reading, 'the cable companies might in the long run have been forced to succumb to the competition of their successful rivals,' with the danger that 'if the Eastern Telegraph and its associated companies were to be forced into voluntary liquidation portions of the system might pass into foreign control.' The conference had evolved 'a bold and statesmanlike solution', a privately owned communications company subject to Government supervision. The plan seemed admirably designed to secure the two chief advantages which should be enjoyed by a joint imperial utility undertaking – the economy of private management and the security of public control.

The Bill received the royal assent on 5 February 1929.

A month later (on 6 March) the man who had initiated the negotiations on behalf of Eastern Telegraph, Sir John Denison-Pender, its chairman, died at his London home, 6 Grosvenor Crescent, at the age of 74. The funeral took place at the little village of Slaugham where he had had an estate up to the death of his wife in 1920. At 12.15 pm Greenwich time that Saturday 9 March 1929 all the telegraph circuits of the Eastern Associated companies worldwide stopped for one minute to allow the staff on duty to stand in silence at about the time of the committal of their chairman's body to his Sussex grave.

Another month passed and the two new companies created by the merger were registered – 8 April 1929, which *The Zodiac* chronicled as A New Date In English History. On that day the new 'Court of Directors' (à la Bank of England) consisting of the Boards of the two groups, met for the first time in the third floor Board Room of Electra House, Moorgate, which, since Eastern Telegraph was the dominating partner, became the Head Office of the new combine.

The combined Eastern/Marconi 'Court' consisted of twenty-two members and included Senatore Guglielmo Marconi. As laid down in the conference report, they were the same for both the Merger and the Communications company. If Sir John Denison-Pender had lived, he and Lord Inverforth would have been joint-Presidents of the Merger Company, which it was decided to call 'Cables and Wireless Limited'. But in the event 63-year-old Inverforth was appointed sole President, and Sir John's 46-year-old son John Cuthbert, 'Governor' and joint managing director with Fred Kellaway, who was also a Deputy Governor with the Earl of Clarendon. A Management Committee was formed to conduct the day-to-day business of what was no more than an investment house, with which the Imperial Communications Advisory Committee had no concern.

The Communications company was called 'Imperial and International Communications Limited'. Its shares were held not by individuals but by the cable companies and the Marconi company, whose ordinary shares in turn were owned by Cables and Wireless (whose shares alone the public could buy). Electra House nominated and, after consultation with Dominion governments, HMG approved as chairman the distinguished economist 46-year-old Sir Basil Blackett who from 1904 to 1922 had been at the Treasury, latterly as Controller of Finance, after which he was Finance Member of the Executive Council of the Governor General of India, and a Director of the Bank of England. He was also made a director of Cables and Wireless. In his first public speech after his appointment he made it clear that Imperial and International had been formed not primarily to make profits but to give service to the governments and people of the empire.

J C Denison-Pender and Fred Kellaway were appointed joint managing directors; Edward Wilshaw became General Manager and Secretary of both companies; but general policy was formulated by a Management Committee.

Obtaining the consent of the Dominion governments to the terms of the agreements reached by their representatives with the various wireless and cable undertakings, so that the recommendations of the conference could be carried out, was far from a foregone conclusion.

With only two days to go (on 8 May), Australia waived its stipulation that Imperial and International should first come to a revised agreement with Amalgamated Wireless, and agreed to sign the Merger Agreement. But signing for the sake of unity did not remove Australia's objections, which remained over the years and became the platform for a second attempt to wrest back unfettered control of Australia's own communications when circumstances became more favourable.

Although New Zealand too had signed the Merger Agreement, her Prime Minister Sir Joseph Ward, speaking in the House of Representatives on 27 August, said he thought the British Government had made a great mistake in giving up its superior interest in the beam system to enable Eastern Telegraph to become part proprietors. After the announcement of the merger there was a phenomenal rise in the shares, he said, and a lot of money was made. 'The whole thing is a very regrettable transaction. The Merger company is now able to dictate its terms. That the British Government agreed to let us all go was an unfortunate and a wrong thing from the point of view of the Pacific Cable and the Empire's future. I am sorry New Zealand has no opportunity of protesting against the sale of the Pacific Cable.'

The Times thought this outburst was based on a misapprehension. The merger was recommended unanimously after prolonged consideration. Why had not New Zealand's representative protested at the conference? The cable companies were now able to capitalise their position before feeling the effect of wireless competition. The paper welcomed 'the first effort at imperial industrial rationalisation'.

The next hurdle was to persuade shareholders to transfer their shares. At the 95th general meeting of Eastern Extension on 29 May 1929 (the day before the General Election which ended Stanley Baldwin's five year administration and brought Labour back to power) the new chairman, J C Denison-Pender, justified the Board's initiative in seeking a merger. It was not likely, he said, that so high a dividend would have continued if the competition with governments had been allowed to increase. A prolongation of conflicting interests would have impoverished them. Almost the last words his father had said to him were, 'This merger is a great thing and a good thing'.

It was the last general meeting under the old conditions, and similar 'last' meetings took place on the same day of Eastern Telegraph and Western Telegraph, and the pertinent alterations made to their articles of association. In telling shareholders of the Advisory Committee representing HMG and Dominion Governments the chairman observed, 'It is of course quite understood that this will in no way interfere with the commercial operation and control of the two great companies about to commence'.

Lord Inverforth told the general meeting of Marconi's Wireless Telegraph Company on that same 29 May that it was Eastern Telegraph having reserves of £3,730,240 that ensured a rate war would do the cable companies very much less damage than the Marconi company which had no reserves. The Marconi company had agreed to the merger because of HMG's decision to hand over the Marconi-built imperial beam system in the UK to the Post Office. After that the only revenue the company could hope to derive from it was a royalty of $6\frac{1}{4}$ per cent on gross traffic receipts. Where would they be if their traffic revenues came only from their foreign services and investments in their Indian and Dominions companies? The solution reached by the imperial conference had been regarded by every foreign power as a model to be followed in the development of telegraphic communications throughout the world. The greatest interest of all had been shown in the United States where steps had already been taken to form a similar merger. It was a landmark in the history of world communications. Before the merger, empire communications had been in the hands of eight different authorities. No wonder foreign communication interests were beginning to assert themselves in the empire.

The commercial operation and control of these great companies is left entirely free and unfettered, but on general questions of policy there will be an opportunity which the directors will always welcome of consulting the Advisory Committee. That Committee will be a body entirely outside both companies, so as to be able to exercise an impartial judgement on any questions which may come before them. We look forward to working in the friendliest possible spirit with this Advisory Committee.

The chairman went out of his way to explain that Marconi's Wireless Telegraph Company (which had made a profit of £413,000 in 1927) was in no way going out of business. While disposing of its 'traffic assets' (including minority holdings in overseas wireless stations built for government sponsored companies like Amalgamated Wireless (Australasia) Ltd), it would continue to carry on its sales and manufacturing business, and the research essential to the success of these (which included talking pictures, television and the use of microwaves in communications). The 'entertainment side' of the business was sold to The Gramophone Company (which became His Masters Voice).

Also on 29 May was made the key Treasury Agreement (Cmd. 3380) which regulated the relations between the new companies and the imperial governments.

Before Imperial and International could take over, it had to fulfil its undertaking to transfer to its employ the 13,000 staff engaged under varied conditions of service by the old companies, on terms not less favourable than those they already enjoyed. This entailed long discussion and negotiation, but was completed by the summer of 1929. On the whole Eastern Telegraph staff received higher pay, a shorter working week (forty instead of forty-eight hours) and longer holidays than Marconi staff who thus benefited from the merger. The future looked bright and most of the new contracts guaranteed employment for five years.

By 27 September 92 per cent of Marconi shareholders and 98 per cent of Electra House shareholders had given their assent to the transfer of their shares. The Treasury Agreement inherited by the new Labour Government from the outgoing Conservative administration became binding. Shareholders in any company of the Eastern Telegraph group received £299 18s in nominal value of fully paid shares in Cables and Wireless Ltd for every £100 nominal of stock or shares held. Every £100 of Marconi shares were exchanged for £437 of shares in the new company (see Appendix D for details).

On Sunday 29 September the London ends of the 'Imperial' cables from Canada were transferred from the Post Office Telegraph

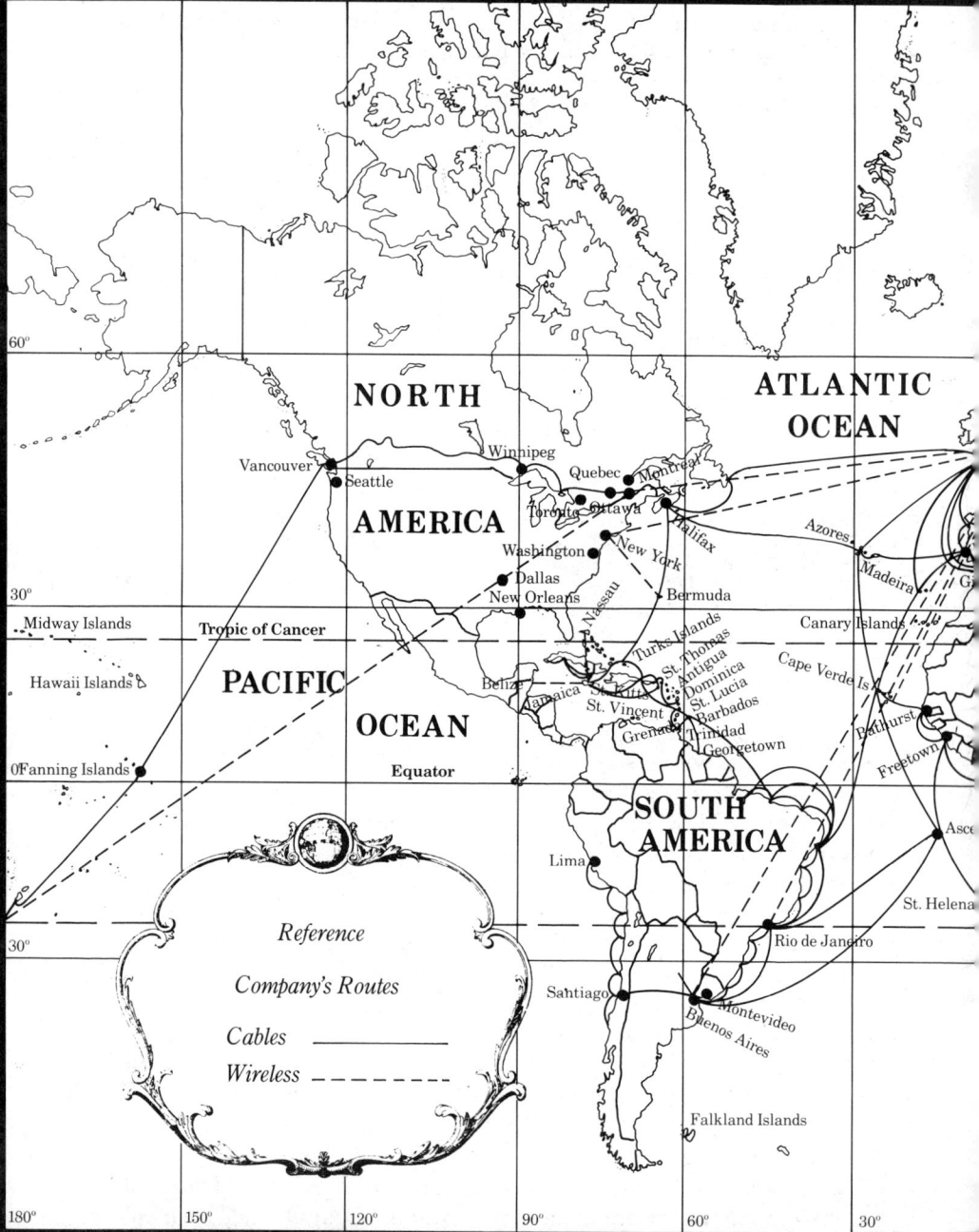

60°

NORTH

ATLANTIC
OCEAN

Vancouver
Seattle
Winnipeg
Quebec Montreal
Toronto Ottawa
Halifax
Azores

AMERICA

Washington New York
Dallas
New Orleans Bermuda
Madeira

30°

Midway Islands — Tropic of Cancer
Canary Islands

Hawaii Islands
PACIFIC
Belize
Turks Islands
St. Thomas
Antigua
Dominica
St. Lucia
Cape Verde Is

OCEAN
Jamaica
St. Vincent
Barbados
Grenada Trinidad
Georgetown
Bathurst

0 Fanning Islands
Equator
Freetown

SOUTH
AMERICA
Asc

Lima
St. Helena

30°
Rio de Janeiro

Reference
Santiago
Montevideo
Buenos Aires

Company's Routes

Cables ————

Wireless — — — —

Falkland Islands

180°   150°   120°   90°   60°   30°

Moscow

ASIA

Berlin

ussel

ROPE

Rome

Teheran

Malta

Cairo

Karachi

Delhi

Peking

Shanghai

Foo-Chow

Tokio

PACIFIC

RICA

Port Sudan

Aden

Bombay

Calcutta

Rangoon

Hong Kong

OCEAN

agos

Madras

Bangkok

Saigon

Philippine Islands

Manila

Maldive
Islands

Colombo

Singapore

Loango

Nairobi

Zanzibar

Seychelles Islands

Keeling Islands

Port Darwin

Fiji Islands

nda

Mozambique

Salisbury

Mauritius

Rodriguez

Tropic of Capricorn

AUSTRALIA

Brisbane

Sydney

Norfolk Is

ape Town

Lourenço Marques

Durban

INDIAN OCEAN

Perth

Canberra

Adelaide

Aukland

Melbourne

Wellington

30°   60°   90°   120°   150°   180°

Office to Electra House, Moorgate, and the shore ends were diverted from Mousehole to Porthcurno where fourteen operational cables now terminated. Imperial and International took over the operation of the UK terminals of the Post Office beam wireless stations at Bodmin, Bridgwater, Skegness and Grimsby working to Australia, India, South Africa and Canada, and the terminals were transferred to Radio House from which the non-imperial wireless stations operated by Marconi (and now by Imperial and International) were already operated. On the same day, although no physical transferring was needed, Imperial and International took over the management of the Pacific Cable and the West Indian Cable and Wireless System of the Pacific Cable Board.

These services, together with the cable system of the Eastern Associated Companies, opened to the public under the banner of Imperial and International Communications Limited on 30 September 1929. This remarkable concentration of world communications was a major technical achievement of the utmost complexity, and that it was carried out without a hitch and without interruption to any client's services in so short a time was a great tribute to the skill and energy of the staff concerned. The new public utility company controlled 164,400 nautical miles of submarine cable – one half of the world's total submarine cable mileage of 325,000 – 13 cable ships, 253 cable and wireless stations and offices valued at £24,857,000, plus a stock of new cable worth £824,000. A list of all the companies whose communications assets it had acquired by 31 December 1929 is given at Appendix E.

Before Imperial and International began operations they acquired an additional company but not in the field of communications. This was the British East African Broadcasting Company of Nairobi, purchased on 18 September.

A company which remained outside the Imperial and International umbrella, though inside Electra House, was the Direct Spanish Telegraph Company which was formed by the first Sir John Pender in 1872 as a private venture under the name Falmouth and Bilbao Submarine Cable Company without ever bringing it into his Eastern Telegraph group. The profits went to the Pender family personally. The India Rubber and Gutta Percha Company laid the cable from Bilbao on the north coast of Spain to Kennack Cove, Cornwall just below Goonhilly Down, whence it was taken to a cable house in Falmouth. The fifteen-mile land line had to cross several fields and was continually being ploughed up by farmers which made maintenance difficult. So in 1921 the shore ends were diverted to Porthcurno (POR-BIL-1) where a second Bilbao cable (POR-BIL-2) was brought in in 1925. Both were 'fast' cables – eighty words a minute. In Spain the cable at Bilbao was carried by land line to Barcelona and thence to Marseilles, and in the 1920s there were through connections to Italy via Italcable. Direct Spanish had its own staff in Bilbao and Barcelona, and representatives in Madrid and Marseilles. The London end was worked from a small back room – Room 8 'Strictly Private' – at Electra House, Moorgate. In 1978 Roy Baker, who joined Direct Spanish as an operator in 1928, remembered the rivalry with Eastern Telegraph who still used stick perforators to punch tape, whereas the operators of Room 8 were trying out a Kleinschmidt electric punching machine. They also experimented with the first gummed tape – previously tape had been stuck to the telegraph form with ordinary glue. All went well until they had to make a copy of a telegram on a wet tissue copier and the tape came unstuck. Eastern Telegraph had a small research laboratory at Electra House and used Direct Spanish staff to test new equipment. Direct Spanish did not become part of Imperial and International until 1933, when it was absorbed in return for a payment of £3,500 a year for twenty years.

Also excluded initially from control by Imperial and International was the Direct West India Cable Company which Sir Campbell stuart had recommended should buy out the Pacific Cable Board's loss-making West Indian System.

But in the event the Pacific Cable Board bought Direct West India with its 8,000 miles of cable and twenty-two stations of Halifax and Bermudas, Cuba Submarine and West India and Panama. Then on 22 July, a fortnight after the conference had announced the merger plan, Eastern Telegraph bought their shares.

The component parts of Direct West India continued to operate under their own name, and outside Imperial and International, in whose first accounts there appears an item of £720,000 for the purchase of 160,000 Direct West India shares – the inherited Eastern Telegraph purchase? West India and Panama became an 'associated' company of Imperial and International who took it over fully in 1932. The Regional Superintendent of West India and Panama, Maurice Petit, was based on St Thomas at this time. The manager of the Barbados station, J R Thomas, retired in 1929 and of the small staff then running the Bridgetown office only Victor Bladon and George Edwards remained. Later the Bridgetown office was closed and, as Jimmy Cozier the distinguished local chronicler of these events records, 'the name of West India and Panama went out of use in Barbados after serving the island for sixty years'.

Meanwhile, in 1928, a hurricane had knocked out the wireless station on Antigua. The aerial was soon re-erected and the station back on the air, but the falling earth screen left a permanent mark on the head of engineman Llewellyn Pereira.

> The then Officer-in-Charge after an exciting day retired to his quarters anticipating a rest, only to find his room swamped with water. The same year P R K Grant had the unique experience of descending through the floor of the operating room. P R K is still sore because the Company at that time did not pay the ciropractor's [?] fee for his injured knee (typed note headed 'Wireless Station Antigua' found in Antigua archives, 1978).

North of the Caribbean, the Bermuda Wireless Coast Station, call sign VRT, began operating from its two 300 ft aerials at Lily Park, St Georges in 1928. The tower masts were a familiar landmark to shipping for years. The station kept in touch with all craft sailing round the island's shores and gave Direction Finding bearings to any approaching it. When on one of its flights from Germany to the United States the *Graf Zeppelin* took a more southerly course than usual and overflew Bermuda, the station's DF system helped her find her way.

On 21 December 1931 the Bermuda Telephone Company opened a radiotelephone service to New York in conjunction with the American Telephone and Telegraph Company. This was the second international radiotelephone circuit ever to be opened, preceded only by the New York-London service.

Far-speaking, impossible through long distance telegraph cables as seen, was undoubtedly a more desirable form of communication than far-writing. As Sir Basil Blackett remarked to a Royal Empire Society audience in November 1930, 'Consider how different the history of the Versailles Conference might have been if President Wilson had been in daily telephonic communication with Washington'. The wide application of overseas wireless telephony was obviously the next stage in the evolution of telecommunications, but by its terms of reference Imperial and International were excluded from developing it. Why? Sir Basil Blackett, who had to execute the Merger Plan, had taken no part in the deliberations which formulated it. He had had no opportunity of raising the question round the conference table, and now there was a 'we' and 'they' situation between his new employers Imperial and International and his old,

HMG, with a gulf between the two which provided the usual opportunities for resentment and misunderstanding, for the clash of personalities and misinterpretation inherent in every 'communications gap'.

It seemed clear to the outsider Sir Basil Blackett that when the imperial conference left overseas telephony to the Post Office and in the same sentence declared that it *'will* agree' (not 'can' or 'should') – that it will agree with the company the terms on which it would have the right to use its wireless stations for this purpose – that the Government would at once arrange to do that very thing. Apart from the explicit nature of the enjoinder, did not the Communications Company 'charter' make it clear that the establishment of empire services which worked in opposition or out of harmony with the scheme deprived it of much of its value and militated against the objects it was seeking to achieve?

The directors' bewilderment was expressed in a peculiarly bitter passage of their annual report of 6 July 1930 which covered the first twenty-one months of their activities, 1 April 1928 to 31 December 1929.

> We did not, of course, assume any necessary implication that this Company should, or would desire to, take over the operation of Imperial wireless telephony, but we did certainly place this interpretation upon it, that the Company's stations which already existed would be used for that purpose, and the matter was therefore taken up with the Government. It was a great disappointment to us that the Government should have refused our offer to co-operate in providing the wireless link for the overseas telephone services. We cannot believe that it is in the true interests of this country and of the Empire that there should be competition between Government overseas telephone services and telegraph services operated by this Company, specially constituted as it is a Public Utility Company under considerable Government control. This is in effect to revert to the undesirable position which had developed in 1928, when Government-operated Beam wireless services were competing with cable services operated both by Governments and by Companies.

It seemed to them scientifically unsound and uneconomic to draw a hard and fast line between radio telegraphy and radio telephony. It was particularly regrettable in regard to research. The Post Office's use of its monopoly had deprived the only British wireless telephone company of the opportunity to operate any wireless telephone services from Britain. Their most formidable competitors, the great American traffic companies, had no such handicap. Imperial and

International were at a considerable disadvantage with them in their struggle to maintain the position in this country and abroad in wireless telephony. The anomaly was emphasised by their associated company conducting the wireless telephony at the Australian end of the British-Australian service; and that was going to be the case too in India and Canada.

Since the first twenty-one months' profit of £2,023,150 (income £11,381,250, expenditure £9,358,100) represented an annual rate well below the standard (twelve months) profit of £1,850,000, there was no setting aside the half of the 'excess profits' above that standard which the charter stated should go to financing rate reductions. In spite of this, under pressure from the Dominions, reductions were made for political reasons, and the company asked, without success, for a rebate on the losses so sustained – reckoned at £79,000 a year in Australia, and £85,000 a year in South Africa.

Other expenditure in that first twenty-one months included large non-recurring items such as the £1,015,000 paid for the Indo-European Telegraph Company whose land line from London to Teheran ceased to operate on 28 February 1931. This, as seen, had been opened sixty years earlier with the object of constructing a direct trans-continental line some 3,800 miles long to link Britain with Russia, Persia and India.

Apart from having to find £720,000 for 160,000 shares in Direct West India, the company inherited the £1,154,100 debt of the Pacific Cable Board to the Government, the outstanding amount from the £2 million borrowed in 1901–3 to lay the Pacific Cable. The company undertook to repay the National Debt Commissioners £77,500 a year to clear off principal and interest by 1949.

Even if traffic had sustained itself at the level of 1928, the boom period when the standard revenue was fixed at its optimistically high figure, the chances of the company at once meeting its obligations were slender. At their final meetings, shareholders of the various companies had been warned to expect lower dividends, but within months circumstances had so changed that management found its quasi-commercial, quasi-state constitution so restrictive as to make progress well nigh impossible.

The world economic crisis heralded by the Wall Street crash of 1929 and set in motion by the financial collapse in Austria and Germany in the summer of 1931, dispersed the optimism of 1928 with unprecedented speed. Britain's exports fell from £839 million in 1929 to £666 million in 1930, to £461 million in 1931. At the beginning of 1929 unemployment in Britain stood at 1,250,000; in December 1930 at 2,500,000. When it rose to 2,900,000 in August

1931 the Labour administration fell and a 'National' Government was formed 'to pull the country back from the edge of the precipice'. In July the Bank of England were losing gold at the rate of £2½ million a day. The raising of tariffs and abandonment of the gold standard did little to revive the nation's flagging trade, and management at Electra House saw traffic falling month by month with no way of stimulating greater usage.

The number of chargeable words carried by the company fell from 244 million in 1929 to 121 million in 1935. Takings in the first six months of 1930 were down 10 per cent, from £3,108,150 in 1929 to £2,807,000 due, said the chairman, to 'dislocation of the world's commerce'. 'It will be readily appreciated that a depression so widespread in its incidence and so grave in its effects carried with it a loss of traffic which is the inevitable result of the decline of that world trade from which the need for international communications arises.' The international difficulties which confronted the nations of the world, he said, underlined the need for co-operation, not competition, in long distance communications. It was not an occasion for illusory self-interest but a pooling of resources of goodwill. The whole system had to be kept going whether the traffic was large or small, unlike a factory, part of which could be closed in a crisis.

The 1929 net profit of £1,050,000 fell to £326,000 in 1930; and then to £75,000 in 1931. There was a slight improvement to £80,000 in 1932; to £209,000 in 1933 and £625,000 in 1934. To pay the 1⅓ per cent dividend in 1930 (£525,000) it was necessary to draw £200,000 out of the General Reserve.

The decline was certainly accelerated by dislocation of the world's commerce, but it derived quite as much from increasing foreign competition which had to be faced with or without an Economic Crisis.

'There is room for both of us,' said Sir Basil Blackett in his Royal Empire Society lecture of 13 November 1930. There was room for Americans and British, as well as for other national systems if everyone worked in co-operation with their eyes not on domination but on service to mankind. It was not an ideal shared, however, by Sosthenes and Hernand Behn of the International Telephone and Telegraph Corporation, formed in 1920 and by now well entrenched in Latin America and a large part of Europe. ITT operated the six Atlantic cables of the Commercial Cable Company and the Commercial Pacific Cable laid in 1902–6 between San Francisco, Honolulu, the Philippines and the Far East, in which Imperial and International held one half of the share capital with the rest divided equally between the Danish Great Northern company and the

American Mackay group. Nor was it the philosophy of Owen D
Young and D Sarnoff of the Radio Corporation of America (RCA),
formed in 1919 by General Electric to take over the American
Marconi company; nor of Newcomb Carlton of Western Union
Telegraph Company, founded 1851, which owned ten transatlantic
cables and leased one of them from the British company, Anglo-
American, for £262,500 a year.

In 1933 ITT began negotiations with RCA to join in the creation
of a world-wide system of direct radio services from the USA with
low tariffs. Electra House saw such a development as jeopardising
Britain's predominance and appealed to the Imperial Communications Advisory Committee for help. Sir Campbell Stuart suggested to
Newcomb Carlton of Western Union that Britain and US could
come to a pooling arrangement over transatlantic cable receipts, and
Denison-Pender brought Behn of ITT and Sarnoff of RCA into the
conversations. But the talks came to nothing; neither did the plans
for an ITT-RCA merger. But on 26 February 1934 President
Roosevelt recommended to Congress the creation of a Federal
Communications Commission. In a Memorandum written on that
day headed 'Projected Merger of Communications Companies in the
USA and its Effect on Empire Communications', the Advisory
Committee pointed out that the possibility of an American com-
munications merger was one of the main reasons given by the
Marconi company to the imperial conference of 1928 for setting up a
British combine which could resist the threatened domination of
world communications by the USA.

> In the present financial position of Imperial and International,
> competition from a strong American combine might lead to disaster
> and might imperil the existing structure of Imperial Communications
> . . . The British group cannot face an American combine unless it has
> put its house in order and is in a position to frame policy in full co-
> operation with the Governments of the Empire which will ensure
> active management and development of its communications system,
> giving the best facilities to the telegraphic user combined with the
> lowest tariffs which can be justified on a commercial basis.

In their Third Report to the Governments of the Empire of 1 May
1934 the Advisory Committee warned that integration of internal
and external communications companies of the US was likely to
follow the establishment of a Federal Commission.

> A merger of American external services based on control of internal
> communications in the USA would be a much more dangerous

opponent of Imperial and International than the present competing American companies. It is possible that such a merger might practically write off its cables and start a programme of active wireless development based on low tariffs, and it would be likely to aim at direct services with British Empire countries.

The Americans were not alone in this. By the Statute of Westminster of 1931, which gave legal form to the Balfour Declaration of 1926, the Dominions became autonomous communities within the British Empire equal in status and in charge of their own external affairs, though united by common allegiance to the Crown. The concept of an imperial 'chain' of communications radiating from the 'Mother Country', with a British company operating both ends, no longer appealed, and from now on each Dominion Government sought to establish its own direct communication with other members of what was now called the British Commonwealth of Nations, and with foreign countries, and preferably by the cheaper and easier medium of wireless.

The Indian Government wanted a direct service to China, the Canadian Government to the Far East and the Australian Government to Japan. Australia's distance from Europe made communications of special significance to her. James Scullin, Australia's Labour Prime Minister, rejected John Denison-Pender's plan to form a Public Utility Board. He wanted complete unification – nationalisation in fact – of all internal and external communications, a policy maintained by subsequent administrations under Lyons, Menzies and Curtin, each of whom tried to gain control of the wireless services as a first step. In August 1933 Denison-Pender reminded Scullin of the words in the 1928 Report demanding 'the whole-hearted co-operation on the part of all governments concerned', and urged him to reconsider the Public Utility Board idea in order to fulfil the 1928 ideal of securing a common basis for the telegraphic communications of the Empire. Archdale Parkhill, the Australian Postmaster-General, replied to Denison-Pender on 25 August stating he gathered from the tone of his letter that he regarded the policy of the Australian Government as being out of harmony with the principles laid down at the 1928 conference. On the contrary, he said, their plan of 'terminal unification' would contribute to those objects. Australia's Labour government did not want a semi-state, semi-private enterprise utility company of the kind Electra House were establishing throughout the Dominions to relieve Imperial and International of the financial burden of maintaining the distant cable terminal. They wanted their external communications system to be wholly state-owned, as their internal system was; and they did not want to be diverted from this objective by agreeing to a halfway scheme by which Amalgamated Wireless, with the agreement of Imperial and International, took over the operation of the Australian end of the cable service.

Though the Australian government had a controlling interest in Amalgamated Wireless, the British private enterprise Cables and Wireless had a minority holding, as they had in all other government-sponsored communications companies in the Commonwealth. Cables and Wireless had a large shareholding in Overseas Communications of South Africa Limited, appointed one director and shared in the receipts under a 'wayleave' scheme. They also had 75 per cent of the share capital of Indian Cable and Radio Communications Company Limited formed in 1932 to amalgamate the country's overseas wireless and cable services. They received payment of a wayleave and had a share in the net profits. They appointed four of its twelve directors.

Outside the empire, throughout the 1930s, Imperial and International had to compete with nearly 450 distinct wireless telegraph services operated by foreign undertakings. In many countries particularly Japan, Italy and Germany – the future 'Axis' powers whose military combination hoped to defeat the world – these were government-financed, and took positive action to divert traffic from the British international cable network. Of the 450, more than 150 were new short wave wireless services which had opened since 1928. Governments were more and more refusing to renew cable concessions on expiry. Foreign traffic with the Empire almost always avoided the cable routes. The French passed cablegrams to New Zealand through the Tahiti wireless. Messages from Japan to South Africa were going via Belgian Congo wireless. Germany and France were seeking to set up direct wireless services to Canada; Germany, Japan and USA to Australia; USA, Holland, Germany, Italy and Japan to South Africa; USA, China, French Indo-China to India. Such direct circuits, the Advisory Committee reckoned, would lose Imperial and International £24,000 a year.

A new threat was Airmail, particularly to the company's cheap Letter Telegram service which constituted more than 20 per cent of all its traffic. These telegrams were delivered one or two mornings after handing in. The Imperial Airways flying boat service took only two days to Canada, two and a half days to India, four and a half days to South Africa, and seven days to Australia. The airmail they carried increased from three million letters in 1928 to 17½ million in 1934. Imperial Airways with its twenty-eight flying boats was subsidised by British and imperial governments to the tune of £750,000 a year. Apart from this, for some countries in Europe, airmail was already quicker than letter telegrams, and in other cases the comparatively slight difference in time was offset by the great saving in cost.

How could the company prevent itself being submerged by the Economic Crisis, by foreign and imperial competition, by airmail?

By adding to its operations within the field permitted by its 'charter', by seeking to have that field extended, by cutting expenditure, by closing unprofitable routes, by seeking compensation for non-commercial liabilities, by adjusting rates (up or down), by improving the management structure in the light of experience, by injecting new blood.

The need for a new or better communications system is not diminished by a trade depression, quite the opposite. There was plenty of new business, but as far as Electra House was concerned there were too many after it. However, the past reputations of Marconi and the Eastern Telegraph Group, and the skill of their negotiators, ensured that orders and concessions came their way, crisis or no. They won a twenty-five year concession from Southern Rhodesia, built a wireless station at Singapore to work Hong Kong, Bangkok and Rangoon (they saved £30,000 a year by closing the Bombay-Calcutta-Rangoon cable), built wireless stations in Turkey and Peru, a combined wireless and cable station at Spitalfields, London.

Development of the new medium of 'photograms' was still within their terms of reference, and during 1930 two picture telegraph installations were ordered for the Australian beam circuit, and experimental services were instituted with the other Dominions.

There was considerable potential in building and operating internal telephone systems abroad, and valuable contracts were obtained in Cyprus, Turkey and elsewhere.* Many such telephone

---

*The service of the China and Japan Telephone Company with fifty subscribers was taken over by the Hong Kong Telephone Company in 1925. The police on Hong Kong Island had been able to telephone their colleagues in the New Territory since 1899, but not the public. When wireless came to Hong Kong in 1922 it was not used for international telephone calls until 1927 when the Government Commercial Radio Service established its short wave link to Manila; but this was not for the general public either. When Hong Kong got its first public Central Telephone Exchange at Lane Crawford House in August 1931, this was connected with the Hong Kong to Canton overhead telephone cable completed in August 1931. For commercial users, the Government Commercial Radio Service established communications with other towns in China, and in Indo-China; and by 1931 to Europe
continued overleaf

systems were already operated by the American combines, but Imperial and International saw to it that the process was not allowed to continue unchecked. Apart from the profit, whoever ran the inland national telephones were able to specify the distribution of international telegraph and telephone traffic.

This was the only aspect of telephony open to the company under its charter. The resentment at not being allowed to operate overseas wireless telephone services from Britain, expressed in their first annual report, stimulated a determined campaign to have the policy reversed and the company's field of operation widened. In 1930 the company published a public relations tract entitled *Imperial Wireless Telephony.* The Marconi short wave beam system was admirably adapted for telephony it said. Not only had the Post Office 'totally ignored' the company in equipping its Rugby station for telephony, but had bought American plant manufactured by ITT trading in the UK as Standard Telephones and Cables (STC). 'The Post Office attitude resulted in damage to British prestige, encouragement of American interests in securing influence and business abroad, and tended to discourage British enterprise and research.'

The booklet recorded how after Denison-Pender had written to the Chancellor of the Exchequer a Cabinet Committee was appointed to investigate the position.† It sat in private and the proceedings were confidential. The company were invited to make statements and its witnesses were admitted to a part of two of the sittings. Their request that they might be allowed to examine and report on the evidence tendered was ignored. Their request for a test of their system of telephony with one of the Dominions was refused. The company never met the experts called to the Cabinet Committee and never discovered the questions put to them. All the company knew was that their system and the Post Office's were both 'probably' equally capable of providing the service.

On 26 February 1930 the Postmaster-General issued a statement in the committee's defence. They wanted to concentrate all transmitters and aerials for both telegraphy and telephony at Rugby, whereas the company offered to build and run a Multiplex System for an annual payment of £11,000 plus 20 per cent of the gross

---

via Colombo with an alternative route via Manila and the USA. By 1937 there was a radio-telephone to Hankow, and to Australia, Macao and Sabah which the public could use on the national telephone network, through the services of the Chinese Government Radio Administration.

† It consisted of Lord Thomson, chairman; Pethwick Lawrence, Financial Secretary to the Treasury; G M Gillett, Secretary to the Overseas Trade Department; and H B Lees-Smith, the Postmaster-General.

British receipts using a number of sites round London; or another method for £33,000 plus 10 per cent by which there was one transmitter for telegraphy and one for telephony both using the same aerial. But the Post Office stood its ground and refused to allow the company to participate.

The Press, ever the champion of private enterprise, gave the booklet wide publicity. The *Daily Mail* of 31 January 1930 carried an article by Sir Basil Blackett trenchantly attacking the Post Office.

> Last summer I told Mr Lees-Smith the PMG that he could, if he wished, give full effect to the slogan 'Telephone Imperially' by Christmas, but his department failed to take advantage of the facilities we offered. Indeed they ignored us and preferred to carry on with what I consider is their inferior copy, developed by their own engineers, of our shortwave beam system. For the last four months they have been trying to communicate with Australia by their system and failed.

Lees-Smith made a statement in the House on 26 February confirming that G W Howe, professor of engineering at Glasgow University and Dr F E Smith, Secretary of the Royal Society, had pronounced both systems as capable of providing the service, but that of the Post Office as more elastic. The Post Office should remain free to try *all* systems and choose the one which suited its requirements best. It had strictly observed the six months' delay before placing an order and had in fact tested the Marconi system; but they had decided to get the equipment from STC because it was superior in design and cheaper.

The company realised that further action would be unproductive. They had to accept the Post Office as the overseas radio-telephone authority* and plan their future round that *fact*.

---

*The British Government's wish to handle overseas telephony was shared by most other governments. As seen, P & T Departments discussed international telegraph problems at periodical international telegraph conferences from which, much to the Penders' annoyance, Eastern Telegraph representatives were excluded. In 1885 the duties of the International Bureau of the Telegraph Administrations set up by the International Telegraph Union (founded 1868) were extended to cover telephony, and from that year to 1932 International Telegraph Conferences dealt with telephony too. However, in 1923 the European Administrations established a Telephone Consultative Committee (CCIF) independent of the Telegraph Union. In 1932 the convention which was adopted at the Madrid Conference provided for separate International Telephone Conferences. At Madrid in 1932 the international organisation of telegraphs, telephones and radio was unified by the creation of the International Telecommunication Union.

9

Depression
1931–1938

Having failed to extend their permissible field of operations, Electra House now sought to increase their profit by cutting their £5,300,000 annual expenditure.

The first savings were in wages and salaries. The new combined company found itself with 8,429 operators. By the end of 1929 these had been reduced to 7,784. The loss of 645 and the saving of £170,000 a year was effected mainly by voluntary retirement of cable staff. The services of forty marine officers were terminated on the sale of six of the fleet of eleven cable ships – and annual saving of £190,000. Within two years and nine months the established staff of 9,494 (all jobs, including operating) was reduced by 1,085 to 8,409, and the unestablished staff was cut by 503. This amounted to a saving of £370,000 a year. The number of directors was reduced from twenty-two to fourteen. Lord Clarendon, the second Government-approved director, resigned on being appointed Governor-General of South Africa and was replaced by Hon George Peel. The seven others who resigned received £10,500 in compensation. Those who remained reduced their fees voluntarily by 20 per cent. The changes, taking effect from 1 January 1931, brought about an annual saving of £35,000.

Admiral Grant, the long serving director and one-time managing director of Eastern Telegraph, stayed on the Board to devote his whole time to problems, human and organisational, arising from having to dismiss so many at a time of unprecedented unemployment resulting from the Economic Crisis. J C Denison-Pender, as the current Father of the company, was personally concerned in seeing that terms of leaving were favourable. The blow of being made redundant by the mechanised 'Regenerator' relays had been softened by the offer of an Extended Furlough Scheme, supplemented by the granting of gratuities to less senior men; and this was the basis of the plan for the new wave of dismissals – known, because of the date/time code at the head of the message announcing it, as the D/N (Dee Bar En) Scheme (4th day, so 'D'; and 1300 hrs, so 'N' – J was omitted).

Admiral Grant's disagreeable task was helped by the existence of two staff associations, who faced the situation stoically and let it be known that they considered the terms offered very generous. The Eastern Cable Staffs' Association was replaced by the Imperial and International Communications Staffs' Association open to all employees (except messengers), including many grades who had not hitherto enjoyed organised representation. In March 1931 it launched an official quarterly organ *The Gazette.* It had 140 in the Women's Section, 534 in the Head Office Section, 540 in the Clerical Assistants Section, and 2,088 in the Stations Section. Alongside it was the Association of Cable and Wireless Telegraphists with almost identical objects, originally formed to protect sea-going operators.

The first 600 to go, as seen, were volunteers who, for various reasons, were content enough to leave the service at that moment in their career; but by 1931 operators did so because of their fear of what the future might bring, and fear that the offer might be withdrawn before they applied. A man of 40 in the fifth grade was offered a gratuity of about £788 plus refund of his own pension contributions. A man of 45 (the eligible age) was offered Extended Furlough payment of £164 a year till September 1934, the end of the five years from 1929 to which staff taken over were guaranteed employment under the Merger Plan. Most of those who applied would have been over 30 and therefore married with families. Trained in the blind-alley occupation of telegraphist the prospect of finding other employment was extremely bleak. Service in the tropics made them unfitted for training in another craft. Few had managed to accumulate any savings for the rainy day with which they were now confronted.

Admiral Grant's announcement that the Extended Furlough and Gratuity offers were being withdrawn on 31 August 1931 brought a rush of applications, and by the end of the year some 2,500 had left and another 700 went in the first six months of 1932. But the economies were still not enough. At the beginning of 1932 J C Denison-Pender made a personal appeal to the staff 'to make a contribution for the remainder of the year on a defined scale towards the revenue of the company' and to consent to an increase in their hours of duty by one hour a day, and a reduction in payment for overtime and Sunday duty. He told shareholders at the third annual meeting of Cables and Wireless on 21 July 1932 that the response to this appeal had been 'instantaneous and overwhelming'. Over nine months the saving had been at the rate of £200,000 a year. It involved a scaled contribution of 15 per cent downwards, and came from the highest officials to the messenger boys whose voluntary contribution of tenpence a week was not the least valued part of 'this gift'. The directors were proud of the evidence of loyalty and appreciation of the difficult times both by the old staff and the new entry in making sacrifices and contributing to the common cause. The response, he said, which was voluntary, had not been 100 per cent, but the majority had showed their concern for the welfare of the company. Working expenses at the stations had been reduced by £171,400 in 1932, by £450,000 in 1933. The appeal received the full support of the Imperial and International Communications Staffs' Association who balloted members many of whom, however, wrote to express dissatisfaction. Shortly afterwards the Association was dissolved.

The situation was eased by coinciding with a transition from the multiplex system to the teleprinter system, which the Post Office, following a commission to the USA in 1929, adopted in 1932. This was the beginning of 'Telex'. Hitherto 75 per cent of the cost of the telegram service had been in operating, but the teleprinter doubled output as compared with manual morse and halved the line costs. The invention of F G Creed, the one-time telegraphist with the Central and South America Cable Company in Chile, had arrived.

Electra House lost no time in trying to shed what they considered 'non-commercial liabilities', and asked the Advisory Committee to allow them to abandon the cable systems between the Mother Country, India, South Africa and Australia, or have the Government contribute towards keeping such unremunerative cables for strategic purposes. They sought permission too to discard potentially commercial routes which were not proving profitable. But the key to profitability, as ever, lay in a proper adjustment of the rates. Any changes had to be referred to the Advisory Committee, and in March 1931 they applied for an increase designed to raise revenue by £75,000 a year. Fearing that the main incidence of this would fall within the British Empire, the Advisory Committee thought it right, before agreeing, to set up an Enquiry into Imperial and International's position generally, even though only two years had passed since the sessions of the imperial conference which had brought it into being. The Advisory Committee informed the Partner Governments of the company's application to increase rates, and of its own decision to hold an enquiry, on 24 March. It then wrote to tell

Imperial and International – on 7 April. This unexpected turn of events was immediately discussed at a meeting of the Management Committee at Electra House at which they agreed to accept the holding of an enquiry in principle. On 15 April, Sir Basil Blackett told the Advisory Committee that the matter would come before the Court of Directors on 21 April. At that meeting, he warned, they might not consider that the company's proposal for an immediate increase in rates was the best method of dealing with the company's difficulties, 'and that a decision might be taken to approach the Governments with a view to the Communications Company being taken over by the Governments'*.

Sir Campbell Stuart pointed out to Sir Basil that an immediate increase in rates had been the company's idea not the Committee's, and that he had already told the Partner Governments. 'Having regard to these facts,' said Sir Campbell, 'should the proposal now be withdrawn, the manner in which the serious financial situation facing the company is being handled by those responsible for its direction and conduct would not create a very favourable impression.' It seemed to Sir Campbell, according to his note of his meeting with Sir Basil, that an increase in rates had been suggested in the interests of the shareholders. It would appear too that when Blackett referred to 'the company's' proposal, he meant 'the Management Committee's'.

Denison-Pender, Blackett, Kellaway, Peel and Wilshaw met Campbell Stuart, with Sir Norman Leslie, the secretary of the Committee in attendance, on 28 April, in order to discuss the terms of reference and constitution of the proposed Committee of Enquiry.

Denison-Pender said he was glad the Advisory Committee was sympathetic to the Court of Directors in their difficulties, for some of his colleagues had felt rather hurt at the tone of the Committee's last letter. The main subject of enquiry for the Committee, suggested Blackett, would be the discrepancy between the Standard Revenue and the present position. An enquiry into the direction and conduct of the company's affairs might be read to mean an enquiry into the individual competence of those engaged in its management. That was a matter for the shareholders, and they should be represented on the Committee. Kellaway agreed. The proposed terms of reference indicated a lack of appreciation of the real gravity of the situation and possible a reflection on the Court of Directors. The company must agree the names of those composing the Committee.

---

*Sir Campbell Stuart's version of what Sir Basil Blackett said – Campbell Stuart Papers, Folder 9, Royal Commonwealth Society Library.

On 29 July Denison-Pender told shareholders that largely thanks
to Sir Montagu Norman, Governor of the Bank of England, they had
reached an agreement that an Investigating Committee should be
appointed consisting of three distinguished gentlemen, 'whose
names will inspire confidence throughout the Empire,' to enquire
into the company's present position. On hearing this, a shareholder
said the company had been formed in opposition to the present
members of the Government,* and they had obstructed it at every
turn. 'In fact, they said that not only did they not like the company,
but they hated it (laughter), and would do all they possibly could to
make life as difficult as they could for it.'

The chairman of this 'Imperial Communications Inquiry Committee' was Wilfrid Greene KC, the future Master of the Rolls, and
the other two members were Lord Ashfield (Albert Stanley),
chairman of the London Underground group of companies which in
two years' time were to be part of the merger which formed the
London Passenger Transport Board, and Lawrence Holt, chairman
of the shipping line Lamport and Holt. Their terms of reference were

> To inquire into the position of Imperial and International Com-
> munications Limited and all matters related thereto and bearing
> thereon, including the causes and consequences of the great discre-
> pancy between the Standard Revenue and the present earnings, and
> to suggest measures which will remedy the present situation.

The Greene Committee issued its report on 21 November 1931,
and the Advisory Committee sent copies of it to the two companies
on 7 December with the request that it should be read only by the
directors and not senior officials (to which the company objected).
In setting out their reaction to it in a letter of 17 December, the
Court of Directors stated they regarded the findings in the main as
'helpful and far-seeing'. When the principal recommendations were
carried out, the Court were satisfied that the Empire's communications would be placed on a stronger foundation.

The Greene Committee thought the requirement that the chairman and one of the directors should be approved by the Government
subjected the two to a dual responsibility that was undesirable. In
law a director could only be responsible to his company and its

---

*Ramsay MacDonald, the Labour Prime Minister, resigned on 24 August 1931 and
formed a National Government the next day, which was reconstituted in November
with J H Thomas as Dominions Secretary, Sir Philip Cunliffe-Lister, Colonial
Secretary, and Sir Kingsley Wood, Postmaster-General. MacDonald remained
Prime Minister.

proprietors.* They wanted to see an improvement in the company's publicity, and the substitution of wireless for cables wherever a clear economic advantage could be shown. They thought the system of administration at head office threw too great a concentration of business on the General Manager/Secretary. With all this the Court of Directors agreed. But they still saw no reason why directors should not hold other paid appointments, and objected to the proposal that the company should change its name, which would be injurious and costly. They were in no way satisfied that it was in the company's interest to reduce its capital.

The Advisory Committee referred economic matters of this sort to accountants Sir William McLintock, Alan Rae Smith of Deloittes, and Sir Gilbert Garnsey of Price Waterhouse. This 'Committee of Three' opposed a reduction in capital, but on the grounds that it was impossible to estimate 'normal revenue'. They also agreed with the company that the only realistic way of calculating a subsidy for maintaining non-commercial strategic cables was to wait until each section of cable was proposed for supersession and permission to do so was withheld. In the event the Advisory Committee sought authority to abandon 17,532 miles of cable and the Government sanctioned 13,570 miles. Of the 36,492 miles for which sanction might be sought in the future 19,010 were scheduled as required for strategic purposes.

On the crucial matter of rates which had provoked the investigation, the Committee of Three agreed with the company's proposal to discontinue all reduced-rate traffic (particularly the Letter-Telegram at one-fourth rate) and substitute a new category at one-third 'ordinary' rate and to increase the minimum message from twenty to twenty-five words (and thus the minimum charge). The new tariff was introduced in 1932. If it had been in effect in 1931, said the Committee of Three, the company's revenue would have been £178,500 the greater.

But the company had also proposed a wholesale increase in the full rate – tantamount to raising the charge of 1s 4d a word to 2s 4d on all traffic outward of UK. This was condemned out of hand in the Greene Report and never got as far as the Committee of Three.

But the Greene Committee's main task was to decide whether the Communications Company as constituted by the imperial conference was in fact suitable for dealing with the new situation which had developed since 1928. Evidence was given to them for the transfer of the undertaking to some form of state ownership,

---

*Though 'approved by' is hardly the same as 'responsible to'.

preferably a Public Utility Board. But for their part, stated the Greene Committee, they did not believe that the dual obligations laid upon the company to earn a profit for its shareholders, and to conduct communications under a measure of governmental control, were irreconcilable. If so important a public service as the conduct of imperial communications was entrusted to a public company some measure of governmental control was inevitable. Moreover there were many public companies in being at the present time in which the interests of the user and of the shareholder were fairly and effectively adjusted. The main causes of the decline in the company's revenue had been the commercial ones of trade depression and competition. In their opinion a public company was in the present conditions far better suited to contend with such conditions than a Public Board. In contending with the difficult conditions, the common interest and close relationship which the company shared with all the imperial governments should be a help rather than a hindrance. The difficulties in meeting the obstacles had their origin in the company's system of direction and management rather than in its nature, or the control to which it was subject.

> The system whereby management is vested in a committee of four equals is open to grave objections, and is probably largely responsible for the extent of the difficulties in which the company finds itself. We recommend that the 'Management Committee' be abolished forthwith.

In their comments of 17 December the Court of Directors insisted that the Management Committee had been the only solution, as management by the chairman would give it to a man with dual Government/Company responsibility. But in a dissenting note Sir Basil Blackett said he agreed with the requirement that the chairman should be approved by the governments. Moreover, in his view, unity of control could only be secured by making him managing director as well, and also Governor of Cables and Wireless and managing director of all other companies in the merger including the Marconi company. 'What has been lacking, and still is, and will be lacking, is a single chairman and managing director of the Communications Company with adequate powers, devoting himself entirely to the business of communications.' He agreed the Management Committee system had been largely responsible for the difficulties.

When asked by the Advisory Committee to expand on his dissenting note, Blackett said (30 December 1931) the question at issue was not the form of organisation but the personnel. In his view, the chief executive should not hold other appointments. While a Government-approved chief was the ideal arrangement, its success in meeting the needs of the case depended entirely on the choice of the right individual. Given adequate powers his position would be strengthened not weakened by the fact that his appointment had received the approval of the Partner Governments. But in view of the history of events since the company was formed, the Governments should not now insist on that condition. 'While not indefensible, the requirement of Governmental approval for the appointment of the chairman of the Communications Company has from the outset been used by my colleagues as an argument for not entrusting him with adequate powers.' He thought it preferable in the circumstances to leave the choice of chairman to the unfettered judgement of the shareholders.

The Greene Committee also pronounced on another matter vital to Imperial and International: overseas wireless telephony. The partial rationalisation of communications services in the UK brought about by the 1928 conference could with advantage, it said, be made complete by the transfer to Imperial and International of all external communications (telegraph and telephone) now operated by the General Post Office. Any hope that such a recommendation might be acted on in the immediate future was dimmed by its relegation to the Committee of Enquiry under Lord Bridgeman which had just been appointed to investigate the running of the Post Office, to which the company would have an opportunity of putting their case (which they did, but to no avail).

J C Denison-Pender informed the Advisory Committee that the Board agreed the usefulness of the Management Committee was outlived; they had abolished it. On 19 April 1932 Sir Basil Blackett resigned as chairman of Imperial and International but remained an ordinary director of that company and of Cables and Wireless. The Court of Directors then nominated J C Denison-Pender who was already Governor of Cables and Wireless for approval by HMG as chairman in Blackett's place. In informing the Advisory Committee of these moves in a letter of 3 May 1932, Denison-Pender said the company had already acted on many of its recommendations. Some 2,830 men had been 'retrenched' since the inception of the merger, and 9,965 staff out of 11,411 – some 87 per cent – had accepted his scheme for voluntarily reducing their salaries by £203,000 a year. All were safeguarded by the five year agreement which expired in September 1934. They had set their house in order, he said, and they could not but feel that their actions merited recognition. The Advisory Committee seemed satisfied on the whole, and on its advice the Government approved the appointment of Denison-Pender as chairman of Imperial and International the week before the annual general meeting in July. But, with the Greene Committee, it regretted that agreements for joint working of cables and wireless had still not yet been reached in many parts of the empire. But the leopard could not change its spots. Electra House would never be other than cable-orientated so long as its management was Eastern Telegraph dominated. The Advisory Committee considered it was a very broad assumption for the company to make that in the majority of Empire points communications were adequate, and that no improvement or development of facilities were needed. 'Further, on a long view, it must be assumed that international communications traffic will increase to many times its present volume. It seems improbable that new long distance cables will be laid in future for telegraphy alone, on grounds of capital cost, so that the tendency must be, in the end, for wireless to supersede the cable system. Can we contemplate leaving foreign countries such as the USA to the mastery of the radio art, while the British group remains tied to what may become an outworn cable system?' (1933 Annual Report to Partner Governments).

The Advisory Committee considered there had been no significant development of the company's policy on the lines contemplated by the Greene Report. 'It seems to us to be doubtful whether the need for an active and aggressive policy is fully recognised and whether the company is at present disposed to show that proper balance between the cable and wireless systems which is necessary if an American combine is to be squarely faced.'

To the public in Britain at any rate, wireless and cable services seemed to be united. The different routes 'Via Marconi', 'Via Eastern', 'Via Empiradio', 'Via Pacific', 'Via Western' and the rest, were done away with, and every telegram out of Britain went 'Via Imperial'. Though internal delivery of the company's telegrams by the Post Office had been discussed by the Greene Committee, there was still a large and smartly uniformed corps of messenger boys at Electra House and branch offices in London and in the provinces in 1932 who took them by hand, using public transport, direct to the door of the addressee.

One of those who became a 'Via Imperial' messenger boy in London in that year, as a lad of 14 just left school, was Frank Maynard, who retired in 1978. Normally, telegrams for Buckingham Palace were delivered from the company's branch in Parliament

Street, but on the day of King George V and Queen Mary's Jubilee
they were so busy they asked Head Office to help them out. A bag of
congratulatory telegrams for their Majesties under his arm, young
Frank, smartly uniformed, was soon in Birdcage Walk but took time
to penetrate the crowd before reaching the palace gate through
which he had to pass in order to make his royal delivery. The
policeman held back the throng, pushed the gate open and let him
through. He marched proudly across the courtyard and half-way he
heard cheers from the crowd behind. They were not for him,
however, but for the addressees who had appeared on the balcony at
that moment. He was to remember the occasion when, some forty
years later, he made the same walk to receive his MBE from Queen
Elizabeth II in the New Year Honours of 1978.

The date of Frank Maynard's entry into Imperial and International employ is significant, because it was in 1932 that recruitment
of staff began again after an almost total stoppage for ten years. The
D/N Scheme was considered closed by March 1933 when more than
4,000, a third of the combined workforce of 1929, had been retired.
In addition, a large number of Eastern Telegraph-trained men had
been transferred to the newly constituted overseas companies with
whom they would stay for the rest of their careers instead of being
available for company managerial posts anywhere in the world. So
the pool of 'management material' was running low.

London Training School at Hampstead was re-opened in 1932
with F Freathy as Manager (appointed June 1931). A member of the
first course of some ten 16-year-olds that autumn of 1932 was,
appropriately enough, the grandson of William Ash, the long-
serving principal of the Porthcurno Training School now closed,
John Ash. His father, Percy Ash, had joined the Foreign Service of
Eastern Telegraph in 1903, served in Suez, Rodriguez and
elsewhere, and from 1914 been Engineer, London Station, Moorgate. After eighteen months, training at Hampstead, John Ash was
also posted to London Station with its three floors – Cable Circuits,
the Beams, and the HF Galleries – working a forty-two hour week,
six days a week with 'Early Off' on Saturdays if traffic was slack. He
was then back at Hampstead for another two years on the
Engineering Course from which he graduated as an Assistant Cable
Engineer in 1936.

Training the telecommunications managers of the future took time
and money.* The Depression had by no means lifted and traffic

* The training of Eastern Telegraph probationers was paid for by their parents, but
after the 1929 merger they paid £1 a week which was refunded to them at the end of
training so they had some £80 or so to take with them on their first posting.

remained low. Gross revenue of Imperial and International fell from
£4,946,800 in 1932 to £4,576,500 in 1933. But management were
determined not to be caught unawares when world trade revived. It
had not helped that the company had not attained as full a revision
of its 'charter' as it would have liked. For instance there had been no
re-assessment, in te light of Depression conditions, of the £250,000
beam rental which had been based on the actual and potential
earning power when the service had been in the hands of the Post
Office. However, they had no need to wait on the Government or the
Advisory Committee to carry out their plan, conceived in 1929, to
give the two new companies, created in that year, an administrative
headquarters in a building of their own separate from Electra House.
In 1929 they purchased a site on Victoria Embankment for that very
purpose. They demolished the 1879 London School Board's building
which occupied it – it had become the office of the London County
Council's Tramway Department (about to be merged into the new
London Transport) – but kept Arundel House next door at the
corner of Arundel Street. They had Sir Herbert Baker design them a
new building to their own specification and it took four years to
complete. Then, for some reason, the Court of Directors decided it
was unsuitable. They tried to sell it, but could find no one willing to
take it off their hands. So they had to move in after all.

They missed the opportunity – presumably intentionally – of
demonstrating Imperial and International's independence of the
past by giving the building a name dissociated with any of the
company's component parts, and unbelievably called it 'Electra
House', retaining the name also for the old Eastern Telegraph
'London Station' at Moorgate. So there were two buildings in
London with the same name – Electra House, Victoria Embankment, and Electra House, Moorgate.

Mrs J C Denison-Pender opened Electra House, Victoria Embankment, with much ceremony on 11 May 1933 following a
luncheon in the hall of the Society of Accountants and Auditors next
door (once the house of Lord Astor who died in 1919). At the time of
the merger the staffs of the various companies had occupied seven
different buildings, so centralising them in this fine new building
(which still stands in 1979) was a great improvement.

In the entrance hall there was a list of all the companies in the
group. The London Platino-Brazilian Cable Company was liquidated in 1933 but that was the year Direct Spanish was taken over, as
seen, and a new company formed called The Eastern Telegraph
Company (France) Ltd. When Eastern Telegraph's French concession expired in 1931 the French Government refused to recognise

Imperial and International Communications so the concession had to be renewed under the old title.

The move to Victoria Embankment was accompanied by important administrative changes. Edward Wilshaw, who had now been with the company forty years, relinquished the dual position of General Manager and Secretary, was elected a director and given the single appointment of 'Chief General Manager'. S G (Sidney) Farmer, who had seen thirty-two years' service, took over as Secretary. Fred Kellaway died in 1933 and his place in the Court of Directors was taken by H A White who had been associated with the Marconi company for fourteen years. All of these refused any increase in their emoluments. The Greene Committee hoped to see a better relationship between the Advisory Committee and the company, and two of its members also joined the Court of Directors at this time at the request of the British Government. One of these was Sir Courtauld Thomson, chairman of the Advisory Committee. Commenting on these appointment in *Opportunity Knocks Once,* Sir Campbell Stuart observed

> I saw the wisdom of this decision, but was somewhat surprised to be told by Sir Warren Fisher [of the Treasury] that the company desired the services of Sir Norman Leslie, our secretary, as their managing director. Sir Norman Leslie was an ideal secretary, but hitherto had not engaged in business.

Sir Courtauld Thomson and Sir Norman Leslie, who became joint managing director with Denison-Pender, took the place of Charles Hambro and Admiral of the Fleet Lord Wester Wemyss, both of whom had died. 'It is very satisfactory that after three years working with the advisory committee,' Denison-Pender told the annual meeting on 25 May 1933 'our relations are so cordial that they accepted our invitation to join our board'.

John Cuthbert's 26-year-old eldest son, John Jocelyn Denison-Pender, had entered Eastern Telegraph just before the merger. In 1930 he had become Assistant Secretary of Imperial and International, and in 1933 he was appointed Wilshaw's Deputy Chief General Manager.

The following year there was a further re-shuffle with the appointment of Edward Wilshaw as Joint Managing Director of the communications company which, however, no longer bore the name Imperial and International.

The Greene Report was never published, so no one knew of the Court of Directors' opposition to the Investigating Committee's proposal that the name of the company should be altered. But in

1934 the company, reacting to another of the report's recommendations, appointed their first Publicity Manager, H L Morrow, and it
may have been he who persuaded them to change their minds. Be
that as it may, J C Denison-Pender had no qualms about telling
shareholders on 24 May 1934 – the meeting was on Empire Day
every year – that in their publicity Imperial and International were
stressing the telegraphic nature of the company's services by cable
and wireless. 'The reason is,' he went on, 'that we have for some time
realised that in this respect the lengthy title of the Communications
Company was not helpful.' The name was originally chosen, he said,
to express the various communications undertakings of which it was
composed, and indicating the imperial and international nature of
the activities of the combined service. However, though expressing
an admirable sentiment, it did not constitute an effective name for
their business. Shareholders were asked at an Extraordinary Meeting that followed the general meeting to change the name of Cables
and Wireless Limited to 'Cable and Wireless (Holding) Limited'
and Imperial and International Communications Company Limited
to 'Cable and Wireless Limited'.

One of the first announcements made by the company under its
new name was the opening in October 1934 of a facsimile picture
transmission service between London and Melbourne.

By 1935 the total volume of Britain's annual export trade had shrunk
by £800 million. As world commerce slowly revived so did use of the
world's telecommunications of which Cable and Wireless had so
large but diminishing a part. The company made a net profit
(investments and receipts) of £209,000 in 1933, £625,000 in 1934 and
topped the million in 1935 – after only £75,000 in 1931. Up to 1934
the dividend on the preference shares was kept to $2\frac{3}{4}$ per cent, but in
1934 it was $4\frac{1}{8}$ per cent and in 1935, $5\frac{1}{2}$ per cent. However, Britain's
semi-state, semi-private communications company was not alone in
seeking to benefit from the revival of traffic. In May 1934,
J C Denison-Pender was deploring the decline of what he called
international telegraphy, the result of each country wanting its own
overseas telegraph system working direct to its own more important
trading areas and so cutting into the economic structure of
international telegraphy. A concession which in John Pender's day
would have been taken for granted was now highly prized – as that
secured from Saudi Arabia for both wireless and cable in 1935.

The move towards telegraphic nationalism was a manifestation of
a more deep-rooted urge for self-determination which threatened to

pitch a world recovering from economic turmoil into another kind of crisis. In March 1935, Adolf Hitler's National Socialist government in Germany repudiated the disarmament clauses of the Versailles Treaty; in October, Benito Mussolini's Fascist government in Italy invaded Abyssinia. In telecommunications circles there were more and more references to the importance not only of imperial unity but *strength.* The submarine cable telegraph which yesterday, it seemed, was on the point of being phased out, regained importance along with the now far-from-academic word 'strategic'.

In such circumstances it was natural that the focus should be more on Cable and Wireless as an instrument of government than an earner of dividends, that the Imperial Communications Advisory Committee should become more master than watchdog – a situation calculated to prove irksome to the long-serving, commercial-minded, cable-oriented management at Electra House, now divided into two cliques, the Denison-Pender/Courtauld Thomson/Norman Leslie camp and the Lord Inverforth/Edward Wilshaw camp. Each had their own solution to the company's current problems, and to forestall a clash with the Advisory Committee, the former group came up with a plan to dispense with it altogether. On 21 February 1935, Sir Norman Leslie told a meeting composed of himself, Denison-Pender and Colonel Sir Donald Banks that the management had under consideration a complete re-organisation of the company's structure so as to include all the new Dominions telecommunications companies into one great imperial company in London. The various empire governments would hold shares in it proportionate to their interests, and be directly represented on the Board by their High Commissioners. The Advisory Committee would be abolished. Colonel Banks got the impression that Denison-Pender was not as enthusiastic about this plan as Leslie, and the colonel himself told the meeting he did not think it was practical politics; if the Advisory Committee were to go, liaison with the Dominions would disappear. Nothing further was heard of this plan.

Denison-Pender, as Government-approved chairman, had no hesitation in making a direct approach to Britain's Dominions Secretary, J H Thomas, to whom in the spring of 1935 he poured out the complete catalogue of items on which he sought help from the Government – beam rental reduction, co-operation with the Post Office over telephony, a subsidy for strategic cables. When he saw Campbell Stuart on 7 March he said Thomas had supported the company's position and had arranged for him to have an interview with Neville Chamberlain, the Chancellor of the Exchequer, at which Thomas would be present. Denison-Pender also told Stuart that he felt some of his staff deserved honours in the coming Honours List.

In Sir Basil Blackett's view, as seen, what was needed was not so much a new company structure as new faces. Sir Basil was killed in a motor accident in Germany in August but, in any event, as the victim of the clash of personalities which left him an ordinary director, he would not have been in a position to countenance a palace revolution. Others were – and did.

On 30 October 1935 there took place at Campbell Stuart's London house a meeting attended by himself, Sir William McLintock, the accountant, E St J Bamford a Treasury official, Sir Alexander Roger and Viscount Hinchingbrooke. The plan discussed at this meeting centred round Sir Alexander Roger whose £12 million group of telephone companies was headed by British Insulated Cables and included the International Automatic Telephone Company and the Telephone and General Trust. A leading figure in the group was the American A F Adams, leader of the independent telephone companies in the USA in opposition to ATT.

A Note of what occurred at this meeting was written by Bamford and is among Campbell Stuart's papers in the Royal Commonwealth Society.

> Sir Campbell Stuart said that the meeting had been arranged to discuss, informally and without any commitments, the proposition of the inter-marriage between Sir Alexander Roger's Telephone group and the Cable and Wireless group on the basis of an exchange of Directors, which would leave Sir Alexander Roger as Chairman and Managing Director of the Cable and Wireless operating company, with a working Board of not more than twelve, of which one-half would be nominees of Sir Alexander Roger . . .
>
> Sir William McLintock said that they were all agreed that the present direction and management of C & W Ltd was hopelessly ineffective. The right way was to get rid of the whole gang and even if they had to be compensated on a five years' basis it would be a profitable investment to the Company since none of them need be replaced. At the same time he could not see any reason for Sir Alexander Roger to take on his own Boards any of the present directors who would obviously be useless to him. In his view, the best way of handling the business would be a root and branch reorganisation by which the shares of C & W Ltd, the operating company, could be distributed to the present shareholders of C & W (H) Ltd, giving them at the same time a share in the unwanted cash reserves in the group.
>
> Sir Campbell Stuart pointed out that wholesale compensation to the existing Board on these lines might cost anything up to £250,000,

and that he understood Sir Alexander Roger would not find it difficult in dealing with the rest of the Board if the king-pins, Denison-Pender and Iverforth, were knocked out.

Nothing is known of any subsequent meetings or what led the conspirators to drop the plan. But no 'coup d'état' on the lines envisaged on 30 October ever took place, and presumably no one ever came to know that such radical steps had been contemplated outside the group and Sir Warren Fisher of the Treasury. Denison-Pender certainly never knew of them. He thought he could circumvent the 'interfering' Advisory Committee, whom he was currently boycotting, by renewing the request, first made in July 1935, that the Government should call an Imperial Wireless and Cable Conference with the specific purpose of reviewing the recommendations of 1928. On 31 January 1936, Major Tryon, the PMG, told him he was not prepared to do this, though both of them would have been aware that an opportunity would present itself when the Premiers met at the coming coronation of Edward VIII who had succeeded to the throne on the death of his father George V on 20 January.

But official patience with Denison-Pender was running out, and a few days before Easter, Sir Warren Fisher asked him to call at the Treasury. He told him 'quite definitely' that the Government had decided not to renew Government approval of himself as chairman, or of George Peel as a director, when their term expired; that whatever the strict wording of the documents might be* the Government intended to approve another chairman selected by itself from outside the company, and two other directors would be added to the Board, one approved by the Government. Warren Fisher's Note of this meeting continued: 'Reference was made to the question of Mr Denison-Pender being recommended for a baronetcy which it was emphasised had nothing to do with the present discussions "since the Government decision with regard to the changes in the Board was 'immovable'"'.

Fisher told Denison-Pender that the British Government had decided to maintain the present machinery of the Advisory Committee and was not prepared to discuss the matter with the imperial governments; that the governments had decided not to give the company any easement in the matter of beam rental, strategic cables etc., but if the company wished these matters to be ventilated

---

*The 1928 imperial conference had recommended that a candidate for chairman of the Communications Company should first be nominated by that company and then approved by the Government.

it could raise them through the proper channel, which was the Advisory Committee with which it must re-establish contact.

Denison-Pender reported this to the Court of Directors who, he told Fisher on 21 April, were unwilling to take any new members. Moreover they thought that for him to be retired would be contrary to his personal prestige and be construed as interference by the Government. At a meeting with Campbell Stuart the same afternoon Denison-Pender asked for the Advisory Committee's help in getting imperial communications put on the agenda of the coming Imperial Conference. He hoped too that the Committee would ask the Dominions to stop setting up any more direct wireless services.

He was particularly worried by Australia's intention to establish a direct wireless circuit to Japan; and Stuart said he would discuss the matter with Sir Clive Baillieu, the Australian representative on the Advisory Committee. Baillieu put the Australian Government's case in a letter to Denison-Pender which the recipient considered 'insulting'. He told Stuart on 23 April it ought to be withdrawn. Baillieu had put a pistol to the company's head, and the Board would not stand for it. The Australian PMG had told the Eastern Telegraph manager in Australia that they had no intention of starting a competitive wireless service. When Stuart suggested Denison-Pender should put the company's case in writing to the Advisory Committee, the latter said that was not possible. The Advisory Committee spelt out their view of the general situation in a letter of 29 April which made the Court of Directors 'very annoyed'. They were particularly upset by the absence of any offer to reduce the beam rental. They thought the time had come 'to force publication of the Greene Report'. To compensate for the contemplated competitive wireless services, they wanted a Government subsidy.

The following evening Lord Inverforth telephoned Campbell Stuart. He said he was 'very troubled about recent developments' and made an appointment to see Stuart the next day. From that meeting Stuart gathered that Inverforth thought Warren Fisher of the Treasury was hostile to him, but that maybe he (Stuart) could be captured for the 'I & W Camp' (Inverforth and Wilshaw). Inverforth felt John Cuthbert and his son John Jocelyn should go, along with Norman Leslie and Courtauld Thomson. He and Sir Charles Barrie were the only businessmen in the show. He could run the business profitably if A H Ginman was brought back from Canada; he would work with him, together with Wilshaw and White. 'CS tried to find out from Inverforth what DP had told the C & W Board. Inverforth said he understood DP was to have an honour, perhaps a peerage, and is not to be approved as chairman.

... Inverforth said DP would give up both jobs [C & W and C & W Holding] for a peerage, and that Leslie would not mind being Chairman under DP and had promised to protect young DP. Inverforth said young DP was not good enough for General Manager under a weak Chairman.'

Stuart told Inverforth such matters had nothing to do with the Advisory Committee, and that he was sure that the question of an honour did not come into the picture. Denison-Pender, said Inverforth, would attack the Government in his chairman's speech at the coming annual meeting unless the Advisory Committee or Warren Fisher could assure him the beam rental would be reduced. But if he won on that point, he would 'go out gracefully'. Inverforth was worried by the fact that Denison-Pender did not tell the same story all the time, and met directors individually and not collectively.

When Warren Fisher saw Denison-Pender at the Treasury on 8 May he made it quite plain that the Treasury Agreement was not going to be torn up by the Cable and Wireless Board. The Government wanted new blood on the directorate. When Denison-Pender asked him whether their differences could go to arbitration, Warren Fisher told him bluntly, 'The Treasury does not arbitrate'. The matter could not be put to the Government again; the decision was immovable.

Campbell Stuart saw Inverforth, Wilshaw and White on 9 May from whom he gathered that Courtauld Thomson and Norman Leslie had been foisted on the company by Warren Fisher. Stuart's brief handwritten notes on this meeting ran

> C & W had no commercial policy
> Wilshaw = chair?

The Treasury now had a quiverful of bargaining points and when Warren Fisher met Denison-Pender on 15 May he offered to modify details in return for the chairman's co-operation in 'facilitating' the essence of the Government's programme of reform. So long as it was not Sir Norman Leslie, who must vacate his managing directorship at once, he would consider approving a new chairman *selected by the company* and not the Government, as he had said in March. Imperial communications would be on the Imperial Conference agenda so long as the company agreed to an investigation by the Advisory Committee, whose report would be the subject of discussion at the conference. Pending this investigation and completion of the report, the Committee would recommend that imperial governments held their hands on direct wireless services and rate reductions. If this was not agreeable to the company, communications would not be raised at the conference at all. If he accepted the Government's terms, Denison-Pender could tell his shareholders on 22 May – Empire Day was on a Sunday – that he had asked for the company's 'case' to be put to an imperial conference and HMG had graciously agreed that this should be done.

The company then formally presented the Advisory Committee with Six Points on which it requested a ruling by the imperial conference:

(i)   no new direct wireless services to be allowed between Empire and foreign points save with compensation by governments for any losses;

(ii)   no rate reductions save with compensation;

(iii)   compensation for upkeep of strategic cables;

(iv)   abolition of £250,000 a year beam rental to Post Office;

(v)   abolition of £77,545 a year Pacific Cable debt;

(vi)   operation of overseas telephone services conducted by Post Office should be in same hands as the overseas telegrams.

Cable and Wireless had no alternative but to accept the Government's formula, but of course HMG was no longer the Mother Country which dictated policy to her imperial underlings. No decision in Whitehall could make Canada or Australia act against their democratic will. On the matter of independent direct wireless services, the Canadian Government said it could not reverse its decision to license a trans-Pacific service from Vancouver. The Australian Government 'preferred not to give a formal standstill undertaking', but agreed not to re-open the matter except after consultation with the Advisory Committee. All other Dominions governments gave specific undertakings of the kind HMG requested, but the loop-hole left by the two most nationalist of the Dominions was never closed. And in its annual report of 19 May 1936, the Advisory Committee felt obliged to state: 'The Governments of the Empire cannot easily forgo for an extended period their rights to license new wireless services'.

John Cuthbert Denison-Pender duly vacated the chairmanship of the Communications Company on 24 June. He ceased to be the third joint managing director but remained an ordinary director. This left Edward Wilshaw and Sir Norman Leslie as joint managing directors of Cable and Wireless, while Denison-Pender remained Governor and joint managing director of Cable and Wireless Holding. Then Sir Norman Leslie resigned from the Court of Directors and relinquished all connection with the group. That left Edward Wilshaw as sole managing director of Cable and Wireless, and he was thereupon appointed, in addition, Government-approved chairman. John Jocelyn Denison-Pender was appointed General Manager.

HMG were as good as their word, and in 1937 His Majesty created J C Denison-Pender the first Baron Pender of Porthcurno.

Wilshaw was now 56, one year over the retiring age. He had his pension, and at first, it is said that he was reluctant to take over the post, advocated by Sir Basil Blackett, of sole chief executive. He had other activities in view. He was a Justice of the Peace and had aspirations to emulate Sir Harry Twyford, once an Eastern Telegraph employee, by becoming Lord Mayor of London. But he was finally persuaded by his colleagues not to desert what seemed in 1936 to be 'a sinking ship'. Above all, it was now his responsibility to make good the company's strained relations with the Advisory Committee with which its future must be closely linked, so long as the company was constituted as it was.

He knew that the Committee's investigation of the company was pending, and on 6 July he had the first of a series of sessions with Sir Campbell Stuart, chairman of the Advisory Committee and the representative of both the British and Canadian governments. He told Stuart he was prepared to close all the company's offices and use the Post Office to collect their traffic if Western Union and Commercial Cable could be got rid of. Stuart told him Cable and Wireless must not expect to be spoon-fed by the British Government. He should reach a commercial arrangement with the American companies which would make it worth their while to retire from the UK. What about Cable and Wireless running the Western Union cables? As a bait they could offer to sell their South American business to Colonel Behn. Wilshaw said he had thought of this, but his colleagues on the Board were against it. Reconstruction of the group was outside the Committee's terms of reference for the investigation, said Stuart; whereupon Wilshaw confessed that the company had always been divided, with himself, Inverforth, Sir Charles Barrie and H A White on one side, and JCD-P, Sir Norman Leslie and Sir Courtauld Thomson on the other. Sherlock was between the two. He agreed with Stuart that re-capitalisation should have taken place years ago.* He thought directors' fees were

* In 1935 the capital of the holding company, Cables and Wireless, was reduced from £53,700,000 to £47,246,600. At the annual meeting of the operating company on 24 May 1937 shareholders carried unanimously a resolution for capital reconstruction involving the cancellation of the arrears of Preference stock dividends and the reduction of the issued amount of A and B ordinary stocks by £22,500,000 to £6,880,000 and amalgamation into one class of ordinary stock. Preference stockholders were given new funded stock at 4 per cent representing what they were owed in arrears. The directors said it was a plan to liquidate the past and provide for the future. It was not expected that the National Defence Contribution which the Chancellor announced in his April budget would affect the capital reduction plan.

excessive. Denison-Pender drew £6,400 a year from Cable and Wireless, £2,000 from Marconi Mercantile Marine and £2,000 from Cable Investment Trust. Wilshaw had been offered £6,400 and asked Stuart if he should accept it. That was up to him, he was told. He intended bringing back A H Ginman from Canada who would have Jocelyn Denison-Pender under him. He spoke highly of J J Munro's ability but he was not liked by outside parties.

It was a good beginning, and it seemed that the two men had established a new personal and official relationship in place of the recent antagonisms.

Though 1936 made for peaceful relations between Treasury and Electra House, it was war conditions for the staff of Direct Spanish at Bilbao. At 6 am on Sunday 21 July, their office on the third floor of the P & T building was broken into by some twenty armed members of the local Popular Front Militia. They were ensuring the communications centre was not seized by army detachments taking orders from General Franco whose revolt against the newly elected left-wing government on 18 July began the Spanish Civil War. On being told the third floor was foreign property the men withdrew. They returned later to post guards armed with sub-machine guns. The Direct Spanish staff were issued with passes signed by the Civil Governor and countersigned by the Frente Popular. On 25 July HMS *Verity* and HMS *Wishart* arrived in the harbour and took off the ladies of the British colony. A month later the rebel destroyer *Velasco* shelled the oil storage tanks by the harbour and rebel planes attacked the government aerodrome at Lamaico. But no cables were hit and Direct Spanish carried on as best they could. The British vice-consul at Algeciras, a Franco stronghold, where the consulate received a direct hit from the government warship *Jaime Primero,* was an ex-Eastern Telegraph manager with twenty-five years' service, E G Beckingsale who, with his wife, luckily escaped unhurt. Bilbao escaped serious damage of this sort but, being blockaded by land and sea, business dwindled and telegraph traffic virtually stopped. The town was evacuated on 18 September.

Spain's constitution was overthrown by force and within three years the rebel government's control was recognised by HMG. Hopes of postponing a world conflict were buoyed by something called Collective Security, but unilateral preparing for the worst continued in the cabinet offices and board rooms of Europe, not least in Whitehall and Victoria Embankment. Procrastination and the creaking democratic process had led to the British Government's communications being found wanting on the eve of the First World War. Those responsible for Planning – the 'in' word of the thirties – were bent on making sure that circumstances did not repeat themselves. A committee had been set up in the Admiralty in 1931 to determine policy on the laying and maintaining of submarine cables in time of war, and it met for the first time in five years in July 1936 under Admiral Sweeney, the officer in charge of lighthouses, with the new chairman of Cable and Wireless as a member.

Edward Wilshaw had been a 34-year-old executive of Eastern Telegraph in 1914 and seen the effect of the shortcomings at first hand. In his talks with Sir Campbell Stuart throughout 1936 he was determined that achievable recommendations should be ready to lay before the imperial Premiers when they met in London in May 1937 for the Coronation of what was now going to be the Duke of York and not the Prince of Wales whose brief reign as Edward VIII had ended in abdication. But there was one strictly commercial matter which could be considered separately – rates, and the effect on them of the terminal and transit charges which empire governments still imposed. Wilshaw chaired an Empire Rates Committee composed of two representatives of Cable and Wireless, two of the Advisory Committee and the permanent heads of empire governments' postal administrations. They lost no time in placing charges on a new, simple and coherent basis, and in their report of July 1937 recommended a uniform rate of 1s 3d a word in sterling territories.

If this ensured future prosperity, the company's staff would benefit too. For at this time, at Wilshaw's instigation, the Directors approved a profit sharing scheme which gave staff a direct interest in the undertaking. It included the large messenger staff for whom for some years now there was also a part-time education scheme for which they were excused one day a week.

But new rates and profit sharing would not help Britain face the Nazis. In 1936, as in 1913, the company's development of wireless was far behind what it should have been. This was due, said Electra House, to their obligation under the Treasury Agreement of 1928 to maintain the cable system for strategic purposes which they reckoned cost them £450,000 a year. If Britain went to war, the Dominions and Colonies would go with her as before. Communications strategy would be essentially imperial; it would depend to a large extent on 'secure' cables but it would have to take full account of the wireless circuits which could never be cut or diverted. However, in the twenty years since the end of the Great War, unity of control and direction of communications within the Empire using both mediums had been only partially realised. There was still no joint working of cable and wireless in Canada or Australia. In their annual report of 31 January 1937, the Advisory Committee suggested that 'a possible solution of some of these difficulties might be sought in the acquisition of the system by the State'. The problem of co-ordinating telephone, telegraph and air mail could be more easily handled in that way. Responsibility for the upkeep would fall on the taxpayer and not the private shareholder; it would provide an easier way of meeting the inroads of foreign competition. The terms of acquisition could be settled by arbitration, which might involve the governments in heavy payments for compensation. Operation would not necessarily be more efficient and might be more expensive. Paragraph 61 of their report, which of course was 'secret' and never published, set out another possibility.

> 61. A possible form of State ownership would be to organise communications in the hands of an Imperial Public Utility Board jointly constituted by the Imperial Governments with a complete monopoly, on the lines considered by the Greene Committee. There would then be complete participation by the Dominions and India in the Imperial scheme, thus giving real unity of control and other advantages. We agree with the Greene Committee, however, that such a scheme, with the relinquishment of sovereignty by the Imperial Government which it implies, is outside practical politics.

Paragraph 62 gave the Committee's reasons for sticking to the present constitution.

> The basis of the 1928 merger scheme was the free association of the Imperial Governments in support of a system of overseas telegraph services owned and operated by private enterprise. Notwithstanding the catastrophic world conditions under which it has had to operate, this scheme has so far achieved its main objects. It is still faced with grave difficulties, but these difficulties do not appear insuperable, and we can see no practicable alternative which offers a better prospect of success. We conclude, therefore, that this basis of the 1928 scheme should be retained. If this system is to operate successfully, however, it must be so organised that it can offer in every country of the Empire a rapid and sufficient service at a competitive tariff. For this purpose co-operation between C & W Ltd and the other undertakings and Governments concerned throughout the Empire should be improved. C & W Ltd should be given the greatest possible freedom to operate on a commercial basis, and 'non-commercial' obligations and charges imposed by the Governments should be reduced. Whenever

practicable the Governments should facilitate the acquisition of further services by C & W Ltd which may complete the unity of the Imperial telegraph scheme. Every effort should be made to avoid the introduction of new routes or services which may withdraw traffic from the Company's system.

They concluded that if practical steps could be taken to achieve those objects the company could reasonably hope to consolidate their position and maintain the essential imperial service without recourse to Government ownership or Government subsidy.

But the tide was running in the opposite direction. In fact very little effort was made to avoid the introduction of new services which might withdraw traffic from the company's system. In the spring of 1937, Cable and Wireless heard that the Director General of Indian Telegraphs was once again pressing his associated company Indian Radio and Cable Communications to open a direct wireless circuit between India and China. On 16 April, Sidney Farmer pointed out to the Advisory Committee that this would mean 'a considerable loss' to the company; could they persuade India to postpone the project? On 16 June, Farmer was asking the Advisory Committee if they had heard of the wireless tests taking place between the administrations in Malaya and Burma. Would the Advisory Committee see that the inauguration of a direct wireless telegraph between the two was deferred? For Cable and Wireless proposed replacing the land line between Burma and Malaya via Bangkok by a direct Government wireless circuit between Rangoon and Singapore at 15 annas a word, the same as the land telegraph.

In the year of its inventor's death – Guglielmo Marconi died 20 July 1937 – it was evident that wireless was still as divisive an element in the field of communications as it had ever been, and Electra House was only just becoming reconciled to it.

The most serious threat to the unity of an imperial system centred on London still came from Canada and Australia. Cable and Wireless had to operate its cable service between these Dominions and Britain at both ends, and at the Dominion end alongside a local wireless company. In 1937 it was proposed that Wilshaw should go to Canada to rationalise the position, but he never went. Later, however, Cable and Wireless acquired a controlling interest in the Canadian Marconi Company whose chairman, A H Ginman, was also the Cable and Wireless manager in Canada and was eventually given a seat on the London board.

There was what Campbell Stuart considered 'an unbridgeable gap' between the ideas of Cable and Wireless and Amalgamated Wireless Australasia on every aspect of the merger. The matter of collection and delivery of messages in Australia was a standing source of grievance.

Canada and Australia had agreed and then attacked the 1928 Merger Plan. Their representatives had helped formulate the proposals resulting from the Advisory Committee's investigation of 1938, and the optimists hoped that this time their governments had authorised them to do so because they fully intended to carry them out.

The heads of agreement reached by the empire governments with the aid of the Advisory Committee were set out in a White Paper (Cmd. 5716) presented to Parliament on 8 April 1938. The proposed settlement, which was subject to approval by Parliament, called for financial sacrifices by the British Government, by Dominions governments and by Cable and Wireless, but they would be offset by the benefits, particularly to users of the system. The settlement provided for immediate and substantial reductions in overseas telegraphy rates and relieved the strain on the company's finances. The third purpose was to expose the negative attitude of the Dominions governments to the system which they had established in 1928, and to plead with them to start giving it their support and co-operation so that the company could plan for the future with greater confidence.

The Empire Flat Rate Scheme of 1s 3d a word for full rate, plain language telegrams to and from the UK, and between Dominions and colonies, was designed to save the public £500,000 a year, and would start on 25 April. It meant 5d a word for letter-telegrams, $7\frac{1}{2}$d a word for deferred telegrams, and 10d a word for code. The charge which the British Government had on the company's revenues through the Beam Rental was exchanged for an interest in the company's equity. The twenty-five year Beam Rental of £250,000 a year (ending 1953) was cancelled from 1 March 1938; the Government agreed to hand over the freehold of the beam wireless stations and take 2,600,000 £1 Cable and Wireless shares in return. Insofar as the annual sum earned in dividends from these shares was unlikely to equal the £250,000 a year in rental, the 'surrender' of the latter represented 'a substantial sacrifice' on the Government's part. Moreover 'in view of the wider public issues involved,' the net consequential loss in the operation of the beam stations over £100,000 a year would be borne by the Exchequer. The cost of this concession, stated the White Paper, was outweighed by the great importance alike on political, commercial and strategic grounds of maintaining intact the company's system of imperial communications.

Under a separate agreement, confirmed in a letter from Wilshaw to the Secretary of the Treasury of 27 May 1938, HMG, now a shareholder in Cable and Wireless, became entitled, after consultation with the chairman, to appoint a director to represent the Government's interests – this time 'responsible to' not merely 'approved by'. This director would hold office at the pleasure of the Treasury and would not be required to submit himself for approval or re-election by shareholders. He would not be a member of the Advisory Committee. A change in the company's articles of association would be required. But the Treasury would not exercise this right unless they were satisfied it was in the public interest. Pending such an appointment the company agreed to give to anyone the Treasury nominated 'all such information as the Treasury may reasonably require to satisfy themselves that the public interest is being safeguarded'. In fact, the Treasury chose to suspend the right indefinitely, and rely instead on Sir Campbell Stuart, chairman of the Advisory Committee, as their 'representative' for as long as the good relations continued to exist between him and Wilshaw by which Wilshaw sought his advice and collaboration whenever he saw fit, bearing in mind the interest HMG now had in the company.

The existence of those relations had given great satisfaction to the Treasury, and it was their earnest desire that they should continue unchanged. The new arrangement would not involve anything more than a continuation and strengthening of Wilshaw's present relations with Stuart.*

The White Paper of 8 April 1938 also announced the reduction of the Standard Revenue fixed in 1928 at 6 per cent of the capital, or £1,865,000, which had never been operative, to 4 per cent of capital or £1,200,000 (Cable and Wireless made a profit of £1,253,919 in 1937). The Anglo-Continental telegraph services operated by the Post Office and the company were to be co-ordinated under a Joint Purse; and the licence which the Post Office granted the company in 1929 due to expire in 1953 was extended for twenty-five years from 1 January 1938 (to 1963).

Though there was no reference to overseas telephony, no cancellation of the Pacific Cable annuity, and no hint of a direct subsidy for operating strategic cables,* Electra House were well

---

*On 21 September 1936 Stuart told Wilshaw in confidence of his plan to have the representatives of the Dominions and Colonies on the Board of the Communications Company (increased to 15) instead of the Advisory Committee, which would be abolished. It was Stuart's intention that Wilshaw should be appointed chairman 'with suitable protection for some years'. Stuart saw himself as the director representing Canada. The plan came to nothing and Stuart was appointed Government-approved Liaison Officer between the Treasury and the company.

satisfied. They had got their new 'charter', and it had left the structure of the company as a private, profit-earning corporation intact. In their annual report presented to the meeting of 16 May 1938, the directors felt assured 'of the goodwill, support and co-operation of all the Empire administrations who will give Cable and Wireless free publicity facilities in all their offices and publications'. They hoped that the settlement, particularly the Empire Flat Rate Scheme, would not only stimulate traffic but have wide imperial consequences in view of its influence on social and commercial intercourse within the Empire.

The heads of agreement were embodied in the Imperial Telegraphs Act 1938 just in time for the Munich Crisis which gave Britain and her allies respite, but confirmed to most that it could only be a postponement of a second world war which those in telecommunications knew, ironically, would stimulate telegraph traffic and heal the company's financial troubles with very much greater assurance than any parliamentary bill or tariff revision.

---

*A paragraph in the White Paper explained that the company's responsibility in this regard had been taken into consideration in framing the settlement.

PART III
1939–1979

10

Wilshaw's Way

1939–1942

What effect did the changes of the late 1930s have on life in the company's branches overseas?

By 1938 Cable and Wireless Limited and its wholly owned subsidiary companies were operating in 146 locations around the world from Accra to Zanzibar. In Hong Kong it was more than a change of name; it was on 1 January 1938 that Cable and Wireless first became responsible for the fixed wireless services of Hong Kong, using the existing Peak wireless receiving station and the Cape D'Aguilar transmitting station. The company soon came to a temporary arrangement with the Hong Kong Telephone Company over the provision and connection of external telephone services for the national network.

Under its old name, the company had taken over communications at Bahrain in 1931. They established a wireless telegraph link with Sharjah in 1933 and with Aden in 1936; and by the end of 1938 with Saudi Arabia, Kuwait and ships at sea. A telegraph cable was laid to Bushire in 1935, and some six years later to Fao. These cables were brought ashore at Jabalat-Mani on the north coast of Bahrain island two miles from Manama, and at Hoora.

In the western hemisphere the old names persisted – and seemingly the way of life that had put Western Telegraph at the bottom of the Eastern Associated league. On 14 July 1938, H S Ramsay, manager of the Rio Grande branch in Brazil, where the group's operations were still carried on under the name of The Western Telegraph Company, wrote to his regional manager in Rio de Janeiro to report that his offices were in need of a clean. The inside was very dirty and had not been decorated since 1924. Outside, the building was conspicuous for its shabby appearance; it had not been painted since 1931. The Casino cable house also needed attention.

> A great deal of wilful damage has been done recently by persons so far unknown. Practically all the windows have been smashed by revolver shots, and several window frames have been practically broken away. The water pump standing on the verandah has been deliberately smashed.

Ramsay said he had personally paid a local man to keep watch out but he suggested the time had come for the company to employ a full-time watchman and build a small shack for him to live in.

Only Ramsay's good personal relations with the Inspector of Police saved the company from becoming even more conspicuous, if not notorious, when later that year the police threatened to take action against Western Telegraph 'for acquiescing in the perpetration of an illicit business, inasmuch as they were accepting and transmitting a telegram known to convey illegal information for the purpose of Jogo de Bicho'. The latter was a local numbers game, conducted by letter and telegram and banned by the State. A resident of Regalia, Rio Grande, Senor del Mauro, devised an elaborate scheme to scoop the pool. Unfortunately it involved the Revising Clerk at Western Telegraph, one Nelson Franca and his assistant Ferreira. All the suspects were arrested and then released. Why? asked Ramsay of his friend the inspector. With Jogo de Bicho being illegal, said the policeman, a denouncement would lead to a spectacular disclosure of numerous persons being involved, with Western Telegraph at the centre. In view of this he was going to recommend that the investigation was 'archived', and that Western Telegraph exercised stricter supervision in future.

> The inspector then took me into his confidence (wrote Ramsay in his account of the affair to his superior in Rio de Janeiro) after first making it clear that he was speaking as Senor Carvalho to Senor Ramsay, and not as Inspector Chefe da Policia to the Gerente da Western, and told me Senor Ferreira had made a written confession that he falsified this telegram aided by Messrs Nelson Franca and Octavio Franca.

Ferreira gave the inspector his confession in return for a promise that he would not be denounced, and the inspector asked Ramsay not to use this information as a reason for dismissing him. He had confessed to falsifying other telegrams. But Ramsay had to confront Ferreira, who cockily told him he knew he would never be denounced as it would implicate the company. Ramsay pointed out to H S Sieyes, Brazil regional manager, that Rio Grande being so small it was difficult to find a sound lawyer to advise him on how to handle the culprit. They were now in the invidious position of knowing who was responsible, but being unable to punish the guilty or exonerate the rest of the staff who had felt the disgrace intensely.

Ramsay had roughed it at Porthcurno with the rest of them, but for anyone used to European standards, service with Western Telegraph in Brazil in the 1930s must have been somewhat of a trial. Life on the Rio Grande was far from romantic and not only on land. It took him nine days on *ss Itaquatia* for the 800 mile voyage accompanied by his wife from Rio de Janeiro to Rio Grande to return from a business visit to Sieyes. It was far from a pleasure cruise.

> For new arrivals in Brazil with no knowledge of the language or customs of the people (he wrote to his regional manager when he got back), a voyage like this would be almost a torture. I do not recommend this type of boat for general transport of staff, but in an emergency, as in my case, the voyage can be tolerated. You definitely need a sense of humour and a good stock of patience, also the ability to adapt yourself to conditions which in no way conform with your general ideas of diet and hygiene. The ladies' bathroom was impossible and the ladies' lavatory disgusting. The tit-bit of the whole trip however was when the table steward wiped my plate on the seat of his trousers before handing it to me.

The voyage out to Brazil from England was always broken by a call at Sao Vicente in the Cape Verde Islands where, as Arthur Grass had done before them, new recruits were still stopping off to spend their six months 'consolidation' after successfully passing the course at Hampstead and before proceeding to their first proper posting as fully qualified Assistant Engineers. When John Ash went there in the 1930s he found there was no longer the big Mess of earlier days, though the building remained and its tennis court. There was a staff of twenty but most were married and only six were bachelors. After six months at St Vincent, and another four months on the full staff, he reported to F Hannigan and Watty Edmonds of the Staff Office at Electra House to be told where he was to go next. He knew it was no use expressing a preference. Anyone who did so was inevitably sent in the opposite direction. So in 1938 he found himself in Buenos Aires where again there was nothing resembling the 1920s Mess, with only fifteen or so expatriates and the rest Argentinians. For him it was the beginning of a typical Cable and Wireless Foreign Service (F.1) career which took him from South

America to West Africa (Lagos), to the Atlantic Ocean (Bermuda), to Peru, Hong Kong and Chile and then back (thirty years later) to Sao Vicente where as Honorary British Vice-Consul he had the satisfaction of renewing his own passport.

Marine staff had a different circuit determined by the various cable ship bases, the most popular of which in the 1930s was at Singapore – the 'plum station' as Bill Cross recollected in retirement in Vigo in 1978. He joined the cable ship *Recorder* in 1938, built in 1902 as the *Iris* for the Pacific Cable Board; moved to the Indian Ocean, Fiji, Gibraltar, and ended up in Aden.

In the West Indies, Western Telegraph and the Pacific Cable Board were fused in 1929 as Imperial and International, with West India Direct (West India and Panama, Cuba Submarine, and Halifax and Bermudas) joining in 1932. They assumed the name Cable and Wireless in 1934 but from 21 August 1936 the business was conducted in the name of The West India and Panama Telegraph Company Ltd which, so far as local governments and the public were concerned, became the working company of Cable and Wireless in the West Indies. This followed an agreement between Cable and Wireless and West India and Panama of 6 September 1935. But eighteen months later they changed their mind and in March 1938 the name was altered to 'Cable and Wireless (West Indies) Limited'.

Unlike most other parts of the world for which operators and managers had to be found in Britain, the West Indies produced a large number of staff locally. Thus there were no Messes for young expatriate bachelors as there were in the Far East and Mediterranean. Junior staff at West Indian branches lived and ate at home; some Barbadians were posted to the other islands. They had no inclination for the collective living and playing so loved by the English public schoolboy. Barbados had dominated the British West Indies ever since political and religious refugees from Britain settled there at the beginning of the seventeenth century, whose descendants, unlike the inhabitants of the other islands, were able to maintain a continuity of commercial and political activity uninterrupted by foreign occupation. Between the wars most of the staff of telegraph stations in the West Indies came from Barbados, like Johnny Bourne who in 1923 joined the town office of Western Telegraph in Bridgetown fresh from school and was put to taking down messages on the telephone received at the cable station at Dover, St Lawrence. There was no formal training; he did whatever F C Walker, the manager, told him to do. After he had been there six years he was in charge of accounts and had the task of training the department in the new methods demanded by Imperial and International. It took three years to make the change. Two years before the amalgamation of the three telegraph companies operating in the West Indies, he was joined at the Western Telegraph office in Bridgetown by a fellow Barbadian, Edward Stoute (1927). In 1978 he recollected how the new masters came from England to inspect their premises and equipment, and 'condemned' the Western Telegraph building at Dover. They kept it in use, however, and even added to it, right up to 1970 when it was replaced by the new Wildey building.

The D/N Scheme applied to all staff, whether they had been with the Pacific Cable Board, Western Telegraph or West India and Panama, and within months of becoming employees of the combined company they were being asked to make voluntary cuts in salary and work longer hours. The job of a cable operator was in any event a very demanding one in the West Indies where it was more important than in most places to maintain the critical balance which allowed simultaneous sending and receiving by the duplex system. As soon as the reader of incoming messages noticed an excess of 'fat dots' (which could be a dash), 'drop outs' or 'splits' (three dots or a dash?) he had to call the engineer to restore the balance.

For Hubert Montplaisir (the 'r' is silent) who joined West India and Panama in Castries, St Lucia, in 1920, it was hand-sending right up to 1957 when he retired as manager of the station. When the Pacific Cable Board came to St Lucia in 1928 and set up a wireless service to compete with his company's cable telegraph, he had little contact with the newcomers as they had no town office and ran their telegraph from the wireless station on the Morne. Apart from the manager, West India and Panama only had three operators, and once, when one of them went ill, Montplaisir was on duty for two nights and three days at a stretch. At his request the Divisional Manager at Barbados sent a replacement whose introduction to St Lucia was to be put at once on a twenty-four hour shift, from one o'clock the afternoon he arrived to one o'clock the next day.

Not only long hours but long service characterised the Barbadian staffs who manned the Caribbean wireless and cable stations of those days. Hubert Montplaisir's thirty-seven years from 1920 to 1957 were beaten by Ainslie Skeete who joined the West India and Panama station in St Lucia in 1918 as a Learner Operator of 18, and only retired in 1962 after forty-four years' service. He became a qualified member of the staff at St Lucia two years after joining. As a young man of 22 he once found himself in charge of the St Vincent office with only an even younger probationer, Galbraith Cropper, to help him, when he had to relieve the chief clerk while he took a month's holiday. The headquarters of West India and Panama were on the island of St Thomas (now part of the American Virgin Islands), and for a strange period of his early career he found himself staying there at the Grand Hotel where his superintendent, Maurice Petit, also lived. He was not on leave, yet he had no duties. The days passed and the hotel bills, which he knew he could not afford, were running up. So one evening in the hotel lounge he steeled himself to ask Petit why he had been sent to St Thomas. Making sure he was not being overheard, his manager told him in conspiratorial tones that his stay was all part of the procedure necessary to get him to the Puerto Rico office, the island next to St Thomas. This was United States territory, and the Americans were refusing to issue any more Englishmen with work permits. They allowed tourists to land on a day-trip, and the plot was for the Mr Skeete who had been staying as a tourist at the Grand Hotel for several days to take a day-trip to Puerto Rico and then 'miss' the boat back, make straight for an arranged hotel and next morning report to the West India and Panama office in the town. In this way the increase to the branch complement by one clerk would never be noticed. But Ainslie Skeete was fated never to take part in this elaborate piece of play-acting, for he contracted a serious bone disease in his ankle and was sent back to St Lucia before Day-Trip Day arrived.

He became 'Responsible Officer', as it was called, at St Lucia at the age of 27 in 1927; and when it was known that West India and Panama were to be merged with the two other companies, Maurice Petit sent Skeete a personal note reassuring him that the St Lucia office would not close and he was in no danger of losing his job. By picking up the various techniques required of him, from cable testing to operating, he made himself a jack-of-all-trades and later took a course at the Barbados Training College, established in the 1930s, when the New Zealander, A B Hathaway, was wireless training officer.

Already fully qualified as a mechanical and wireless engineer/operator before joining Imperial and International in 1934, in British Guiana, was Guyanan Frank Savoury. That was the year the company took over the naval wireless station at Georgetown to which he had been apprenticed. In anticipation of the merger, he taught himself touch-typing on a portable machine he bought for the purpose. As a wireless man he took the cable course at the Barbados School in 1937, and returned the following year to Guiana where Grenadan Lloyd Smith, ex-Pacific Cable Board, had been installed as manager for Imperial and International.

Another way of becoming a qualified telegraphist was to join the Pacific Cable Board at Barbados as an Abstract Clerk, which was what 16-year-old Barbadian D Courtney Frost did in October 1928 at a salary of $24 a month – £5. Soon he was a Probationer Operator learning to send Morse at twelve words a minute at $30 a month; and shortly a Proficient Operator at $40. Most of the senior operating staff were retired naval officers. But the man who had entered as a clerk in the Accounts Department made sure he could do the work as well as they. He remained for forty years and retired in 1968 as branch manager of St Vincent, a post he held for ten years.

Looking back on those days in 1978 'Frostie' remembered that the main objective had been to make their more sophisticated wireless service quicker and more efficient than the West India and Panama cable service which charged about the same rates. Though professional rivalry was tense, relations between the staffs of the three companies who would circulate throughout the West Indian islands were personally cordial – to be a Telegraphist generated loyalties greater than belonging to any one telegraph company. There were also frequent inter-company postings. Barbadian Galbraith Cropper, who retired in 1959 after thirty-nine years, transferred from West India and Panama to the Pacific Cable Board quite early in his career. At the Pacific Cable Board office he found fellow Barbadian Jimmy Cozier (whose mother's family had come to Barbados after the Battle of Sedgemoor). Cozier had joined in 1931 at 16 as a wireless operator at Deep Water Harbour, doing six hour shifts. His career as a telegraphist has been described as 'short and stormy' and he was one who did not make it his life. He became a newspaper owner, columnist and distinguished author/historian. Jimmy Cozier and Edward Stoute between them know more about Barbados history than any others alive.

The telegraph companies in the Caribbean as a requirement under the terms of their concessions had to provide an information service to the community in the form of international news and weather forecasts. A daily news bulletin was received in morse from an agency and was transcribed and typed out in the form of a legible news sheet, with one copy for the Governor, one for the local newspapers and another posted on the company's noticeboard for any passers-by to read.

In 1933 the Pacific Cable Board's news system was superseded by a re-modelled up-to-date service compiled by the Canadian Press agency (Canapress), 1,500 words a day of world news, two-thirds of which, however, was British. Pinning it up outside the company's

front door was an easy enough operation, but there were times on
Antigua when getting the copy through to Government House
presented difficulties. On 3 June 1934 the Officer in Charge of the
Pacific Cable Board, Antigua, sent the Inspector of Police this well-
observed account by his messenger of his encounter with PC 35.

> I beg to report that I was sent to deliver a Government cable
> addressed to H.E. Governor. On arrival at Government House the
> sentry was not at his post, but looking over the fence. I rang the bell
> and the sentry called to me to 'come this side'. I said 'I have a
> Government cable. You must come to the gate and take it'. He said
> 'You are damn forward. If I call you, you must come to this side.' I
> waited five minutes for him to come to me, then rang the bell a second
> time. The butler came and took the cable. Whilst I was waiting for the
> receipt, the sentry came to the gate and threatened, among other
> things, to arrest me and lock me up. After receiving the receipt, I told
> the sentry I was going to report the matter to my Chief, and he replied
> 'To hell with your chief!'

A substantial investment income, which gave a misleading picture of
the company's buoyancy, had postponed the earlier introduction of
measures such as the Empire Flat Rate Scheme (from 25 April
1938), the Flat Press Rate (from 15 April 1939) and the Empire
Greetings Telegram (5s for thirteen words from 1 May 1939). The
latter, known as a GLT, was introduced with a flourish by Wilshaw
allowing everyone to send a free greetings message on the opening
day. An enterprising newspaper correspondent in Nairobi saw a
story in having 10,000 Kenyans send loyal greetings to King George
in London. He was making plans to line them up at the telegraph
office when fortunately the manager was able to persuade him that
the result would be chaos and call it off. One Scotsman, however,
managed to send his family greetings for 1939, 1940, 1941, 1942 and
1943 before his thirteen words were exhausted.

In the few months left before the war, such inducements at once
demonstrated their effectiveness as a stimulus to the greater use of
overseas telegraphs. There was an immediate growth of traffic
within every part of the Empire (except Ceylon). Given a con-
tinuance of peacetime conditions they would doubtless have made a
considerable contribution to recovery. But they were overtaken by
conditions which made them superfluous, conditions which had
always shown themselves to be a greater stimulator of traffic than
any commercial ploy – war.

In the next five years traffic increased 400 per cent. The number of words sent increased from 44 million in 1938 to 354 million a year in 1944. War took care of the shareholders. In the first full year of hostilities (1940) profits nearly doubled to £2,375,700, and for 1941 were £2,928,350*. Commercial stewardship could take second place. The circumstances of a Britain fighting for survival demanded different thinking, new priorities, new values.

In their 1938 report, presented to the meeting of 8 May 1939, the directors revealed an offer had been made to create a system of wireless telegraph and telephone communication throughout the colonial empire – excluding the Dominions, that is – without cost to the administrations or peoples concerned. Wireless sets would be put at each cable station where no commercial service existed, to enable them to keep in touch with aircraft and ships.

This 'Colonial Wireless Scheme' had, in fact, been mooted in 1937 but it took a long time to be accepted. Each colonial government had to be consulted and discuss it. Agreement was only reached a week before war broke out. The sets had been ordered earlier, however. Dispatched under naval protection, most were working by May 1940. Radio-telephony was somewhat of a novelty, even for the West Indies where a sophisticated wireless telegraphy network had been in operation for many years. Officer in Charge, Antigua, to Divisional Manager, Barbados, 14 March 1940:

Colonial Wireless Set

The interruption of Bar-Aga contact was afterwards traced to a dirty spark plug which prevented the motor from giving an even voltage, and a faulty Main Magnifier (PT.5) valve in the transmitter. These have been attended to and we are now functioning correctly again. Contact was still established with St Kitts, and the editor of the *Magnet* spoke to editors of the St Kitts newspapers. An excellent contact was made between H.E. the Governor, speaking from St Kitts, and The Administrator, Antigua, on 13th. The conversation lasted about twenty minutes, during which not the slightest alteration or adjustment to the set was made.

---

*On the outbreak of war the UK Government introduced an Excess Profits Tax applicable to all commercial undertakings ranging from 60 per cent in the first year to 100 per cent in subsequent years on all profits over a certain level. It was ruled that the revenues of Cable and Wireless were subject to this tax, the Group being assessed as a whole. In the published accounts of 1939, 1940 and 1941 no allowance was shown for EPT; but it was in 1942, when the profit was shown as £1,278,700, still above the Standard Revenue, however. The final assessment of EPT consumed a substantial part of the earnings which would have been reckoned as Excess Revenue available for rate reductions under the 1929 Agreement. From 1939 to 1942 shareholders received the standard four per cent dividend and no more.

There were going to be few opportunities to add to the company's system as constituted at the end of 1939. It was a question of reducing the risk that the system, such as it was, would become even smaller through neglect to take the proper precautions. The accent was on making sure that it was always in a condition to facilitate the highly complicated political and military activities of the Allies, spread all over the world, in their common cause of defeating the enemy. Every eventuality had to be anticipated. In peacetime, concentration of terminals at Moorgate from the cable heads and the eight wireless stations was an operational asset; in wartime, a target which no enemy could fail to overlook.

The 105th meeting of the Imperial Communications Committee of the Committee of Imperial Defence agreed to appoint a sub-committee to prepare a scheme with Cable and Wireless for the establishment of a reserve cable room as alternative to Moorgate, 'and such other measures in the UK as are considered necessary by them for the maintenance of the services of the company in time of war'. Edward Wilshaw had already turned down the idea, discussed with the Post Office, of an emergency station at Twickenham because it was too near the river. But at a meeting of the 'Imperial Communications Committee, Sub-Committee for Consultation with Cable and Wireless Limited' – not to be confused with the Imperial Communications Advisory Committee – which had its first meeting on 16 February 1939, F W Phillips, Director of Telecommunications at the Post Office, told Wilshaw the foreign telegraph companies had already selected provisional sites for emergency stations in northwest and west London. His company's site depended on whether the Government would pay for it, said Wilshaw. They had no surplus equipment; new apparatus would cost £50,000 and take two years to provide. In any event, if Moorgate was put out of action, their Central Telegraph Station would move to one of the western counties – Marlborough or Dorchester had already been considered.

The position of Cable and Wireless, insisted Wilshaw, was altogether different from that of public utilities which could recoup themselves by raising their rates; his company was committed to the Empire Flat Rate Scheme. His company should be treated as a quasi-government department. If 'the authorities' required the company to provide an emergency station the whole cost should be borne by HMG. Of greater concern to him was the safety of Porthcurno. The Admiralty had not thought it necessary to provide the alternative station which he had suggested at the time of the Munich Crisis.

Having spoken about Marlborough and not wanting an

emergency station in London, Wilshaw then told the sub-committee that during the 1938 crisis, on his own initiative, he had set up an emergency operating room at their other Electra House on Victoria Embankment. This was news to Sir Campbell Stuart, who chaired the sub-committee, and to Trentham and Phillips – and to Warren Zambra, the sub-committee's secretary. It brought the meeting to an abrupt close. They decided they could take no action until Cable and Wireless had given the Post Office the fullest information on this.

Wilshaw set out his views on emergency arrangements in a letter to Sir Campbell Stuart dated 22 March. The company's buildings at Moorgate and Victoria Embankment were far enough apart, he said, to make it unlikely that they would both be hit. A bomb-proof and gas-proof station could be built as a second line of retreat at Electra House, Victoria Embankment. As regards the wireless stations, improvisation should be resorted to when one or other of them was destroyed.

Wilshaw had his way and all the more important cable and wireless telegraph circuits at Moorgate were duplicated at Victoria Embankment in a second line, reinforced central telegraph station which became known as The Fortress. A rehearsal move from Moorgate to Victoria Embankment by 1,500 staff in taxis, buses and Underground trains proved that the change-over, if ever it became necessary, was practical. A third line of defence was established in an unassuming suburban villa, 12 Hamilton Road, Ealing. Certain provincial branches could take overall control if London was isolated. All offices were fitted with wireless on which they could talk to Electra House if the cable was cut.

The company's wireless station at Bodmin was taken over in its entirety by the Air Ministry on 3 September. If the enemy put the transmitters at Ongar and Dorchester, and the receivers at Brentwood and Somerton, out of action, what was left of the Cable and Wireless system would be operated from Porthcurno.*

No work was, in fact, done to protect the cable station at PK until after the fall of France, but special constables were on guard duty before the fatal 3 September 1939, and one night soon afterwards some 300 Scottish infantry arrived. They were quartered in the old school and had taken up their positions by dawn. The beach was sealed off with wire and an unscaleable, flood-lit fence. Tank traps were dug to catch armour which landed from the sea, and the area covered by artillery concealed in 'hay stacks' and 'bus stops'.

---

*The GPO cable to Belgium and Germany terminated at Dumpton Gap, and the cable to France at Cuckmere. The Post Office connected these to a central point in Moorgate.

Rocket projectors and a Bofors gun would ward off attack from the air.

Lord Lloyd, whom Winston Churchill appointed chairman of the Imperial Communications Committee as well as Colonial Secretary on becoming Prime Minister in May 1940, gave orders at once for the major work to be carried out at Porthcurno which Wilshaw had advocated in February 1939. Contractors Edmund Nuttall assembled a skilled team of 200 Cornish miners from Truro, supplemented by experts from Ireland and Yorkshire, and they began tunnelling into the cliff rock on 25 June. Their task was to provide an alternative underground, bombproof suite of offices, instrument room, generating station and cable terminal in as quick a time as possible. Pneumatic drills, for which compressors were parked on the tennis courts, opened up the granite for the high explosives whose continuous detonations shook the valley from 30 July, through summer and autumn, to 27 October. In the early days nets protected the surroundings from flying debris; but in the rush to finish the job they were soon dispensed with, and one afternoon a boulder the size of an armchair hurtled through the manager's kitchen window. Luckily neither Mr nor Mrs Bell were at home. The contours of the valley were changed by the mountains of rubble stacked along its sides waiting for clearance up the little railway built for the purpose. Under the stern eye of their Irish foreman, the men worked from the top of the cliff and inside the excavation simultaneously, creating a winding staircase within the cave following the rock's curving strata. In ten months they removed 15,000 tons of granite, making two tunnels each 26 feet wide, 23 feet high and 150 feet long. At the top was an emergency exit up 119 steps (in case the blastproofed doors jammed), and over the entrance was a 70 foot thick granite roof, reinforced with concrete. They were taking no chances.

All the circuits were moved into it, and the chairman's wife opened The Tunnel (as it was, and is, known) on 31 May 1941. It was never hit. Over it flew the new Cable and Wireless house flag which had now replaced the old Eastern Telegraph/East India Company standard since May 1939. It was a red St George's cross superimposed on a golden cross on a dark blue ground. In one corner was the monogram C & W, and in the other a gold crown to symbolise imperial associations.

If HMG had taken possession of the company's stations and cables on the outbreak of war as they were entitled to do under the Treasury Agreement, they would have shouldered the cost of building The Tunnel as a matter of course. In fact the Government decided not to exercise that right and to leave the company in the hands of Edward Wilshaw and his Court of Directors, which included Lords Pender and Inverforth, in the belief, presumably, that they would react to the war situation with an awareness of the nation's needs which would obviate the disruption of a change of control. Tempting though it must have been to make the Government (Post Office) and company one, particularly in view of the obvious pleasure taken by the Cable and Wireless chairman in emphasising the 'we' and 'you' situation, they trusted that the hybrid organisation, whose suitability had been confidently confirmed by the Investigating Committee's report of 31 January 1937, would rise to the occasion. They demonstrated their support of the chairman by having him created (in December 1939) a Knight Commander of the Order of St Michael and St George, of which Sir Campbell Stuart was already a member, but in the same honours lists was promoted to Knight Grand Cross.

Ready to delegate wartime operation of its overseas telegraph service in this way, HMG decided nevertheless on strict supervision of the transmission of messages in the form of censorship. The use of codes was forbidden. Some 250 retired Cable and Wireless employees were recalled and dispatched as censors to the company's branches throughout the UK and overseas. Some fifty went to Aden alone, which presented accommodation problems. Lord Lloyd wanted more than this. In a letter of 12 September 1940, he told Sir Campbell Stuart that he would like to change the terms of reference of the Imperial Communications Committee of the Imperial Defence Committee (of which Stuart had been a member in a personal capacity since May 1939, when he was told not to communicate their documents to the Imperial Communications Advisory Committee of which he was chairman). He wanted to make the ICC the Government's instrument for taking effective decisions on all cable and wireless policy except where reference might be necessary to the War Cabinet. But Sir Kingsley Wood, Chancellor of the Exchequer, saw difficulties in changing a group, created to advise the Imperial Defence Committee and the Government, into an executive body. 'It seems to me,' he wrote to Lord Lloyd on 21 September 1940, 'too much to expect departmental ministers to surrender their responsibility to, or submit to the final authority of, an Official Committee.' Lord Lloyd did not press the point when the committee next met on 1 November.

Protection and deployment of the means of communication available presented comparatively straightforward problems. The 'cable war' took its predicted course. The two German cables from

Emden to the Azores (and New York), and from Emden to Lisbon, were cut within hours of Mr Chamberlain making his historic announcement on 3 September. When Roderick Mann returned to Vigo after leave in 1940 – he had to come up by train from Lisbon as the port was closed – he found the Germans had hired a Spanish trawler and cut the cable to Porthcurno. There was no wireless, so the company closed the station and sent the manager, Gerald Edwards, and his staff home. Mann, however, elected to remain in the peninsula, and was seconded to the Foreign Office attached to the British Embassy in Madrid.

Two of the company's cable ships were in the Mediterranean at the time: the *Mirror* with an all-Spanish crew at Gibraltar, and the *Retriever* at Malta. The Admiralty had accepted war risk liability for both since August. The day Italy joined the Axis against Britain and her Allies (10 June 1940) the captain of the *Mirror* had the ship's blacksmith file a grapnel to a sharp point; had it thrown over the side and dragged across the sea bed where it hooked and cut the Italians' submarine cables to South America and Spain. The Italians had already severed two of the company's five cables between Gibraltar and Malta. At midnight on 10 June the last message was telegraphed across to Italy from Tripoli via Malta, and the cable sealed and handed over to the Chief Censor. Malta remained a vital communications centre throughout the war in spite of the vicious air raids. The company built an emergency station in a tunnel under Fort St Angelo which they shared with Royal Navy Signals.

Of equal importance was the station at Aden which Italian bombers attacked on 11 June 1940 from their base in Eritrea but never succeeded in putting out of action. A month previously the families of the Cable and Wireless staff had been sent to India, but the station never ceased to maintain links with London, Yemen, Nairobi, Colombo, Seychelles, Bombay, Port Sudan. Aden was also the depot for the cable ship *Lady Denison Pender* whose smart white paint overall had to be changed to a less striking grey. Her pitch was along the east coast of Africa and the Arabian Gulf. In 1978 Charles Martin, her chief engineer, remembered the black-out material on the windows raising the temperature of the engine room to 140 degrees Fahrenheit; watches were changed from four hours to three.

When France concluded an armistice with Germany on 22 June 1940 the manager of the company's branch in Marseilles assumed that the whole of territorial France would come under enemy rule. So he dismantled his equipment and took it with him on the last boat to Bône on the Algerian coast opposite. On hearing the Germans intended a partial occupation only and that Marseilles was in the free zone, he returned and put it back. Luckily the search party of the Italian submarine which stopped his freighter in mid-Mediterranean miraculously failed to find the hidden sets.

On the third day of the French Air force's attack on Gibraltar in September 1940, the company closed its circuits in the main office and moved to emergency headquarters inside the Rock. Next time it was confidently expected that the enemy would attempt to capture the headland. Wives were sent home and staff issued with uniform. The station was well below its complement of operators. It was a question of six hours on duty and six hours off. An enormous amount of traffic was handled for the armed services. The attack never came and the branch was still 'standing-by' for invasion on VE-Day.

The Mediterranean was not the only scene of action. Operators at St Lawrence, Barbados, where A G L Douglas was Divisional Manager, West Indies, had an eight hours on, eight hours off shift – two men sharing twenty-four hours. At the St Kitts wireless station the two men chose to work twelve hours on and twelve hours off. Ainslie Skeete claimed he was on twenty-four watch at St Lucia. The war sent traffic rocketing – as elsewhere. As Jimmy Cozier recollects,

Not only was there all this volume of traffic, however. Far more taxing, both to the patience and to the fingers, were the numerous URGENTS, nearly all Government telegrams, which meant clearing the lines and interrupting traffic, forcing the operator to start all over again. In fact merely URGENT lost a lot of its pristine pull, and there were MOST URGENT and then XX, short flash messages of such extreme urgency that they overrode all other traffic in importance and had to be pushed through in split seconds. At no time were XX messages more troublesome than on 12 December, 1939 when they began coming in from Pernambuco [later called Recife] from which line they had to be transferred to the North circuit for flashing through to London. They came in fast and furious, and it was obvious that something was up.

Something certainly was up! The German pocket-battleship *Graf Spee* was engaged in battle with the British ships *Exeter, Ajax* and *Achilles,* commanded by Commodore Harwood. The story of the scuttling of the *Graf Spee* is too well known to bear repetition, but the three days Captain Hans Langsdorff and his men remained in Montevideo harbour meant hectic work for the cable operators at Barbados.

As in World War I, many of the enemy's cables were diverted rather than cut. The Italians' Malaga-Canary cable was diverted to form a Gibraltar-Casablanca cable; and their Lisbon to La Panne

(Belgium) cable taken to Brighton to become a Gibraltar-London link. The Malta-Bône (Algeria) cable became a casualty soon after the capitulation of France in June 1940. When the *Norseman* cut another Italian link, the Italians retaliated by cutting a third Gibraltar-Malta cable. Finally all five cables from Malta to Gibraltar were broken, though the three between Malta and Alexandria were kept working throughout the siege, and all so-called FIL traffic which had to go by cable circulated between London and Malta via Alexandria and Capetown. The company's cable ships were kept busy repairing all the cables around the African continent, often having to do so in areas where U-boats were known to be operating and without naval escort. It was typical of the important and dangerous work which the cable ship fleet and the company's marine staff were required to undertake throughout the war years which Charles Graves has described so dramatically in *The Thin Red Lines.**

But most of the messages between Malta and London had necessarily to go by wireless. The transmitter had been built by A R Tyrer, a brilliant wireless engineer. Much of it was hand-made. Tyrer was tragically killed in February 1941 when he lost his balance and was electrocuted making adjustments to the set which he insisted should not be switched off and so delay the transmission of vital Most Immediate government messages. An urgent signal was sent to the cable ship *Recorder* ordering young Wilfred Davies, who had served two years on the ship as a wireless engineer, to fly out to Malta at once as a replacement radio operator (he stayed until October 1943). The transmitter remained operational throughout the Siege of Malta. The Maltese staff endured the bombing with resolution and great bravery. Despite nearly 3,000 air raids in two and a half years and at least three lengthy blitzes from Luftwaffe bases in Sicily (only sixty miles away), they turned up regularly for duty though weary from sleepless nights, and weak from lack of food – most of the staff lost at least three stone in weight. The fortitude with which Cable and Wireless maintained radio contact with London in such circumstances was recognised – along with much else – when the island was awarded the George Cross by King George VI on 5 April 1942.

The company raised the cable they had laid between Aden and Djibouti, now in Italian hands, which meant the end for the time being of all direct cable communication between Britain and the

---

* His most stirring story is perhaps that of the lonely passage of the cable ship *Cable Enterprise* from St Lucia in the West Indies in 1941 to Singapore via Pernambuco, Rio de Janeiro, Cape Town, Mombasa and Colombo (page 89).

73 In 1955 Cable and Wireless moved its London head office to Mercury House, Theobalds Road (*above*). The building was opened by Lord Reith. The holding company, now an investment trust, remained at Electra House, Victoria Embankment.

75 In August 1959 J. A. Smale (*left*), a former Engineer-in-Chief of Cable and Wireless, opened the new building in Kirby Street, London, which housed the Development and Work-shop Production sections of the Engineer-in-Chief's Department, which was named after him – Smale House. He is seen here at the ceremony talking to R. J. Halsey, Director of Research GPO and part-time director of Cable and Wireless from 1959 (*centre*), and Norman Chapling, Managing Director of Cable and Wireless.

74 That the company was able to partici-pate in the Canadian Trans-Atlantic Telephone project (CANTAT) which gave it recognition in the field of overseas tele-phony, was largely due to the authority and drive of Sir Godfrey Ince (*above*), the Permanent Secretary at the Ministry of Labour, who was appointed Chairman of Cable and Wireless in February 1956.

76 In the 1950s Cable and Wireless had a fleet of eight cable ships based strategically throughout the world to maintain the 150,000-mile British submarine cable network. Here is an 'exploded' drawing of the c.s. *Edward Wilshaw* built by Swan Hunter at Newcastle in 1951.

77 Hauling in the new shore end section of the Barbados/Turks Island cable at Barbados in the West Indies in 1956, from c.s. *Electra* seen anchored out at sea.

78 In 1953 Cable and Wireless had a cable factory in Singapore, and part of it was manned by Malayan boys (*above*), many of whom were the sons of the factory workmen. The caption of this photograph ran: 'As there are no educational facilities for them, they would, if not working, be roaming the streets. There is no absenteeism, since they are enthusiasts in their work and customarily sing while they are doing it.'

79 Inside the instrument room of the key station on Turks Island linking Florida with the West Indies.

80–1 Cable and Wireless 'exiles' posted to remote corners of the world soon acclimatised themselves to working conditions typified by the office at Fao in Iraq (*above*), a vital link in the old Indo-European 'Persian Cable' of the 1860s and still playing its part in 1952 when this photograph was taken. Staff returning to England on leave swapped reminiscences with other members of the F.1. (overseas) family at the Exiles Club (*below*), the house called Meadowbank beside the Thames at Twickenham, originally acquired by Eastern Telegraph in 1920.

82 Major-General Leslie Nicholls who became Chairman of Cable and Wireless in 1951 chats with His Highness Said bin Taimur, the Sultan of Muscat, at Electra House, London, in 1953.

83 On his way back to London from Fiji after officiating at the opening of the Auckland–Suva section of the Commonwealth Pacific Cable, Sir John Macpherson, who became Chairman of Cable and Wireless in 1962 (*left*), visited D. F. Bowie, President of the Canadian Overseas Telecommunications Corporation.

84 In June 1966 H. H. Eggers, who had become sole Managing Director of Cable and Wireless in 1965 (*right*), announced that the company was to participate in satellite communications and had invited seventeen manufacturers to tender for three earth stations. With him at the press conference are (*extreme left*) A. S. Pudner, Engineer-in-Chief, and next to him, E. G. L. Howitt, Deputy Managing Director.

85 Colonel Donald McMillan became Chairman of Cable and Wireless in 1967. Here he is (*left*) with Sir John Carter, Q.C., High Commissioner in London for Guyana, in 1970; E. G. L. Howitt, who succeeded H. H. Eggers as Managing Director in 1969; and A. A. Willett, then Director-Marketing who became 'MD/Ln' in 1973 and resigned in 1977.

86–7 Recruiting, administering and training the huge staff required to operate the services provided by Cable and Wireless has necessarily to be entrusted to men of wide experience and great dedication. With the opening of a new training school for technicians in Hong Kong in February 1966 (*above left*), the company's Far East Communications Centre had three such schools. Responsible more than any other for the smooth transfer of staff in 1950 was Edward (Berry) Mockett who was Staff Manager for fifteen years, seen here in the picture (*right*) with W. G. (Watty) Edmonds, former Staff Manager, at his retirement party in 1962.

88–89 The Cable and Wireless Marine Service with its traditions rooted in the cable laying and repairing exploits of the first submarine telegraph companies of the 1860s still played an essential role in the 1960s. New ships equipped with the latest devices for fault finding and locating kept the fleet abreast of technological advances. Here is c.s. *Cable Enterprise* at her launching ceremony in 1964, with (*below*) Captain P. B. Henderson, her commander, on the bridge.

90–2 Aerials for tropospheric scatter radio systems came in two shapes. (*Above*) curved arrays at Cape d'Aguilar, Hong Kong, in 1968; the smaller, round dish on top of the tower on the right is for a 'line-of-sight' microwave system. (*Below left*) circular dishes in Chalwell on Tortola in the British Virgin Islands in 1966. (*Below right*) part of the control panel of the International Telephone Exchange at Hong Kong in 1966.

Middle East. All cable traffic had to go via Singapore, Australia and Canada. One of the three cables which remained in British hands out of the original five between Malta and Alexandria was handed over to the Services for their exclusive use.

That was one of several which were diverted for the local operational needs of the military authorities. Naturally the company always gave the fullest co-operation on such occasions, and in the laying of new cable and the use of enemy cable in connection with the major strategy of the war. Many hundreds of thousands of pounds were involved in depriving the company of the use of such cables but it was agreed that revenue losses or gains for Cable and Wireless could not be calculated until the war was over.

A 'Note for the Chairman' signed W G R J[acob] gave a good indication of the company's involvement at the beginning of 1941. 'I saw Captain Glover at the Admiralty' he wrote,

> and asked him if he could obtain a decision as to whether the Dakar-Noronha and the Dakar-Brest cables were to be cut by the *Lady Denison Pender* as the ship had now been in Sierra Leone for a couple of days and we had important cables on the West Coast of Africa which wanted repair. Captain Glover said that the Foreign Office had definitely decided not to cut the French cables. On receipt of this decision the Admiralty had said that in that case they would cut the Italian cables between Rio and Noronha. The Foreign Office said they could not do this as these cables run between two points of Brazil and was within the 300 miles limit throughout its length. I then informed him of our information from Pernambuco that the French were going to attempt to repair their own Pernambuco-Noronha cable which, if successful, would give a connection from South America to Germany even if the Italian cable from Rio to Noronha were cut. He said that altered the position again and that he would inform the authorities and let me know the decision in three days' time, and if a decision had not been arrived at within that time he would arrange to let us send the *Lady Denison Pender* down on her own repairs. Commander Bowen was also present at the interview and at one time Captain Glover was quite prepared to allow the *Norseman* to do any cutting that was necessary, leaving the *Lady Denison Pender* free for our use, but he retracted this when Bowen pointed out to him that the *Norseman*'s crew was Brazilian.

There had to be considerable give-and-take, too, over the use of cable ships for urgent repair work. The cable ship of *any* of the Allies – naval, P & T Department, commercial – nearest to a reported break was expected to go and mend it with the greatest expedition. Again the costs could be worked out later. In the spring of 1940 the

Admiralty asked Cable and Wireless to let their cable ship *Cambria* repair a broken cable in mid-Atlantic belonging to the Commercial Cable Company. Could it undertake this on its way back from Halifax to its base in Plymouth? Its restoration was 'a matter of national importance'. They had failed to get a Post Office ship or the Commercial Cable Company's *Marie Louise Mackay* (also based at Plymouth) to go to the scene. The trouble was finding an escort. The company agreed.

The peacetime competitor was now the ally, or the neutral friend, of whom commercial advantage could no longer be taken. The values of the fast receding era known as Before The War (Mark 2) were being modified by the overriding considerations of defeating the enemy. But in some quarters the change was taking time.

For Sir Edward Wilshaw who had left school at 15 and spent the whole of his working life as an executive in a single company in London, the process of adapting came none too easily. Moreover, having been entrusted with supreme power, he was anxious to be seen personally wielding it, not only by the public who were 'his' company's customers, but by the university-trained officials of the Foreign Office and Treasury, and high-ranking officers of the Navy, Army and Air Force who composed the committees he now had to work with; by the members of the Pender family to whom he had been subservient for so long; and by the astute politicians of the National Coalition who had to make the decisions which would win or lose the war. For the first year Wilshaw acted alone; it was not until 1941 that he formed a 'Board of Management' consisting of himself as chairman; Jocelyn Denison-Pender, general manager; Walter Edmonds, staff manager; W G R Jacob, engineer-in-chief; Richard Luff, treasurer; and J J Munro, traffic manager; with J U Burke as the Board of Management's secretary. They met every day. But Wilshaw always handled negotiations with the Post Office personally.

On 1 September 1939 he discussed with Sir Raymond Birchall and F W Phillips the charge to be made for soldiers sending overseas telegrams, and they settled for a flat rate of 5d a word. When Sir Raymond told him the Post Office were abolishing the Urgent rate, Wilshaw thought this would harm Cable and Wireless.

At the beginning of 1940 the company introduced Expeditionary Force Messages (EFMs). These were not free but paid for at the rate of 2s 6d (12½p) for a minimum of twelve words – half the rate suggested on 1 September. A list of 240 standard messages was drawn up, each of which was given a code word. They were based on those actually sent to troops in the Middle East. On receipt of the code words the receiving station had to transcribe them into the phrases of which the most popular was 'Loving Birthday Greetings'. Second most used message was 'All My Love Dearest'. 'Am Well And Fit' and 'My Thoughts Are With You' were close runners-up. It was a wartime service which made little profit, added to the work of over-burdened staff and greatly increased the already inflated volume of Government and Press traffic, but it was greatly appreciated.

It was a surprise to many therefore that when, on 22 July, Wilshaw heard that the Admiralty had decided – without consulting him – to introduce the following week (on 29 July) a scheme whereby naval wireless stations transmitted in code 'short messages of an urgent domestic character' from their families in Britain to commanding officers of bases abroad, he took strong objection. It was to be strictly rationed to two messages a month and was of course free. He only came to hear of the scheme because the naval network was going to be unable to handle the traffic, and the Admiralty asked the Post Office to ask Cable and Wireless if they would handle the surplus by the EFM service. What upset Wilshaw further was the news that the War Office and Air Ministry were planning to do the same.

It was a question of principle. The scheme was to be deprecated as contrary to the provisions of the 1928 Report. In his letter to F W Phillips of 25 July he asked for the matter to be reconsidered, and for all coded telegrams outwards from Britain to naval personnel abroad to be transmitted over the channels of 'my' company. The Postmaster-General thought Wilshaw was wrong, and Sir Raymond Birchall told Jocelyn Denison-Pender on 27 July if there was to be a dispute on a question of principle, the quicker and more definitely it were decided the better. Wilshaw wrote to the Advisory Committee to 'protect the legitimate interests of the company', but of course it was too late to stop the Admiralty starting the service on 29 July.

On 7 August the Post Office told Wilshaw that the Air Ministry and War Office were both starting a similar service on 12 August. On 8 August Wilshaw wrote to the Prime Minister, Winston Churchill:

> Sir, I am informed that the absence of news from home, owing to the present heavy delay in postal correspondence, is considered to be an adverse influence which might affect the morale of His Majesty's Forces serving at certain stations abroad, particularly in the Near East.
>
> In order to help in such a matter I should be happy to institute an

entirely free telegraph service available from and to the Forces in the Near East and their relatives at home, provided that the British Post Office would handle such traffic free of charge to us and to the public in this country.

I hope this would contribute substantially to the maintenance of the morale of the Forces.

The Prime Minister's office sent the letter to Lord Lloyd for consideration by the Imperial Communications Committee, and Sir Campbell Stuart was asked to sort the matter out. Service and government departments were unanimous in the opinion, Stuart told Wilshaw on 2 October, that an entirely free telegraph service to troops in the Near East would be undesirable. But they all wanted a reduced rate homeward service; would Cable and Wireless share receipts with the Post Office on this? Since it threatened to endanger the whole telegraph rate structure, he would do no such thing, he told Stuart on 3 October. Moreover he withdrew his free offer. Lord Lloyd regretted the impasse.

As I understand it, the essence of this alternative proposal is that your Company would take over from the three Services the existing CSN service in the outward direction, while at the same time instituting a homeward service at the same rate, viz. the British Inland Telegram rate; and that the proceeds would be shared as to 50 per cent each between your Company and the British Post Office.

I am bound to say that if this does not commend itself to you the three Services will undoubtedly themselves immediately institute a homeward service as a counterpart to the existing CSN outward service. I feel that, as a matter of courtesy, I should explain this to you, before the Services take action, so that you may be aware of the implications of the present position. The Services regard this as an urgent matter.

I am aware of the letter of the 6th August which you have addressed to the Imperial Communications Advisory Committee, in regard to the institution of the CSN service, but I am bound to point out to you that in war considerations sometimes arise which must over-ride all other considerations (11 October).

The emphasis now swung to incoming traffic. In August 1940 began the series of all-night air raids on London known as the Blitz, the lightning war designed to knock-out the empire's capital at a single stroke. The men at the front were anxiously cabling for news of events at home. At a time when messages had risen to a peak the means of carrying them was severely crippled. When a bomb fell on Hammersmith on 16 August it blasted the land line between

Moorgate and Porthcurno. The Ealing emergency station was switched into circuit at once. Ongar wireless station was hit in September; and on 29 December an incendiary exploded in the lift hall of Electra House, Moorgate, and a high explosive bomb and a landmine fell in nearby Fore Street. The whole area blazed, and in the morning telegraph and telephone wires lay in the gutters. But the building had held together, and as there was no return raid the next day, London Station was patched up to resume its vital role. On the same day the wireless station at Brentwood was hit – and its land lines to London – and for two days a motor cycle dispatch rider had to carry telegrams to Moorgate. In December, too, the line between Moorgate and Porthcurno was again punctured by bombs and landmines. Outgoing telegrams had to be taken to Land's End by courier in a car which brought back incoming messages. The Admiralty and Air Ministry offered to carry telegraph 'slip' by air or sea to Malta for onward transmission by telegraph, but the censors forbade it to go by air. With the total interruption of the Gibraltar-Malta cables, cable traffic to Egypt had to circulate round the Cape, up the East Coast of Africa and through the Red Sea. On 3 January 1941, Wilshaw told Lord Lloyd, of the five cables through the Red Sea, only one was working. 'If anything happens to the remaining cable before the others are repaired, there will be no cable communication with Egypt or the Near East.' The censors were insisting that all messages sent by routes round the Cape or across the Pacific should go by cable. If they went by wireless they might give the enemy information about the order of battle.

With traffic up, and the means of handling it reduced, Wilshaw now had to plead with Lord Lloyd *not* to have to introduce an EFM service for the troops in Egypt. The system could not cope with them. At 2 pm on the previous afternoon, he told him, there was a backlog in London of 62,000 messages yet to be transmitted – some of them handed in four days before. It was 'an appalling state of things'; but any additional messages from Egypt would not only add to the problem but give additional cause for dissatisfaction because of the delay. The position into Britain was the same. New Zealand had 33,000 messages on hand still to transmit; Alexandria and Cairo had 12,500.

However, at a meeting on 9 January, Wilshaw agreed to introduce an EFM service to Egypt from 27 January. The Eighth Army had opened the offensive in North Africa by attacking Sidi Barrani on 9 December, and two days later there arrived in Cairo the new Cable and Wireless Divisional Manager, Middle East, E S Bennett. The afternoon he arrived Bennett had a request from the Combined

Signal Board for Cable and Wireless to lay a cable from Alexandria
along the coast westwards to Tobruk in Libya. Bennett suggested
diverting the cable from Alexandria to Malta, but there was no cable
ship in the area large enough for the job. So it was decided to send
out the *Faraday,* a Siemens ship on charter to Cable and Wireless.
Having loaded enough cable for the Alexandria-Tobruk telegraph at
Woolwich, it sailed for South Wales 'for bunkers' *en route* for the
Clyde to join a slow convoy to the Middle East. However, it got no
further than Milford Haven where it was sunk by a lone Dornier
bomber, which after releasing its bombs was hit by a single shot
which the chief cook of the *Faraday* managed to fire from the cable
ship's 3·7 gun. The enemy plane crashed into the sea.

After his meeting on 9 January, Wilshaw told Bennett of the
decision to introduce EFMs from Egypt. 'I would like General
Wavell and others to know,' wrote the chairman, 'that the introduction of these cheap messages for Forces was my idea, and introduced
of my own volition. I am most anxious to see it made available for all
Forces everywhere.' But the cutting of the Mediterranean cables
explained the capacity difficulty, apart from the opposition of the
Army authorities to sending by wireless.

It seems that the company's unpopularity among the troops was
the exclusion of Egypt from the EFM scheme, and there was no
company representative on the spot to explain things. Now they
knew the difficulties, Bennett told Wilshaw on 17 January, and the
service had been introduced, all was well. 'Feel now that best policy
here, having obtained strongest goodwill and cordial personal
relation, is to bury the past and start afresh. Can see no useful
purpose raising questions which likely rankle, and trust you will
support me in this attitude.'

With the Allies taking the war aggressively into enemy territory,
the risk that Germany and Italy would retaliate by invading
England grew less of a fiction writer's fantasy and became a
contingency for which practical plans had to be worked out well in
advance of 'occupation'. On 6 January 1941, Sir Edward Wilshaw
wrote to Lord Lloyd:

> It has occurred to me that in the improbable event of a German
> invasion of Cornwall immeasurable damage would be done if our
> Porthcurno Cable Station was captured and used for the dissemina-
> tion of false news to the Empire and abroad. I suggest therefore that
> consideration be given, in conjunction with the military authorities,
> so that the station could be destroyed rather than allowed to fall into
> the enemy's hands.
> In order to safeguard the position still further, I could arrange that,

in the event of an emergency, Porthcurno could warn branches abroad, by the use of a pre-arranged code word, that any messages received thereafter should be disregarded unless another pre-arranged code word has been sent at intervals of, say, a quarter of an hour. I should, of course, have to notify in advance the branches abroad of this latter proposal, but I think any suggestion of panic could be avoided by suitably preambling the message describing the arrangement.

Lord Lloyd asked the War Office to work out a plan with the company and Wilshaw discussed the matter with Brigadier G G Rawson on 20 January. At a meeting on 30 January, they decided that rather than have the station completely destroyed it would be 'rendered inoperative' and that similar arrangements should be made for the American cables landing at Sennen Cove, Weston-super-Mare and Dartmouth. Wireless stations too would be put out of action if any of the enemy were seen approaching. Lord Moyne, who succeeded Lord Lloyd as Colonial Secretary and chairman of the Imperial Communications Committee, wrote to Wilshaw on 12 March to say he felt sure that it was wiser to 'immobilise' the stations than destroy them permanently – 'a course which might inflict irretrievable damage on our communications if the order were given prematurely'. The memorised warning code word would be sent to the company's stations at Carcavelos (Lisbon), Gibraltar, Capetown, Fayal, St Vincent (Verde Islands), Ascension, St Helena, Madeira, Harbour Grace, Halifax, Montreal, Rio de Janeiro, Pernambuco and Buenos Aires. Only the manager would know both the 'danger' code word and the 'release' word.

Flame-throwing apparatus was installed at Porthcurno – a 150 foot long pipe with buried nozzles stretching across the bottleneck of the beach 160 feet in front of the cable hut. The pipe was fed by gravity with 'inflammable liquid' from two tanks on the hill above and when lit would discharge at a rate of sixty gallons a minute and throw a flame of thirty feet. There were fears that unburnt fuel would seep through the sand and rot the cables' gutta percha insulation – they had no lead covering. Jocelyn Denison-Pender wrote to Bell, manager at PK, on 28 January saying the company thought 'the use of the apparatus might prove extremely dangerous to our communications'.

There was a test burn, but the flame barrage was never operated in earnest. There was never an opportunity to implement the immobilisation procedure at PK except as part of an 'exercise', but for the manager of the Cable and Wireless station in Athens which the Germans entered in April 1941 it was 'for real'.* He had his wireless station (opened in February 1940) keep transmitting to Somerton as long as it could, which was until half past nine the following morning (25 April) when the English station received the dramatic message 'The Germans are now entering the building. We are closing down. Goodbye. Goodbye'. As Charles Graves reports in *The Thin Red Lines,* 'at Somerton the emotional impact of this message left the operators speechless'. A B Smith, the Athens manager, was trapped in his office soon after this final transmission, and the Germans who entered the building arrested him. He took the decision not to immobilise the wireless station. He was allowed to return a few days later to collect the Greek employees' pay which was in the safe. The sets were handed over to the German Telegraph Company who re-tuned them to Vienna.

Mr Jones, the company's manager at Candia had to make the same decisions when a few days later the Luftwaffe dropped bombs on Crete. He removed his wireless sets to the cave where the British Army had established their headquarters, destroyed his code books and escaped inland. He managed to trek nine miles, but after ten days was captured.

There was no parachute invasion of the company's Central Telegraph Office at Moorgate, but on the night of 10 May 1941, a shower of incendiaries descended on Electra House and Tower Chambers which soon made them a burning furnace. The heavy bomb which hit next door Salisbury House delivered the *coup de grâce.* A number of synchronisers and keyboard perforators were salvaged, but the Tower Chambers fuel store with 2,000 gallons of paraffin, oil and petrol prevented the fire brigade from entering the main building. By ten o'clock next morning it was obvious that Central Telegraph Station was inoperable and would have to be abandoned.

There were no human casualties, and operational staff retreated as planned to operate the circuits installed in The Fortress, the more cramped, alternative London Station, at Electra House on Victoria Embankment. It had also had to find room on the top floor for Mr Smith (The Prawn), the principal of the London Training School with his staff and students, recently moved to Moorgate when the Hampstead building was requisitioned by the GPO as a sorting office. Sir Campbell Stuart also had his office in Electra House, and

---

*'Reliance in the case of overseas branches' stated the instructions drawn up in February 1941, 'should be placed on the initiative and resource of the local managers to gauge the position at the branches with which they are in direct communication, and in the case of the branches invaded on the ability of the man in charge of the cable station at least to give warning that an invasion is taking place.'

not only for his work as chairman of the Imperial Communications Advisory Committee. He was the head of the secret department which conducted 'black propaganda' to enemy countries for the Foreign Office, the Political Warfare Executive.

No one had thought of preparing a place of withdrawal for the company's station on Hong Kong island because, along with everyone else from the Governor and army commander downwards, they were confident the colony would never fall to Japanese troops which were believed to be ill-disciplined and badly trained, and no match for the Scots and Canadians who formed the garrison. It was generally believed that everyone would see the war out in a colony where life would continue much as usual, changing somewhat probably when it became the base for the Allies to mount the campaign which eventually they would have to fight in the China Seas. But with the Royal Navy and Royal Air Force too busy elsewhere to send the island succour, the order went out from Whitehall to delay the inevitable as long as possible by a defence which the War Office recognised could never succeed but would impress the Allies – and the Axis – and tie down a Japanese force which otherwise might be adding to the serious position in Malaya.

When aircraft flew over Hong Kong on 8 December 1941 most took them for RAF planes on another of their exercises, only flying rather lower than usual. When bombs started falling from them, they thought it must be some terrible mistake. The Japanese had dispensed with the niceties of declaring war when, the day before, they bombed the neutral United States fleet as it lay in its Hawaiian base of Pearl Harbor. For the next seventeen days they lay siege to Hong Kong. They made their first penetration through the 'Gin Drinkers Line' across the land from China into the New Territories and then stormed the island.

For four days after 8 December, J Sefton Jenkins (Jenks), the Cable and Wireless manager, and his staff were able to transmit Government messages telling London and the rest of the world what was happening. But on 12 December the Peak Wireless Station, of which H S Rees was in charge, was heavily shelled, eventually knocking it out altogether. Cable and Wireless staff wrecked the Cape d'Aguilar transmitter only days before the Japanese made a landing less than a mile away. A new SWB 8 Major transmitter had been in commission for ten days only, but they threw the Franklin oscillator into the sea and filled the transformers with sulphuric acid. Looters later removed the motor generators. In the office in the town the staff made sure that all the equipment was rendered useless.

Hong Kong surrendered to the Japanese at three o'clock on the afternoon of Christmas Day 1941. Herbert Ascough, the chief engineer who had joined Eastern Telegraph in 1914, was taken hostage and driven to the building where the Japanese had set up their high command. Here, in a tense interview, Ascough refused to 'hire' his services to the Japanese admiral who questioned him. So he and his wife, who during the seventeen-day Battle of Hong Kong had been working in a hospital, were led away to the civil prison at Stanley which the Japanese had made a Military Internment Camp. They joined the other seven British members of the Cable and Wireless Hong Kong branch to spend the rest of the war as 'internees' in the cramped Room 10, Block 14. The Chinese employees went free, and the fifty or so Portuguese and Indian staff withdrew down the coast to the Portuguese island of Macao which the Japanese never occupied – a British Consulate continued to operate there – but used as an army recreation centre. Eastern Extension, by which name the company still operated, had a branch in Macao under Clemente Demée.

The company employed a few Irishmen who, as neutrals, were also not interned. One of these was Joe Leonard whose father was an Irish sailor whose four-masted sailing ship had put in to Hong Kong at the turn of the century. He had liked the place and decided to stay. He married a girl from Hong Kong and joined the Hong Kong police. Joe Leonard who had joined the company in 1919 (and retired in 1964) was on leave at his home in Kowloon when the Japanese units crossed the Chinese frontier. He went to Macao with the other non-internees for the first week. Here the Hong Kong office accountant, a Portuguese, Fideles Rosario, had taken over the branch and been given the responsibility by London head office of paying every member of the staff employed on 8 December one-third of their wages until the war ended. Head Office also allowed him to spend company funds in any way he saw fit to relieve the suffering of the interned expatriates. Joe Leonard returned to his home in Kowloon, and Fideles Rosario arranged to send him 'military yen' (the Japanese Occupation currency) through money-changers who regularly took the ferry to Hong Kong, with which to buy food to supplement the rations of the eight British staff, and Mrs Ascough, at Stanley. Joe received telephone calls from strange rice dealers whom he invited to tea, during which the money was handed over. Internees were allowed one parcel a month, and they sent their requirements to Leonard on a postcard – peanut butter, corned beef, flour, bacon, egg powder, prawns, the solid slab of molasses known as wongtong which acted as sugar. With the help at first of F Silva,

Joe bought the items in the market, took them to Stanley and handed them over. The diet in the camp – mostly rice – was atrocious and became worse as the years went by. There was widespread malnutrition and illness. Those parcels probably saved their lives. Joe's only anxious moments were when he was asked where he got the money to pay for all those parcels. He told them he was working as office boy/coolie in the office of one of the Japanese army commands which was true, and was earning good money, which was not.

Inside Stanley, Jenkins and Ascough plotted to smuggle out a message to London giving head office details of the damage they had done to the apparatus in the town office, and what should be sent to Colombo or Chungking in order to repair it when the war was over. Herbert Ascough wrote a thousand word description on a piece of toilet paper which he gave to a Lieutenant Colonel Doughty, a Canadian Immigration Officer from Calgary who was secretary of the Hong Kong Rotary Club of which Ascough was a member. Doughty was being repatriated along with some Americans with diplomatic immunity. At the suggestion of Jock Frazer, who had been Minister of Defence in the Hong Kong Government (and was later executed by the Japanese along with H S Rees), Doughty concealed the message in the heel of his shoe. He survived the thorough examination they gave him before boarding the ss *Hirano Maru* en route for Goa where the party were handed over to the Swedish ship *Gripsholm*. They thought this was going to Cape Town and Ascough had addressed his message to Loverstock, the company's Divisional Manager, South Africa. But instead the *Gripsholm* sailed to Rio de Janeiro where C J Ellis, the Western Union manager in Brazil, took the shoe, read the message and immediately telegraphed the essence of it to London. The long list of equipment needed to repair the sets followed by airmail – and at the required moment four years later it was ready waiting for shipment.

Japanese tactics at Hong Kong made little impression on the defenders of Singapore where any attempt at capture was also expected from the sea. Penang was bombed from the air soon after the attack on Pearl Harbor. The Cable and Wireless station carried on mainly with Chinese staff. When the office was demolished by a bomb on 16 December, the cables were 'joined through' and the European staff went to Singapore where they prepared for the worst. 'Japanese near. All staff on duty. No orders re destruction gear' read the message from the company's manager in Singapore received at the new London Station, Victoria Embankment, on 11 February 1942. Two days later came: 'Now closing down. Goodbye. Most unlikely able evacuate. Please inform wives'. Later that evening: 'Closing down permanently. All staff well'. The island city surrendered to the Japanese on 15 February, and with it the company's main cable junction to the Far East, Australia and New Zealand. Preparations were well advanced to establish a new wireless station at Singapore to augment the existing facilities, and when it fell it was decided to change the site of the proposed complex to the island of Barbados in the West Indies.

'The staff at Dover,' recollected Jimmy Cozier, 'suspected something was up when within weeks (at least so it seems in retrospect), a long coded service message came in from Head Office for DM/BAR. The code groups were twice interrupted by obvious positional figures, showing latitudes and longitudes. Curious staff members quickly got out their maps to study these figures. Sure enough, they were in central Barbados, and pinpointed after some debate in St Philip and St George valley. Before the end of 1942, lands had been acquired and shiploads of equipment, all addressed to Singapore, had arrived in the West Indies. Mr K D Coombes reported as Project Engineer and construction work began. Within a year the new stations at Carrington, St Philip, and Boarded Hall, St George were operational. A hundred Barbadian artisans worked feverishly to get the building erected on time, and the technicians, some seconded from among the operators, kept pace with them installing equipment almost before the concrete foundations were dried out.'

Communication between London and Melbourne via the Barbados relay station was established in October 1943. The receiving station at Carrington was officially opened by Mrs A G L Douglas, wife of the Divisional Manager, West Indies, on the second anniversary of the fall of Singapore; and on the same day the transmitting installation at Boarded Hall was opened by the Governor of Barbados, Sir Grattan Bushe, who in his speech referred to the company as the silent service which was continuously at work to guard and extend 'those essential communications without which the courage and exertions of our men at arms would be of much less avail. By land and sea, in battle-threatened territories and in perilous waters, the men and women of Cable and Wireless are carrying on their vital task, often in the face of constant danger and with little public recognition'. A similar relay station for counteracting fading was created at Colombo.

Staff on Cyprus, Cocos, Bermuda, Ascension Island, St Helena and Mauritius were luckier than their colleagues on Singapore. A Japanese warship shelled Cocos, which was defended by a company of Ceylon Light Infantry, and hit the cable station, but no one was hurt. The ship hurried away in the belief it had delivered a knock-out blow. But communications were re-established with Durban by next morning. The enemy spotter plane which flew over to confirm Tokyo radio's report that the island had been split in two, was deceived by the fake shell holes which the Cable and Wireless men painted on the roof to maintain the fiction that they were incommunicado.

As a result of the Japanese successes, eleven of the company's stations and 18,000 miles of cable route remained in enemy hands from 1942 until the defeat of Japan in 1945. Some of the traffic had to be shunted via Canada and the Pacific, and the rest by the west coast of Africa, Cape Town and Cocos Island, which became the company's most eastern outpost.

11

Truncation

1942–1946

The unforeseen attack on Pearl Harbor which brought America into the war on the side of the Allies accelerated the process, initiated in the Thirties as seen, whereby the United States Government was allowed to establish direct wireless communication with the Dominions, India and the colonies. It was contrary to the policy endorsed by the Empire governments in 1928 and 1938, but now that the United States was an ally directly concerned with the Empire's war strategy she had every reason to demand access to wherever she would be sending troops, ships and supplies, particularly the exclusive world within a world which on the map was still painted red. The British Government had no hesitation in acceding to their powerful, new co-belligerent's request – but, alas, in their hurry they forgot to notify Electra House. It came as 'a profound shock' to Sir Edward Wilshaw to learn from the letter he received from the chairman of the Imperial Communications Advisory Committee on 4 February 1942, that HMG had agreed with the US Government the principle that they should have direct wireless telegraph circuits to any British colony for defence reasons for the duration of the war. 'The seriousness of the position in my view cannot be exaggerated,' he told the Permanent Secretary of the Treasury on 27 February. 'I cannot believe that full consideration was given to all the implications arising from such a decision before it was made. I should also have thought that, as Government-approved chairman of this company, some intimation might have been given to me before the decision was finally made.' He was personally affronted. Over five pages of foolscap he pointed out how, since his appointment, the company at his instigation, and with some success and satisfaction to all, had carried out all the Government's policy as laid down in the various documents and White Papers. 'And I have been able to bring to bear on all the company's problems the full experience of nearly fifty years in the overseas telegraph service.' Even in wartime the Government should give the Government-approved chairman information on suggested changes in HMG policy 'otherwise the position of any Government-approved chairman may well become untenable'.

By now Sir Edward appreciated that war considerations justified much that would not have been tolerated before 1939, but he feared that they might propel unscrupulous operators into positions of vantage from which it would be extremely difficult – and 'unpatriotic' – to dislodge them once peace was restored.

In a review of his visits to the US which Sir Campbell Stuart set out in a Memorandum dated 13 February 1945, he noted that the American commercial wireless interests which were in competition with each other saw the opportunity of enlarging their field of activities. Fearing a Government-imposed merger of all commercial companies, they each put pressure on the State Department to secure concessions in the British Empire. Sir Campbell discussed the matter with Sir Edward Wilshaw 'who felt the Advisory Committee should adopt a *non possumus* attitude, despite the changed world situation and our very much closer relations in all fields with the USA due to the war alliance'.

In the event the Advisory Committee agreed to the direct wireless circuits on condition they charged the same rates as those of Cable and Wireless for carrying messages from London by cable, though lower rates could have been charged by the direct wireless routes. This was a cause of ill feeling in America who considered the condition was imposed purely in order to protect the revenues of Cable and Wireless. There was a general complaint of excessive delays and abnormally high rates.

When Stuart went again to North America in May 1942 a scheme was before Congress for the merger of the competitive internal telegraph companies, and there was pressure to follow this up with a merger of the American overseas telegraph interests.

> It was being represented that the British interests with government backing were thwarting the rights of the American people to cheap, direct communication with the world, and that America must therefore take steps to ensure that she would emerge from the war in a position of supremacy over the British.

Before they entered the war the US Government had been refused permission to establish a commercial wireless station in the UK.

After Pearl Harbor the request was renewed but again rejected. In a letter of 29 June 1942, Lord Cranborne, the Colonial Secretary, told Sir Edward Wilshaw that in view of the large number of US armed forces in the British Isles the matter of a Services wireless link with America had become acute. So that they should not have to erect their own Services wireless station, the Post Office were placing four channels at their disposal from 1 July. At the same time Sir Raymond Birchall told Sir Edward that in view of Lease Lend they were not raising any charge against the US Forces and that the service would be operated initially by GPO telegraphists. It would not be a commercial service, and there would be no room for social telegrams like EFMs.

By October, American troops were spread out in all parts of the British Empire, and the requests for an EFM service from such points to North America became more and more pressing. On 2 October, Sir Alan Barlow of the Treasury told Wilshaw he had had another telegram from Lord Halifax stressing the urgency of the problem, and asking Cable and Wireless to introduce a service on the lines suggested by the Americans (2s 6d or 60 cents for twelve words). Receipts would be paid into a common pool. On 7 October, the Treasury sent Washington a cable saying it was physically impossible for Cable and Wireless to carry all EFM traffic by cable, particularly over the Pacific and Middle East routes. There must be freedom to route EFM messages by wireless. If that was conceded, and the traffic was confined to messages originating in the US or a Dominion, the company would introduce the service at the requested rate of 2s 6d/60 cents for the duration of the war and six months after it, but other parties in the Dominions would have to agree.

The Dominion with which the United States now had the closest association was the autonomous Commonwealth of Australia whose ties with the Mother Country prompted immediate support for the stand which she and the rest of the Empire had unilaterally decided to take against the dictatorships. When America joined the fight, however, it was natural that politicians in Australia should look to her rather than to Britain. Up to 1941 Australia had adhered strictly to the undertakings it had given in 1928 and 1937, and 'for reasons of national policy' resisted American pressure to issue licences for more direct wireless services. Amalgamated Wireless Australasia had negotiated an agreement with Colonel Sarnoff of RCA but after the 1928 conference, much to RCA's disgust, the Australian Government had refused to sanction its opening. But circumstances altered with the outbreak of war in the Pacific, and the postponed Sydney-

San Francisco service was inaugurated on 26 December 1941. The Australian Government had no wish for Cable and Wireless to be associated with this service and no intention of withdrawing it on the conclusion of hostilities. Early in 1942 the US Government asked for two more US-Australia direct wireless services, and approval was given for a Sydney (AWA) to San Francisco (Mackay Radio) service which opened on 13 April 1942. The opening of a Sydney-New York (RCA) link was only delayed through the inability to secure plant. As D McVey was to say a couple of years later* these services were of 'incalculable value' to Australia. It would be extremely embarrassing to discontinue them if the US Government, 'on whose assistance the preservation of Australia from invasion largely depended', pressed for their retention. The enemy had already seized or interrupted the submarine cables connecting Darwin, Java, Sumatra, Singapore, Labuan, the Philippines, Hong Kong, Shanghai, Penang, Madras and Colombo. Moreover, Australia could not pay Cable and Wireless 'compensation', as that would establish a precedent for doing so on every occasion that, 'in the exercise of its sovereign rights', the Australian Government deemed it necessary to license additional circuits.

It was evident to John Curtin, Australia's Labour Prime Minister, that Wilshaw misunderstood Australia's motives, and he invited Wilshaw to come to Australia so that he could explain the thinking which lay behind the kind of communications company they planned for Australia. He hoped that the conference which he proposed calling in Canberra to review the whole new situation which had developed since 1928, and in particular since 1941, would also be attended by Sir Campbell Stuart.

Stuart had already arranged to go to America, and Wilshaw declined the invitation because, as he wrote to Stuart on 31 August 1942, they felt it was an inopportune moment to enter into long and complicated negotiations of this nature.

> The Court of Directors consider the proposals regarding the composition of this proposed company constitute such a departure from precedent that they could not authorise any representative to commit them to a new company on the lines indicated, which go much further than the provision of the control which every Government rightly possesses over the communications companies serving their respective countries.

---

*On 4 April 1944 at the second session of the conference held in London by the Commonwealth Communications Council.

The company was amateurishly conceived and over-capitalised, claimed Wilshaw. It would give the Government too much control. In any event such a radical change should not be carried out in wartime and under the current emergency conditions. Perhaps Mr Curtin would stop off in Britain on return from his visit to Mr Roosevelt in Washington, and the communications position could be explained to *him.*

When Stuart returned from America he told Wilshaw 'I suppose we ought to think about going to Australia'; but Wilshaw said he thought the matter was dead. It was agreed that a revision of rates was impracticable in wartime, but Curtin was still anxious for a general discussion of the Empire's policy on communications as a whole and how the obligations of the governments to Cable and Wireless were to be interpreted in the present conditions. So he went ahead and called a meeting of heads of the Telegraph Departments of the Empire which sat in Canberra. Sir Campbell Stuart was asked by the British Treasury to go to Australia as its representative and to chair the conference. When the Treasury asked Wilshaw if he was willing that while Stuart was in Australia he should negotiate over the new combined wireless and cable company, the chairman of Cable and Wireless said he did not think that was the way such a matter should be handled.

His absence from the conference was greatly deplored. He submitted certain points in writing, seeking assurances that the direct wireless circuits which Australia and New Zealand had opened with the USA – and India and other colonies were known to be planning – would be limited to the duration of the war and to terminal military traffic. He also raised the question of compensation, but the delegates decided they could not consider these points without Sir Edward being present. 'We are bound to say' stated the report of this 1942 Australian Telegraph Conference dated 22 December, 'that the absence of a representative of the company endowed with plenary powers is a matter for much regret and has been a serious handicap to us in our labours (paragraph 12).'

However, HM Treasury issued a Memorandum stating that the agreements of the 1928 Conference and the 1937 Empire Rates Conference were not such as to justify the company claiming for compensation on legal grounds. They did not constitute the grant to Cable and Wireless of a monopoly of telegraph communications. To impose such an obligation on the Partner Governments would be ultra vires, as not within the power of any government to seek to fetter the rights and duties of the Executive in the future. But apart from the legal position, the opening of direct wireless telegraph circuits had involved a material departure from the principles so far pursued, and Cable and Wireless were entitled to security of tenure, which alone enabled them to make improvements in their services and organisation.

The conference considered there was considerable force in this view. They would ill serve the Empire, they said in their report, if they adopted a policy which had a crippling effect financially on the company. It was essential to maintain a strong and virile Cable and Wireless, not because of the financial interests of the company, but because such a company would best serve the common needs and widest interests of the Empire. The Partner Governments had a moral obligation to Cable and Wireless.

Nonetheless the conference recognised the need for war-time wireless circuits with the USA and their full exploitation. 'We cannot resist the conclusion,' stated the report, 'that our present system is related in too great a degree to the expensive cable system built up over many years, and that this may have led to the British Commonwealth failing to maintain the pre-eminent position in radio which it had attained a few years ago.' The US were establishing radio circuits all over the world and the achievements of the British Commonwealth in this field lagged behind.

The general consensus of those attending the Canberra Conference was that the problem related less to compensation than getting the management of Cable and Wireless to shed its Eastern Telegraph/imperial thinking, and its rigid adherence to its 1928 'charter', which inhibited efficient co-ordination of cable *and* wireless. The forward looking policy adopted by the Imperial Communications Advisory Committee – the 'standing committee' which maintained continuity of thinking between conferences – had only been carried out at the expense of serious disagreement with Cable and Wireless. They recommended that the Committee be re-constituted as The Commonwealth Communications Council with its chairman and secretariat in London but with its individual members residing in their own countries.

Cable and Wireless protested that this 'unilateral change' was unworkable and would lead to friction. With its members so dispersed it could never function as a committee, and the channels between governments and company would be disrupted. Wilshaw's main objection, however, was that it gave too much power to Sir Campbell Stuart, as chairman of the new Council. But there was nothing he could do to prevent it coming into being. Indeed, a copy of the conference report to Commonwealth Governments was not sent to Electra House.*

Far removed from the squabbles at base, Cable and Wireless staff never hesitated to take action on behalf of any of the Allies, and for any of the Services, in order to further military operations in aid of the common cause – very little of which was allowed for in the charters of 1928 and 1938.

Just before the landings in North Africa on 8 November 1942, the company sent a large wireless transmitter to Gibraltar where the Combined Operations 'Africa Command' continued to control the operation for some time after General Eisenhower and Admiral Cunningham had moved their headquarters to Algiers. The company also sent ex-Direct Spanish staff to Gibraltar to assist the Royal Corps of Signals in the tunnel station in the Rock. The old Eastern Telegraph staff quarters had been closed after so many had gone home following the D/N Scheme, but much of the furnishing had been put in store and was now retrieved. The carpets, as Roy Baker recollected in 1978, were alive with moths and chewed to pieces. When they opened the library/billiard room which had been closed for ten years, half a glass of solidified Guinness was standing on the dusty mantelpiece. The billiard table cushions were perished; the tennis court was overgrown with grass. Roy and his colleagues got down to weeding it and mending the net, but on the day of the planned opening game the Army erected a huge marquee on it for a sergeants' Mess.

When the landings began and a Royal Navy transmitter developed a fault, the company lent them a set to maintain contact with the troopships as they moved in. E S Bennett and M T Kempson were flown to Algiers on 16 November and by the following day had established communication from Allied Force Head Quarters (AFHQ) to Gibraltar by cable laid by the company's cable ship *Mirror*. By 2 January 1943, the company had laid a cable from Gibraltar to Casablanca. The telegraph office in Algiers was manned by a mixture of nationalities. Americans handled the receivers, Royal Signals men prepared the punched tape, Cable and

*At the annual general meeting of 29 June 1944, Sir Edward Wilshaw told shareholders about the 1942 Conference 'as a result of which *we understand* that a report was presented to the Commonwealth governments! They were subsequently advised of an alteration in the constitution of the Advisory Committee, he said. If support for the empire policy which the company was created to implement was withdrawn, and its financial resources diverted into other hands, he said, empire communications would suffer disastrously.

Wireless staff operated the transmitters, men of Eastern Telegraph (France) looked after the local lines to the French Post Office. The supervisor was an American lieutenant with a mania for tidiness but no knowledge of telegraphy, and, as Roy Baker recollects, spent most of his time writing home to Mom and Pop.

The 'civvy' status of Cable and Wireless staff operating in battle areas now gave rise to anxiety. It was pointed out that telegraphists in flannel bags and coloured shirts who accompanied the forward troops invading Sicily and Italy in July 1943 could be shot as *franc tireurs* if they were captured. There was also embarrassment over their use of service canteens, accommodation and transport. As 'civvies,' army types tended to look down on them. So after some hesitation, the Army Council agreed that Cable and Wireless staff should be enrolled in a uniformed organisation designated 'Telcom' which gave them military status equivalent to that granted war correspondents. Quite a number of female operating staff in London volunteered for service overseas in Telcom. They all wore khaki battledress, and had access to Officers' Messes and travel facilities. There were no actual 'ranks' but the company's appointment of Divisional Manager was equal to a Lieutenant Colonel, a Manager a Captain and a Senior Operator a Lieutenant. A full account of Telcom's exploits is given by Charles Graves in *The Thin Red Lines.*

Oliver Stanley, the Colonial Secretary, paid a tribute to the work the company's staff were doing at a luncheon in London on 26 January 1943.

> When the end comes (he said), when victory is won, then history will begin to assess the merit. We shall all of us be searching our conscience. We shall be discussing who succeeded and who failed . . . I have no doubt at all, when we come to discuss the part that Cable and Wireless has played, what the verdict of the nation will be – "Well done, thou good and faithful servant!"

But the end had not come yet, and the 63-year-old Sir Edward Wilshaw who, at this time of petrol rationing had taken to riding to his appointments in London in a hired, yellow hansom cab, was beginning to feel the strain. That spring he had a meeting with the man who, as Sir John Reith, had been Director-General of the BBC from 1922 to 1938, became chairman of Imperial Airways, an MP on the outbreak of war and made Minister of Information, then Minister of Transport; created a peer as First Commissioner of Works, and then in 1942 been attached to the Admiralty in the rank of Lieutenant-Commander to undertake various special jobs. Reith was with the Admiralty on 22 April 1943, the day his diary entry read:

> Lunched with Wilshaw who asked if I would like to join the Cable and Wireless Board, and who also said he thought he ought to retire soon [as Chairman] and leave someone else to put to rights the anomalous position in which that concern is.

The anomaly was Sir Edward insisting that the Treasury Agreement must still be the company's charter, in spite of the fact that the Treasury and every Partner Government, as demonstrated by the Australian Telegraph Conference of 1942, were anxious to free him from it. Each departure from what he regarded as holy writ, he saw as sabotage and retrogression; they saw as release and progress.

To be fair, Wilshaw had not seen the report of the 1942 conference, as noted, and on 13 September 1943, Clement Attlee, the Dominions Secretary, and L S Amery, Secretary of State for India, sent Sir Campbell Stuart a telegram saying, 'It will now be necessary to inform Cable and Wireless of the findings of the Conference [at Canberra] which we propose to do. Certain of these recommendations call for immediate action, others await the meeting of the [Commonwealth Communications] Council. We do not propose to furnish Cable and Wireless Limited with a copy of the report which in view of its confidential character should remain a secret document'. Two days later (15 September) Attlee wrote to Wilshaw telling him that the Australian Conference had greatly regretted his inability to be present. However, immediate action was required on certain of the decisions it had taken, one of which was that the Partner Governments had agreed that, as regards the direct wireless circuits which had been established in the war and would stay operating after it, transit traffic should be admitted at each government's discretion. However, the new Commonwealth Communications Council (CCC) was to meet on 1 March 1944, 'and you are asked to be available at that time' when general problems would be discussed and a complete review made of the whole communications system of the British Commonwealth, including the claims of Cable and Wireless for compensation.

In his reply (of 15 October) Wilshaw said he reckoned the latter would amount to £1,000,000 a year. If it became a general rule that the Partner Governments were allowed to admit transit traffic on their direct wireless services it might have grave repercussions. 'The practical effect might well be that New York would inevitably tend to usurp London's present position as the hub of the communications world.'

Wilshaw's wish to retire, confided over luncheon to Lord Reith in April, was not to be granted; but he succeeded in winning that formidable character as a firm friend and ally. 'I had known Sir Edward Wilshaw of Cable and Wireless for many years,' wrote Reith in his autobiography *Into The Wind*.*

> He seemed to have an admiration for the BBC; occasionally we had lunched together and talked about problems of organisation and administration. He had once or twice thrown a fly over me. In December 1943 he had called on me in Admiralty to tell me that at a special meeting of his directors that day he had been authorised to invite me to join the boards of Cable and Wireless and associated companies. Little time would be required of me; he hoped very much I would accept; it would be a great satisfaction to him personally.
>
> Cable and Wireless was a concern of high importance and interest; on this score, apart from any possibility of giving more time to it later on, the invitation was attractive; I appreciated it greatly. But, so driven in Admiralty, I could not undertake more than the minimum required of a director by way of attendance at boards and occasional committees. Though that was all that Wilshaw required, it was what I had always disliked; if a director is to be of real use to a business he must give much more to it than that. I did not know what answer to give.

He consulted Sir John Anderson, the Chancellor of the Exchequer, who told him the Government was much concerned about the company and its future, about its relations with the Dominions and the relations between Wilshaw and Campbell Stuart. It was agreed that he should retain his Admiralty appointment and accept a place on the Cable and Wireless Court of Directors. Reith was pleased at the prospect. 'At Electra House it was obvious that matters were blowing up to a crisis' he wrote in *Into The Wind*. (p 497)

> The directors felt that the change in composition of the advisory body put too much power in Stuart's hands; they had been at odds with him for a long time. In any event they felt they should have been consulted about the change, since they had a contractual relationship with the Commonwealth Governments in respect of the advisory committee. The Treasury, however, appeared to have accepted Stuart's view of Wilshaw and his management of the company; even of his desire by some means or other to have Wilshaw replaced.

Four other directors were appointed at the same time: Sir John Wardlaw-Milne MP in London; E G Brooke, Director Resident in

---

*Hodder & Stoughton, 1949

Australia; Brigadier H J Lenton, one-time Postmaster-General, Director Resident in South Africa; and (at last) A H Ginman, Director Resident in Canada.

With the war seemingly about to end, speculation on the shape of international communications in the new world that lay ahead acquired an air of reality hitherto lacking. It was everyone's confident hope, Wilshaw told shareholders at the 1944 AGM, that their next meeting would be set in calmer days. In the urgent preoccupation with wartime activities they had not been unmindful of the tasks that lay ahead.

What were the company's plans for the future? asked Campbell Stuart of Cable and Wireless on 23 February 1944. Broadly speaking, replied E K Jenkins in a letter of 31 March, the company considered the cable and wireless communications system existing prior to the war more than adequately secured the requirements of the Empire. 'It sees no reason to doubt that, with the restoration of the cable system and the large extension of wireless which has recently taken place the position will be equally secure after the war.' Stuart doubtless could have done with less complacency and more imagination, but will have been encouraged by the decision to establish a Public Relations Office. Wilshaw had been worried by mounting criticism from Fleet Street of delays, due to a large extent to the censors, and when he heard that Ivor Fraser was retiring as Chief Commercial Manager of London Transport, he asked him to come to Electra House to advise him on handling press enquiries. When Wilshaw had been in the Cable Ship Personnel Department he had become friends with Fraser who joined the Western Telegraph Company in 1905 and served on three cable ships, the *Norseman, Mirror* and *Pender.* Fraser had been General Manager of the *Morning Post* for a time, and at 55 Broadway had been in overall charge of London Transport's big Press and Public Relations Department. He established a similar department at Electra House, choosing Colonel J W Wellingham, an ex-employee, to run a Press Liaison Department staffed by telegraphists to supervise the company's communications for the Press, and Harold J Wilson, an experienced journalist then on the *Daily Telegraph,* to head a Public Relations Office. Charged with improving the company's public image, the new advisers stopped the lavish wine and food provided at press conferences which made so bad an impression at a time of rationing, and tactfully persuaded the chairman to stop driving about in his hansom cab which was making him the laughing stock of London and encouraging editors to write of the 'horse-age mind' of Electra House. Instead Wilshaw had the company buy him a new

Rolls-Royce each year which he personally chose at the Motor Show.

The Americans had been giving thought to possible post-war international conventions,* but for the members of the new Commonwealth Communications Council which held nineteen sessions between 3 April and 23 May 1944, the priority was to set their own house in order – the rules of the club within the club. Sir Campbell Stuart was chairman and UK representative; there were individual representatives of Canada, Australia, New Zealand, South Africa and India; and one representative of the colonial territories. Sir Raymond Birchall of the British Post Office, though not officially a member of the Council, was 'also present' at all their meetings. It was up to members to state, Stuart told them at the first session, whether they were prepared to close down their circuits with America after the war. The Americans had objected to the introduction of the Empire Flat Rate which had resulted in 'much ill feeling'.

'The question of our control over Cable and Wireless will also be present to your minds. You have seen what power over their operations we possess, and the problem is, if we possess more power, to exercise it in such a way as to give effect to our wishes and at the same time allow the company to live . . . It will be for you to consider whether any moral obligations we have towards the company can be translated into practical terms.' There was wide dissatisfaction over the service which the company was giving to the Press. Some took the view it would be better to create a British Press Wireless of their own divorced from Cable and Wireless.

At the first session of the CCC on 3 April, a paper was circulated by Sir Ernest Fisk, the chairman of Amalgamated Wireless Aus-

---

*In a report 'Special Committee on Communications – Peace Terms' which President Roosevelt showed to Winston Churchill in Washington in the summer of 1943, US Secretary of State Cordell Hull hoped that a new international telecommunications organisation would be established 'and empowered to systematize world-wide telecommunication'. It would not be of a punitive nature, all countries including enemy powers would have membership on an equal basis; but the United Nations with the largest telecommunications interests would have some reasonable form of veto power or other control. (Secret File, British Embassy Washington no G.285/1; document reference FO 115 3571, Public Record Office, Kew.) The UK Government was a signatory to the International Telecommunication Convention (Madrid 1932) and of the International Telegraph, Telephone and Radiocommunication Regulations (Cairo 1938), and Cable and Wireless had to observe their provisions and regulations. But the company could not vary its rates, or discontinue any route or service, or start a new one, without the concurrence of the Advisory Committee, now the Commonwealth Communications Council. In wartime it had to obey also all directions prescribed by Government Censorship. The chairman of the CCC had direct relations with the appropriate committees of the War Cabinet, the Directors of Signals at the Admiralty, War Office and Air Ministry, and with the heads of the Foreign Office, Treasury, Post Office and other UK civil departments.

tralasia, a Marconi-trained wireless engineer, entitled *Design For A Network of Wireless Communications Linking Together The Various Parts of the British Commonwealth.* The Council appointed a sub-committee under D McVey, the Australian representative, to examine this and report to it, with special regard to the effect it would have on Cable and Wireless. Sir Stanley Angwin, Post Office Engineer-in-Chief, was asked to advise it. Addressing the Council on 12 April, Sir Ernest said the British Empire should be provided with the best communication service possible regardless of cost. Subsidising a network for £3 million a year would be well worth it.

McVey's sub-committee were not altogether convinced that the Fisk Scheme would fulfil its object, and recommended that an alternative scheme should be prepared by consultation with Post Office engineers. The Council agreed that the expansion of wireless should not wait till the war was over, but thought it unsound to have a network for telegraphs without providing for telephone working. Two days later Lord Reith, now a director of Cable and Wireless, saw Wilshaw and, as his diary records 'tried out on him the idea of an imperial company and found him surprisingly receptive. He said that he would put it to the Board. This is very good; it is the obvious thing to do – just what I had in mind for civil aviation. It would be a tremendous thing to do'.

In his autobiography *Into the Wind* he wrote in explanation of this turn of events.

> Drastic constitutional changes were being mooted, and I had come to the conclusion myself that, without any reflection on Wilshaw or the company's performance, there was some rationality in them. It was rather sickening for me . . . But I could not help feeling that serious consideration should be given to a single Commonwealth public corporation operating directly, or by associated companies, throughout Commonwealth and Empire.

On 15 April, he told Sir Alan Barlow, Second Secretary at HM Treasury, of the idea, who said it was certainly the right thing but he had not expected it to come from Cable and Wireless. On 17 April he 'astonished' Sir Thomas Gardiner, Director General of the Post Office, with his suggestion. The following day Wilshaw attended a CCC meeting for the first time. Complaining of the refusal to allow his company a copy of the report of the Australian Conference, he said 'If this newly formed committee is going to keep the company on the perimeter, treat it without collaboration and issue fiats from time to time, then their object will not be achieved'. He ranged over the whole subject in a way Campbell Stuart had heard so many times before, but he never mentioned Reith's plan, which had now been accepted by the Court of Directors. The next day (19 April) he and Reith, with the other directors' approval, explained the plan to Sir Alan Barlow of the Treasury: a single Corporation for the whole Empire set up under Royal Charter, operating with a fixed 4 per cent interest to shareholders who would exchange their stock for Corporation stock which would be guaranteed by HMG and the Dominion governments. If the pooled revenue was not enough to pay the dividend the governments would make up the deficiency. Sir Alan duly reported this conversation to Campbell Stuart that day. Wilshaw made no mention of the Reith Plan when he again attended a CCC meeting on 21 April (though of course the Council chairman already knew of it) but he did ask the Council to consider the possibility of pooling traffic in Empire direct services, divided on a percentage basis, with 10 per cent put aside for development. On 24 April he set out the Single Empire Corporation Plan in a formal letter to the Treasury, asking whether it commended itself to HMG. On the same day Stuart wrote to Wilshaw saying he had been told of his conversation at the Treasury on 19 April, would he come and tell the CCC about it on 4 May?

On 25 April, which happened to be Anzac Day, the situation changed. The Council, wrote Reith in his diary, 'has adjourned for a week because of a bomb from the Australian and New Zealand members [D McVey and J G Young] who are reporting to their Prime Ministers that they are in favour of national corporations in each country.'

It will have come as less of a bombshell to those who knew of Australia's penchant for nationalisation, and of the intention of the Australian and New Zealand governments, noted in the report of the 1942 conference, to discuss with Campbell Stuart their proposals to establish overseas communication companies in both their countries. McVey and Young's 'Anzac Scheme', as it became known, applied that idea to all the Dominions and to the Mother Country, and recommended setting up six public utility corporations eliminating private shareholding. The respective governments would own the capital.

When the Council met on 4 May, for an exposition of the Reith Scheme, McVey said he found its existence 'very embarrassing'. HMG had known of it several days before he and Young submitted theirs, yet they had not been told of it. Before the Cable and Wireless party joined the meeting Stuart told members that it would be led by 79-year-old Lord Inverforth, its president 'the dominating figure in Cable and Wireless'. Yet when Inverforth was invited to speak first, he said he had come to listen and to *learn* something about the plan, and at once asked Wilshaw to tell everybody what he had in mind. Reith also came but never spoke.

Reith described the occasion as 'unfortunate'.

> The meeting took place early in May and was as unsatisfactory as any I have ever attended. Wilshaw did not make a convincing presentation of the case but, in view of the atmosphere of the meeting and of the relationship between him and Stuart, who was in the chair, it was not surprising. Lord Inverforth and I accompanied him; I said nothing. *(Into The Wind)*

Wilshaw described the scheme as a form of nationalisation without the governments having to find the money. For a great many years it had been his ambition to take over the Commercial Cable Company and Western Union, and he thought Cable and Wireless should do that before the Corporation idea was launched. He wanted to extend the Empire Flat Rate of 1s 3d to the USA which could not then object to the Corporation scheme. In all Empire countries there would soon be a conflict between the interests of shareholders and of the public services. There should be public control over major policy but not public interference in management.*

When the Cable and Wireless party withdrew, Young of New Zealand said the plan was very close to nationalisation. Sir Alan Barlow of the Treasury said both the Anzac Scheme and the Cable and Wireless Plan raised serious political issues in the UK. There would probably be differences of opinion in Parliament on the desirability of replacing private companies by national corporations. But speaking as a civil servant 'it would be an awful pity if a big idea of this sort were prejudiced by the avoidable importation of these tiresome questions of whether you are a Socialist or an Individualist, a Conservative or Labour'.

In their first Interim Report of 10 May, the Council stated it had been decided to appoint a Cabinet Committee to examine the UK

---

*In a letter to Sir James Rae of the Treasury of 15 September 1944, Wilshaw saw fit to reiterate that the company was satisfied with the 1928/1937 organisation which was working well, and that 'it was only when the company discovered that the minds of the CCC were moving in the opposite direction and the common policy [of the Empire governments] was in danger of disintegrating' that the single Empire Corporation scheme had been put forward. To the Select Committee on 6 June 1946, he said the scheme he put to the Treasury on 19 April and confirmed in his letter of 24 April, was put forward 'merely as a counterblast' to another scheme he knew was coming into being. But to the next question 'Are you going to dispute that there was need for a scheme of some sort, either yours or somebody else's?', he answered No.

issues of the two proposals. It held further meetings on 13, 16 and 17 May, reviewing wider issues, the improvement of the facsimile network, and the company's service to the Press among them. Wilshaw, who attended the meeting of 17 May, said it would be up to the Council to decide whether the unremunerative Penny Press Rate was continued after the war. The main issue of the Anzac Scheme and the Reith Plan was considered by the Dominions Office on 18 May.

The Council now hardened in favour of the Anzac Scheme. Reith said the company were informed of this 'three weeks later' (after the 4 May meeting). 'Quite inadequate consideration had, we felt, been given to the company's proposal, and it had been rejected.'

The Council believed the company's plan offered a simplification of the present system. Economies could doubtless be effected and it had advantages in dealing with foreign countries; but it failed to provide an acceptable solution of the general problem because it would not be reconcilable with the sovereignty of the Dominions and India.

At the end of June, the Chancellor of the Exchequer Sir John Anderson told Reith matters had reached an impasse. If he could secure Reith's release from the Admiralty, would he be willing to play a leading part in investigations into the policy and constitutional framework of imperial communications? It would involve delicate negotiations between the UK Government and the Dominions governments. In his diary Reith noted: 'Reply: as there was no reflection on Wilshaw or the company – yes.' And there the matter remained for three months.

On 27 July, the Dominions Office telegraphed the Partner Governments seeking their approval for a statement to be made in the House of Commons that the CCC had recommended the establishment of a series of public utility corporations all financially interlocked, one in Britain, one in each Dominion and one in India. Canada wanted the statement postponed as they thought it would harden the tendency to consolidate international communications into two great rival systems, one Commonwealth and one American.

Britain, too, were conscious of not upsetting the Americans. On 28 July Sir John Anderson wrote a Personal and Confidential letter to Campbell Stuart pointing out that one of the important aspects of the possible formation of a series of public utility corporations was 'the United States aspect'.

> The US Government are taking a lively interest in telecommunications questions, and we have recently learned that they intend to ask

through diplomatic channels at no distant date, for negotiations to be opened with the Commonwealth on the subject. The issues involved are difficult and complex, and will require to be carefully considered by the Commonwealth in consultation before any reply can be given.

Anderson understood that Stuart intended returning to the USA shortly to continue his earlier talks. It was clear that such a visit would be untimely; 'I regret to say, therefore, that I have no option but to ask you to abandon it'.

A statement was made to the Press on 23 August, that the proposal for a series of public utility corporations was now being examined, and to give the governments time to consider it the CCC was adjourning its meetings until October. In addition to Cable and Wireless, the Post Office and the Treasury all gave their views on the scheme, and all opposed it. A group of civil servants were asked to assess all the arguments, for and against, and make their recommendations to a committee of ministers, which Reith bombarded with memorandums pleading the cause of his Single Empire Corporation idea. It was known that India and the Dominions approved the modified Anzac Scheme; would HMG agree with them?

Early in the morning of 24 July, a German flying bomb hit the east wall of Electra House on Victoria Embankment destroying Sir Edward's apartment on the second floor as well as the Accountants Hall. Four hundred men and women were on duty, but miraculously only two were killed. The emergency power supply was brought into action and the circuits inside The Fortress were working again within twenty minutes. Removal to Bristol was contemplated but rejected. But in November at the request of the Government the chairman removed himself and a party of senior officials to the Mediterranean and Middle East for a tour of Cable and Wireless branches and to report in particular on the exchange of social telegrams and the distribution of official war pictures. He took with him Jocelyn Denison-Pender, general manager; W G R Jacob, engineer-in-chief; Ivor Fraser, his Press Adviser; and Dr E Bayley, principal medical officer. They covered 10,000 miles by plane and another 2,000 by car. The tour took them to Marseilles, Naples, Cairo, Jerusalem, Cyprus, Athens, Rome, Malta, Algiers, Gibraltar. They had face-to-face talks with many critics, and heard stirring stories of resourceful action. At Marseilles, Wilshaw learnt that their manager and two of the staff had stood firm throughout the German occupation, keeping the plant spotless though the cables had been cut. When the Germans evacuated the town and threatened to wreck the plant, the manager, who had removed and hidden the relays, said that on instructions from Berlin the relays had been destroyed but nothing else must be touched against the day when the Germans came back.

That summer some Cable and Wireless men had taken a mobile radio station to Algiers – five vans and a trailer with nine operators and three engineers. This moved to Naples after the North Africa campaign to provide a telegraph office for war correspondents at each advanced press camp as the armies advanced up Italy. In June 1944 the station was working to the Post Office in London from Rome. The mobile station was dubbed the 'Blue Train' and was the prototype of several others, going later to Klagenfurt in Austria and Vienna. Three 'Blue Trains' were sent to South East Asia Command – the first at Rangoon – and Burma. Telcom Force, Batavia, occupied a former Japanese gaol.

But the main story of 1944 was the Allied invasion of Europe – the 'D-Day' landings in Normandy on 6 June. More than sixty pictures a day were photo-telegraphed overseas by Cable and Wireless during the first eight weeks of the operation, and more than five and a half million words to the Dominions and India. Two days before D-Day the company agreed to handle all Ministry of Information pictures free.

Two days after Wilshaw left for the Mediterranean, Reith noted in his diary,

> Rae [Treasury] told me that the blighted Ministers' meeting yesterday went very well, so my memoranda have been adopted. What a difference I have made by being in on this business. He said the thing was going to the Cabinet next week. Also he thought the ambassador of the scheme would have to go off round the Empire very soon.

By the end of November it looked as if Reith's sustained representations were having effect and the Government were being won round to the Single Corporation school of thought. When Wilshaw returned from the Mediterranean on 6 December, Reith put him in the picture. The chairman supported the idea of Reith making a tour of the Dominions to sound them out on the Reith/Cable and Wireless Plan. Wilshaw had known of this before he went away. But when told it had been suggested that Reith should become the second Government-approved director, Wilshaw was horrified. Reith too realised it would be an unpopular move.

> My colleagues on the board would hate it; they would never again regard me as one of themselves but as a spy imposed on them. Some of them might withdraw their support of the corporation idea. Whereas my known predilection for corporations and a reputation for autocratic action had made some of them rather chary about inviting me to join them, they were now all delighted to have me there. (*Into the Wind*)

So as not to do anything distasteful to Wilshaw, Reith not only rejected the idea of becoming the Government appointee but decided he would be more free in the conduct of his mission to the Dominions if he was not a director of Cable and Wireless at all, and resigned from the Board. Lord Pender (John Cuthbert) retired, and his son Jocelyn Denison-Pender joined the Court of Directors as joint managing director with Wilshaw. Ivor Fraser also became a director.

Hovering in the background was Lord Beaverbrook whose approval Sir John Anderson thought essential. The Canadian newspaper proprietor turned minister had just finished with civil aviation and was expected to intervene next in either oil or communications. He welcomed the choice of Reith, but other members of Winston Churchill's government, and the Dominions Office, expressed reservations about his suitability – and indeed the need for a mission at all – as did many Dominions ministers, who had considered the matter over and done with. The Cable and Wireless directors, however, were behind Reith to a man and gave him a copy of a colourfully worded resolution they had passed at their last meeting. Wilshaw wrote him a personal note wanting him to know how much he had valued the way in which he had championed the cause of the company and its ideals in Empire communications 'which had been deliberately and maliciously defamed and disrupted. It is largely due to you that the position has been reversed, but the ill done has yet to be remedied'.

Lord Reith and his party set out from England on 22 January 1945, in a converted Liberator bomber of RAF Transport Command with a Commando crew as body-guard. The other members were four senior public servants, including Sir Stanley Angwin, assistant director general of the British Post Office, Sir Edwin Herbert, and John Buckley, an administrative officer, a male secretary and Reith's personal girl secretary, whose presence, for Malcolm MacDonald the British High Commissioner in Ottawa, made the visit to Canada 'a trying interlude'.

But Reith's first stop was Australia, home of the Anzac Scheme, the basis of the plan modified by the CCC which HMG had rejected.

93 The opening ceremony in Sydney, Australia, on 3 December 1963 of the Commonwealth Pacific telephone cable (COMPAC) which linked Canada (Vancouver) with New Zealand and Australia. The connection to London was via the trans-Canada microwave radio system and the Commonwealth Atlantic cable (CANTAT) laid in 1961.

94 The £23 million South-East Asia section (SEACOM) of the Commonwealth Round-the-World telephone system was planned at a Kuala Lumpur conference in 1961, with Cable and Wireless agreeing to contribute 46 per cent. Here is the shore end operation of SEACOM 2 taking place in Deep Water Bay, Hong Kong, in November 1964.

95 The Hong Kong/Guam section of SEACOM was laid in the autumn of 1965. Assisted by the repair ship *Cable Enterprise*, c.s. *Mercury* completed the lay into Guam at the end of October. Here two submarine cable technicians on board *Mercury* make the final splice.

99 A telephone engineer at work in the Bahrain International Telephone Switching Centre.

98 Basil Leighton, long service F.I. man who in 1979 was Regional Director, Arab World, with headquarters in Bahrain where the company was the first to introduce Telex in 1963, then taking it to the rest of the Gulf, and in 1966 opening tropospheric scatter links between Bahrain, Doha (Qatar) and Dubai. A Bahrain earth station was opened in 1969.

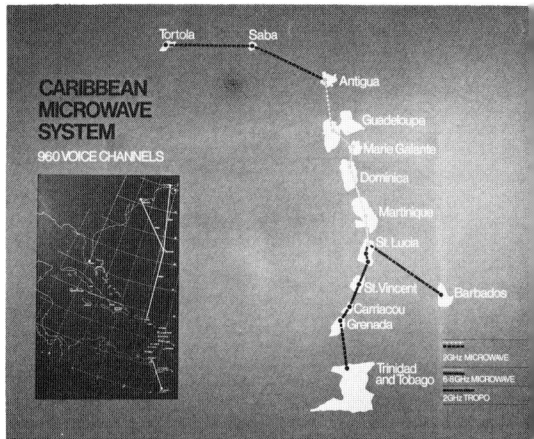

CARIBBEAN
MICROWAVE
SYSTEM
960 VOICE CHANNELS

Tortola · Saba · Antigua · Guadeloupe · Marie Galante · Dominica · Martinique · St. Lucia · St. Vincent · Barbados · Carriacou · Grenada · Trinidad and Tobago

2GHz MICROWAVE
6·8GHz MICROWAVE
2GHz TROPO

100–1 Norman Brooks, in 1979 Regional Chief Executive in the Caribbean, where in 1970 it was decided to upgrade the existing tropospheric scatter system with a SHF microwave radio chain linking the island in the way shown on the map above.

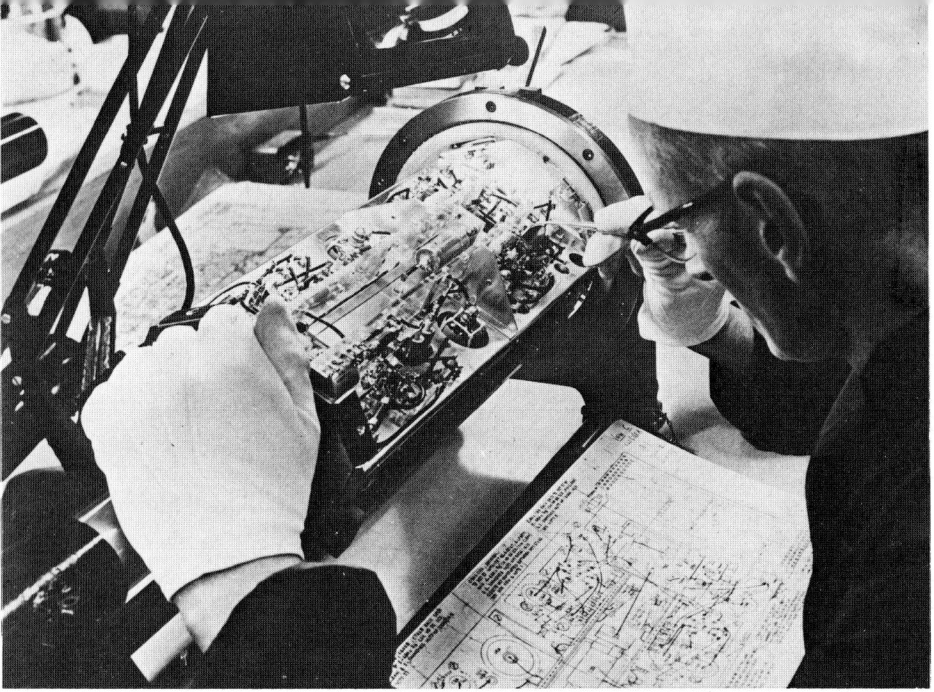

96 The sophisticated interior of the amplifier unit in an STC submerged repeater of the kind used in the Commonwealth Round-the-World Telephone Cable. There were 300 under-sea amplifiers in the Trans-Pacific section between Vancouver, Sydney and Auckland, many at a depth of almost four miles.

97 The problem of paying out a cable which every now and again contained a big bulge – a repeater – was solved by the Linear Cable Laying Gear of the kind seen here on c.s. *Mercury*.

102–3 The clean lines of Cable and Wireless buildings all over the world, created by Fred Hammond, Chief Architect, and the Architectural Services Group, gave the company an image of up-to-date efficiency, adapted to each local scene, wherever they were to be found: (*above*) in the Arabian Gulf; (*below*) in the Caribbean.

104–7 Ever since Edward, Prince of Wales, the future King Edward VII, came to Sir John Pender's reception in Arlington Street 100 years ago, the Royal Family's interest in Commonwealth communications has been unflagging – and greatly appreciated by staff and management in every quarter of the globe. (*Above*) HRH the Duke of Edinburgh watching a repeater being launched from the stern chute of c.s. *Mercury* at Tortola during his 1966 tour of the Caribbean; (*below left*) HRH Princess Anne touring the building after opening the second satellite earth station at Stanley on Hong Kong island in 1971; (*below right*) HRH Princess Margaret talks with H. G. Lillicrap (*left*), who became Chairman of Cable and Wireless in 1970, at the opening ceremony of the Barbados satellite earth station. (*On the opposite page*) HRH Princess Alexandra goes aboard with Lord Glenamara, Chairman since 1976, after naming the latest addition to the fleet, the c.s. *Cable Venture*.

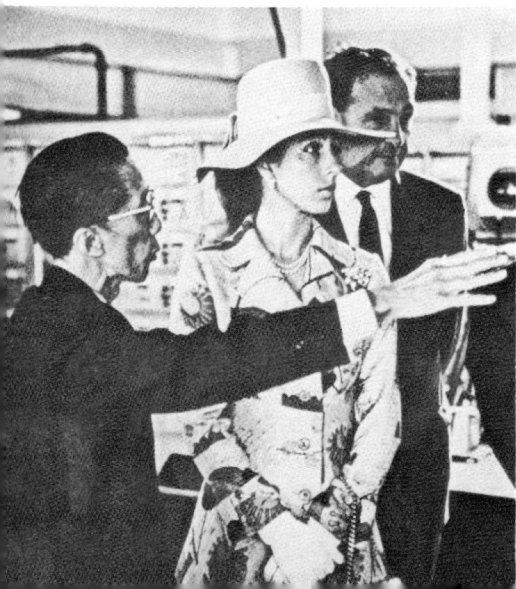

108–9 Within a very few years of moving to Mercury House in Theobalds Road, London, in 1955, Cable and Wireless had extended its activities to such a degree that additional accommodation had to be found for the staff, notably the new Smale House in Southwark (*above*) for Development and Production, and Summit House in Red Lion Square (*below left*) for the Engineer-in-Chief's department.

110 Maurice Bane, in 1979 Regional Director, Philippines.

111 R. A. Rice, Director, Cable and Wireless Ltd.

It was the country which had never abandoned its intention of having sovereign control over all its overseas communication services.

They talked at Canberra for six days. At the suggestion of John Curtin, the Prime Minister, two documents were drafted, one by the Australians championing the multi-corporation idea with maximum central control, one by the British setting out the argument for the single corporation with maximum local autonomy. The two papers were reconciled and submitted to a committee of the Australian Cabinet. This document proposed that each country should have its own corporation, and that there should also be a central organisation with the powers of a holding company. The pooling of sovereignty, pronounced John Curtin tipping right back in his swivel chair so Lord Reith lost sight of him beneath his desk, was no surrender of sovereignty. 'If everybody is to take a purely national view, then it's good night to hopes of peace, leagues of nations, security, fraternity and all the rest of it.'

They took the 'Canberra Proposals', as they became known, to Wellington, New Zealand, where in four days' discussion Prime Minister Fraser approved them with various suggestions for the strengthening of the central control. Then two days in Ceylon and on to Delhi, where the Viceroy warned that the Government of India was not prepared to go further than the original CCC Scheme; but six days later they too had endorsed the Canberra Proposals. In Salisbury the government of Southern Rhodesia did the same. The Prime Minister of South Africa, their next stop, at first insisted that the decisions of the central body must be advisory not mandatory, but in a later letter, Field Marshal Smuts said this qualified acceptance was not to be taken as indicating any reluctance on the part of the Union to be co-operative in seeing the scheme established.

Then via Ascension, Brazil (Natal), and Trinidad to Canada which had received copies of the Canberra Proposals in advance and signalled they were unacceptable. But after five days, Malcolm MacDonald's fears that the Canadian Government would make no concessions proved pessimistic, and they agreed to co-operate on the same terms as South Africa.

The mission's report was a short one – six pages and thirteen of appendices – and according to Reith was 'finished overnight'. After points made by other members of the party had been incorporated, it was printed for circulation seven days after they returned. It recommended an incorporated central body with an independent chairman and a board representing all governments, with local national corporations under obligation to carry out the directive of the central body for the whole system. The central body would receive the net revenue of all the corporations, deduct common expenditure and apportion the balance to the national bodies. The assets and operations not in the territory of any of the partners – under the ocean, in the colonies, in foreign countries – would be owned and operated by a separate UK national body.

The report recognised that dissatisfaction with the existing set-up was widespread and real; it could not with impunity be ignored. Modifications in the constitution of Cable and Wireless would no be sufficient. The plan aimed to eliminate the dominating position of the London commercial company, the substitution of public utility for commercial motive, and the recognition of Dominion sovereign rights. Reith thought the South African and Canadian qualifications amounted only to a difference of form not of substance.

The report was considered by the Cabinet at the end of March 1945, a fortnight after the mission's return. They referred it to a committee under Sir John Anderson. Reith resumed his friendly relations with Electra House and he asked when he could rejoin the Court of Directors. At a lunch to mark his 80th birthday Lord Inverforth told Reith 'Your friends are here, not in the Government'. Conversations he had with Inverforth and Wilshaw made him think the company would take a favourable view of his mission's recommendations. Wilshaw said he wanted to keep in close touch with Reith, and hoped he would attend board meetings to explain points and answer questions. Reith made an appointment to meet the following week. 'Then, despite a telephone reminder, silence; something had obviously happened. I reminded Anderson that I hoped he himself would see Inverforth and Wilshaw soon after they had received the report summary. They were then relatively well-disposed; he might have secured their co-operation before other influence got to work.' (*Into the Wind*)

The Reith Mission report recommended that the UK Government should recognise the Canberra Proposals as a practical basis for reorganisation, and convene an immediate conference in London to review the results of the mission; and, if they agreed to establish the incorporated central body and the other corporations, that they should purchase the assets of Cable and Wireless and help the other Governments purchase the local assets now owned by the company. HMG should explore the 'oceanic assets' proposal.

In spite of Lord Beaverbrook's total disagreement, which nearly jettisoned the whole scheme, in April the Cabinet approved the convening of a conference of governments in London to consider the report's recommendations. Reith was asked to organise it and act as chairman. Beaverbrook circulated a Memorandum ridiculing the achievements of the Reith Mission which in view of his new attitude, was disappointing but not surprising. However, the same week, to Reith's amazement, Cable and Wireless made it known that they too were opposed to the proposals. He knew Sir John Anderson would be mystified and wrote to offer him an explanation. The company apparently now thought the scheme unworkable. Though in 1944 they had been alarmed about the financial future, they were now fully confident they could make the grade if only they were left alone. 'This sudden, sad, complete *bouleversement*' was a great disappointment to Reith.

Wilshaw made the annual general meeting of 29 June the platform for extolling the very considerable services which Cable and Wireless had rendered the Empire during the conflict. In spite of the difficulty of obtaining supplies, forty-seven new wireless circuits had been opened. Four relay stations had been established to overcome fading. The company's cable ships had been continuously at sea protected by the Royal Navy, repairing, laying and diverting cables. The *Retriever* had been sunk by the enemy off Greece.* They were now handling 2,000 pictures a month by photo-telegraph, compared with 100 before the war. Total telegraph traffic was an annual 705 million words in spite of so many cables being lost by enemy action and the call-up of 700 skilled staff. The company had given advice for PLUTO, the famous Pipeline Under The Ocean which had taken petrol to the armies in France. Their engineers were now concentrating on developing the new British plastic, 'Polythene', which would replace Gutta Percha as insulation and make for longer high-speed cable. They had shown that private enterprise in the

---

* In 1978 Eric Barlow, who was the company's cartographer during the war, stated: 'I accept entire responsibility for bringing the cs *Retriever* from the west coast of South America, where she had little to do and could be replaced by chartering the cs *All America*. The cut back of the cable fleet had necessitated a redistribution of the ships. Throughout the war the safety of the cable fleet was paramount in my mind. In this I was ably assisted by Lieutenant-Commander Lace (ex-Eastern Extension and then in service with Admiralty Signals). Far from losing any ships other than cs *Retriever*, we gained two, the cs *Pacific* and the cs *Store Nordiske*. At my request these two ships were moved out of Japanese waters about one week before Pearl Harbor. They were both put to invaluable use in either the Persian Gulf or the Indian Ocean. The cs *Retriever* was not in the Mediterranean when war broke out; she was in the Red Sea. Against my wish she was required by the Admiralty to lay a Tobruk-Alexandria cable which under the circumstances did not eventuate. The Admiralty then took her over – commandeered if you prefer – with the expectancy of diverting the Salonica Cable. She was at wharf in Piraeus, the port of Athens, when the bomb fell. Captain Foy and the crew were killed; but Chief Electrician Eastwood was safe ashore.'

right hands could achieve much. The company's experience in the war had shown that the present system offered the merits of government control without its disabilities, ensuring that the public services were faithfully performed under the vigorous and imaginative direction of private enterprise. 'If left to our own resources and given reasonable encouragement and goodwill by the governments, we and our associated companies overseas can continue to keep abreast of development and offer cheaper and extended telegraph services without adding to the taxpayer's burden.'

The company gave its official reasons for opposing the Canberra Proposals in evidence before the Imperial Communications Conference of partner government representatives which began its sessions in London under Lord Reith's chairmanship on 16 July 1945, two months after the end of the war in Europe with the surrender of the Germans on 7 May, and two weeks before the announcement of the results of the British General Election – a landslide victory for Labour.*

In the view of Electra House, the scheme was impracticable and unworkable. It would make for delay and inefficiency. South Africa and Canada agreeing to come in on a consultative basis only meant the disintegration of the present unified imperial system. The Oceanic Assets Scheme would endanger the company's foreign concessions. If the company came under government ownership foreign countries would immediately take steps to acquire existing concessions and nationalise their overseas communications. That would be disastrous from a strategic point of view and a blow to British prestige.

It all smacked of the fruitless opposition to the Pacific Cable Board of forty years earlier.

As Lord Reith was chairman, the British delegation was not headed by a minister but composed merely of officials who were told to make it clear that the conference would not be allowed to discuss the idea of public ownership. The Cabinet, which was currently 'Caretaker' but might soon become either Tory or Labour, would decide that. Telecommunications policy with the United States would be discussed, however.

Though victory had been won in Europe, the Americans were still

---

* All those concerned in the events of 1943 and 1944 – the dispatch of the Reith Mission and the acceptance of the Canberra Proposals – though members of the wartime, all-party Coalition Government were in fact Conservatives – Sir John Anderson, Lord Lloyd, L S Amery, Lord Cranborne, Oliver Stanley. These became part of the Opposition on 27 July, when Clement Attlee formed his Labour Ministry to which it now fell to tie up an arrangement which in its origins had nothing to do with 'Socialism'.

pressing home their attacks on the Japanese. But two weeks after the new British Government took office, President Truman agreed to drop the first atomic bomb (6 August). Another was dropped on the 9th and on the 14th the Japanese surrendered.

It was the end of World War 2; and for the British members of the Cable and Wireless staff in Hong Kong, interned since 1941 at Stanley, it was the end of an ordeal from which many of them never fully recovered. Herbert Ascough, the station's engineer, liberated himself twelve days before the arrival of the British Fleet, and the official capitulation of the Japanese on the island – the first Englishman to do so. On 10 August, the day after the bomb fell at Nagasaki, he and his wife Dorothy were part of a group of 170 key men and their wives at the Stanley camp whom the Japanese took to the docks and herded into a wooden Japanese coaster which sailed that night for an unknown destination. They were battened down in the stuffy, rat-infested hold while the craft journeyed for three hours through mined waters and a wailing air raid warning as aircraft – probably the RAF who might attack the ship – flew low overhead. They landed on the mainland around 6 am and were taken to a second hutted internment camp, where they stayed until 15 August. When they woke on the morning of the 16th they realised their guards had 'disappeared'. An Indian guard threw a paper over the fence with a front page story reporting the end of Emperor Hirohito's bid to rule the world, and at midday their erstwhile guards were pouring into the internment camp with more food than their shrunken stomachs could cope with. Though Herbert Ascough knew that the English were still technically 'the enemy' to the trigger-happy Japanese soldiers still in occupation of Hong Kong, with two days' good food inside him and feeling reasonably fit he thought it worth risking the confusion to walk out of the camp and try to make his way to the Cable and Wireless station back in Victoria on Hong Kong Island. He found the office full of Japanese and, taking them by surprise, shouted at them that any attempt at sabotage would lead to the British – who he assured them were on the point of arriving – court-martialling the lot of them. In fact Admiral Sir Cecil Harcourt did not sail into Hong Kong harbour with his flagship until 30 August, but the bluff worked and the equipment, which had been working to Canton, was saved from further damage.

On 4 September, A R Harrison and other Cable and Wireless operators had fitted up some CR 100 wireless receivers in a special room for Ceylon Telcom teams at Negombo, twenty-four miles from Colombo. They were told to look out on all frequencies for any sign of life from Hong Kong. One of the Colombo Telcom unit, James

Brophy, was combing his receiver for signals on 6 September, when he heard the familiar call-sign. 'It was like a voice in the wilderness, weak but definitely there.' Someone was sending 'Here Cable and Wireless Hong Kong'. Fifteen minutes later came loud and clear 'Greetings Colombo. Our first contact with the outside world. Ascough.' Harrison immediately telephoned A J Whiteside, the manager at Colombo, who gave London the news – well ahead of the official notification of the liberation of Hong Kong.

Cable and Wireless Hong Kong were truly back in business – though strictly they should not have been. Herbert Ascough admitted to Colonel Gordon, when the R. Signals officer visited the station on 23 September, that he was contravening every military regulation in the book. To which Gordon replied 'And how very effectively too – carry on!' An official War Office message the next day told Ascough that Cable and Wireless were the only essential service in Hong Kong which had not had to ask for army help. From London came a message of greeting from Sir Edward Wilshaw. The manual radio telephone service was re-established with Canton and Macao, but the Governor, Sir Alexander Grantham, was not able to make the inaugural call to Manila until 8 September 1948.

To the other seven Cable and Wireless expatriates in Stanley the three and a half year internment had been a great deal more distressing. Stanley Maunder and C M Newman appeared at Joe Leonard's house in Kowloon in singlet, shorts, and bare feet. He gave them a welcome cup of tea. As their houses had been looted, they were given temporary accommodation by the Government before being sent back to England. The manager, J Sefton Jenkins, was a physical wreck and was returned to his home in Sedlescombe, Sussex, never to work again. Ascough took over as manager.

Preparations had been made for just a situation. Some 150 young men were given training in Britain, Australia and New Zealand to take over the Cable and Wireless stations in the Far East as they were recaptured from the Japanese. In the Press Ship accompanying the British Fleet re-taking Singapore was a telegraph office for Telcom, which was put ashore on the mole of the Telok Ayer Basin together with J D Mackie, the Scots manager of the Singapore office at the time of the collapse who had managed to escape in time. He walked into his old office. There was only one word he had for the Japanese colonel he found sitting in his chair, and it was 'Oot!' Ninety out of the 100 former Chinese staff eventually re-appeared. The transmitter in the Press Ship was soon sending the dramatic stories of eighty war correspondents to news editors round the world. At first they were limited to 250 words each. The instrument rooms in Singapore were out of order, but restored in fourteen days. General Sir Miles Dempsey, GOC British Forces in Malaya, sent the first message – to Sir Edward Wilshaw – expressing appreciation of the speed with which the link had been re-established. Sailing up the Penang Straits from Cocos *en route* for Singapore in the cable ship *Cable Enterprise,* Charles Martin, who was now its engineer, took his turn with one of the three rifles on board potting at mines which had escaped the sweepers in the hope of blowing them up before they hit the ship's bows. On the quay at Singapore he encountered Admiral Lord Mountbatten who asked him which ship he was from.

'*Cable Enterprise,* sir.'

'That's not a ship, that's a boat.'

In London, the Imperial Communications Conference had ended on 3 August, the day the Commonwealth Communications Council met for the last time with Sir Campbell Stuart as chairman. Sir Raymond Birchall of the Post Office, the UK representative, succeeded him as chairman on 31 August. Throughout his twelve year stint as the Treasury Agreement government watchdog, Sir Campbell had frequently expressed misgivings about the way the communications company was organised, as seen. He must have derived a certain satisfaction, therefore, on learning that the report which came out in the month of his withdrawal from the imperial telecommunications scene, declared a fundamental change of organisation was essential. The conference delegates rejected Reith's Oceanic Assets idea which Cable and Wireless had opposed, but had no hesitation in recommending that the governments of the United Kingdom, the Dominions and India should acquire the private shareholdings in the overseas telecommunication services of those countries, and the shareholdings of Cable and Wireless in the Dominion and Indian companies; and that those companies should be re-organised on a uniform basis. So far from being abolished, the Commonwealth Communications Council was to be given wider functions as the Commonwealth Telecommunications Board, still mainly advisory, but sitting in London, and acting as the 'incorporated central body' conceived by the Canberra Proposals. Unlike the CCC its members would not be resident in their own countries. The Board was seen as formulating joint telecommunications policy including rates; co-ordinating and developing the wireless and cable systems of the Commonwealth, all matters affecting the defence of the Commonwealth, and research; and undertaking, when requested, negotiations with foreign governments. It would administer a Central Fund into which the Commonwealth companies would pay their net revenues from their overseas telecommunications services. The submarine cable system would continue to be owned and maintained by the British company, Cable and Wireless, and in certain cases there would be contributions to this from the Central Fund. Such measures were designed to secure unification of Government policy and financial interests so a cable and wireless system could be developed as a whole 'without the artificial routing of traffic to which a divergence of cable and wireless interests naturally tends.'

It was all going to take some time to establish, and in the interim the powers of the Commonwealth Communications Council were extended to cover the Dominion and Indian companies, and many of the functions of the new Board were invested at once in the CCC, of which by a conference resolution Lord Reith was invited to become chairman, with a view to being chairman of the Board when incorporated – which did not take place until 1949. The Partner Governments were asked to see that their companies did not enter into negotiations with foreign telecommunication interests on any major points without the agreement of the CCC; and to take immediate steps to acquire the ownership of Cable and Wireless interests in their countries.

On 1 November 1945, the new Labour Chancellor of the Exchequer, Hugh Dalton, announced in the House of Commons that the Government had accepted the recommendations of the conference report, and subject to the agreement of Parliament the telecommunications services operated by Cable and Wireless would be transferred to public ownership. The latter decision, Dalton noted in his diary of that day,* came from a Cabinet Committee chaired by himself, 'but the policy had been unanimously recommended by a Committee presided over by Lord Reith on which all Commonwealth Governments were represented. It means taking over Cable and Wireless and getting rid of Wilshaw who I hear is very angry. Tom Macpherson, by pre-arrangement with me, rose after my statement and asked whether I appreciated that it would give great pleasure everywhere that His Majesty's Government were following a United Empire Socialist Policy'. Dalton's diary entry for 1 November began: 'This afternoon two new instalments of Socialism were announced after Questions' – the other was Civil Aviation.

Sir Edward was indeed angry. That same day he sent a 500 word letter to the 75,000 stockholders of Cable and Wireless (Holding)

---

*Quoted in his memoirs 1945–1960, *High Tide and After* (Frederick Muller, 1962).

Ltd going over the same arguments once again . . . 'gravest apprehensions' . . . 'impracticable and unworkable' . . . 'anxious to co-operate' but 'bound to make the strongest representations'. Since the Government had accepted the conference recommendations, 'bearing in mind the interest of the stockholders', the directors had no option but to try informally to reach an agreement on the price of the operating company's shares.

For of course the Government were not acquiring the holding company (capital £23,650,000), only the thirty million £1 shares of the operating company, less the two and a half million which they already owned. Neither were they acquiring the operating company's assets – its 155,000 miles of submarine cable, its five cable ships, its five wireless stations in the UK and 200 abroad. As in the case of the Bank of England and the coal mines, the Government proposed making compensation in Government stock not cash. If, in their informal talks, the Government and the company could not agree on the value of the shares, the Treasury stated that the figure should be fixed by an arbitration tribunal on the net maintainable revenue with a multiplier (the number of years' purchase). The transfer was to be made on a day to be appointed by the Treasury. Agreement to put the amount of compensation to arbitration in the event of the company failing to accept the Government's offer was not conceded in the case of the Bank or the coal industry. It was, in fact, unique. The shareholders of Cable and Wireless were greatly in their chairman's debt for persuading the Treasury to make this concession which resulted in the final purchase price being, it is said, very much higher than the first offer.

For Wilshaw, however, the prime task was to ensure that there *was* no transfer of shares. On 9 November, he sent a nine page 'confidential document' to leaders of the Tory Opposition, with copies to the managers of 'all British Empire branches'. 'Very desirable that your Press should acquaint Public of implications involved by Government proposals for acquisition of shares and control' stated the covering cable. He hoped managers would give the document the widest publicity 'without quoting source'. The set-up proposed would perpetuate the weakness of the old League of Nations, said Wilshaw, because one dissentient voice could upset agreement of any proposal. He claimed 'Cable and Wireless Ltd was not invited to participate' in the 1942 Australian Conference – which was far from the truth. To him the Government's decision appeared to be based not on the merits of the case but on political expediency, and had been taken without any evidence that there was a desire for it by the public, the Press, shareholders or employees.

As far as shareholders and staff were concerned, Wilshaw could only assume a majority did not desire it. He never obtained the shareholders' consent to spend the large amount of company funds needed to mount the campaign opposing the transfer; and he never knew what employees thought as he had denied them a voice when he dissolved the staff association with its short-lived *Gazette,* at the first sign of resistance to the 10 per cent pay cuts in the economy drive.*

There was some resentment among senior staff when, as part of his campaign, Wilshaw took it upon himself to tell the Government that heads of departments would not wish to serve under the state. None did, in fact, resign on those grounds. Though not a Pender, and so without the charisma of a Father Eastern, Wilshaw was nonetheless the reliable upholder of the tradition that 'the company looked after you'. On the other hand, anyone who dared to contest his judgement or question his decision – and all major policy decisions were his – soon found himself out of the company altogether or in a position where his voice would have no effect.

In his brief for the Opposition, Wilshaw suggested that one reason for the Government wishing to acquire their shares was the belief that responsibility to the shareholders worked against cheap and efficient service. How could that be so, he asked, when the dividend

---

*In the course of an off-the-record address which he gave to a meeting of the Established Staff of the London Station on 9 April 1942, following their presentation to him of a petition seeking the setting up of a Staff Association, Sir Edward said he was not opposed to the principle but asked whether that was the time for such a move. However, he was reported to have continued: 'I am going to blow my own trumpet and say I have done a lot for the good of the staff, and I feel it would be better not to have an Association. I regard the picture of the shoulder-to-shoulder picture of staff organisation as somewhat academic. This is my forty-ninth year in the Telegraph world and I know more about it than probably any other person in this room. You must excuse me blowing my own trumpet. One of my main considerations during my association with the company has been the welfare of the staff. I leave you to judge whether I have achieved this object. We have had one Staff Association, the ECSA, and I doubt whether there is a man here who could get up and say that he considered it had ever done much good for the staff. Did it achieve anything? It created an atmosphere of passive resistance on the part of the company towards the staff regarding their rights.' When their Profit Sharing project was put forward to the governments of other countries they were not prepared to agree to divide the surplus revenue until they had been convinced that the Cable and Wireless rates of pay were not extravagent. Wilshaw told the meeting he preferred to meet their demands individually rather than collectively; he was not a dictator but preferred to anticipate their eequests. Any pressure by the staff might kill the bonus. 'In Government circles the forming of an Association would have a very adverse effect on the prosperity of the company and of the staff . . . The desire for organised effort is very much in the air today; some people call it the Russian Virus, a fashion now in favour. In this company I think it is no good and not needed.' He proposed there was no discussion on what he had said, but that they met again in six months.

was only 4 per cent, averaging 1½ per cent since the company's inception? 'The Government's view is that, although they are purchasing the undertaking it will not be run as a Government Department, nor will the staff be civil servants and that therefore no real change in status has been made. This argument is obviously specious and could not be maintained in view of the fact that the Government would control the policy and finance, and in effect would also control the management.'

The Press, almost entirely Tory owned, felt obliged to comment adversely on the Labour government's decision. Wilshaw cabled overseas managers a selection of comment from papers, most of whom used the emotive word 'nationalisation' which, in the sense that the coal mines and the Bank of England were 'nationalised' by Act of Parliament, was not what was going to happen to Cable and Wireless. The British Government buying its shares implied that a day could come when they sold them again – to private investors. Cable and Wireless was not to become a statutory body. The appointed day for the transfer of shares was to bring with it no change in the company's legal status, since it remained the limited liability company to which so many foreign governments had granted concessions. Contractual relations with third parties were to be unaffected by a change in ownership. With one small exception it was to remain subject to the regulations and restrictions imposed by the UK Companies Act. The exception was exemption from the regulation which required a public company to have at least seven members. Apart from that, it would enjoy no special privileges, and all the laws which applied to any ordinary private enterprise company would apply with equal force to a Government-owned Cable and Wireless. It was a good example of the fact that a limited liability company has a separate legal existence entirely distinct from that of its shareholders. A great benefit of such an arrangement was that action could be taken against the company which might have caused a diplomatic incident were it to have been taken against the British Government. The proper term for the Press to have used was therefore 'Government-owned' not 'nationalised.'

Wilshaw was on firmer ground in his concern about the effect the change would have on American competition. Whereas the CCC conference had advocated allowing the United States to operate direct wireless circuits with Empire countries on an unrestricted basis, the 1945 London Conference supported Wilshaw's view that these should only be admitted where the volume of traffic justified them and should be confined to terminal traffic only. It was a separate question, and a separate conference took place in Bermuda between representatives of the British, Dominion and American governments to examine it in detail and come to an agreement.

On 13 October, Sir Raymond Birchall, Director-General of the Post Office, wrote to Wilshaw saying they regarded it as most desirable that representatives of Cable and Wireless should attend the Bermuda Conference as advisers to the UK Government Delegation, and he hoped Wilshaw shared that view. The American companies would doubtless attend in that capacity. It might even be possible to make Jocelyn Denison-Pender a member of the delegation.

Wilshaw replied on 19 October, that his colleagues on the Board thought it would not be proper for the company to send a representative to Bermuda while negotiations were taking place regarding the company's future. He offered to send two senior executives in the Traffic Department to provide the delegation with facts and figures.*

It was a characteristic decision and the last expression of a phobia which had done so much harm to Anglo-American telecommunications relations in the past. The Bermuda Conference which opened on 21 November, was designed to reach an Anglo-American agreement which gave the British Government a practical working compromise between American preference for wireless working and British insistence on a balanced wireless and cable system. As such, it was of the greatest importance and high level representation of Cable and Wireless would have made for maximum good will.

The Bermuda Agreement, signed on 4 December 1945, set out the principles to be adopted in determining whether a direct wireless telegraph circuit should be established, and recommended that they should be considered by the next International Telecommunications Conference. The proposal was that they should only be opened when justified by traffic or service needs. The existing wireless telegraph circuits between Britain and America would be retained, and the signatory governments

> shall neither support nor approve efforts by telecommunication companies subject to their respective jurisdictions to prevent or obstruct the establishment of direct circuits between the United States and British Commonwealth points and other countries, and will take such steps as may be appropriate to discourage any such efforts.

---

* The twenty-seven page printed brief which the company's Public Relations Officer sent to the Press in October 1946, stated (p 16) that the company was never invited to take part in the Bermuda Conference.

'At a stroke' the direct US wireless services controversy which had caused so much friction was settled – in America's favour. Each party was to inform the others of intended changes in rates; tariffs were to be in dollars and sterling (at $4·03 to the £). Ceiling rates between US and the Commonwealth were to be 30 cents or 1s 6d a word (later revised to 40 cents and 2s); with 20 cents or 1s for code telegrams (later raised to 26⅔ cents or 1s 4d) and Press rates of 6½ cents or 4d. With the ceiling of Imperial Preference rates at 1s 3d, the arrangement still maintained a margin. In any case Cable and Wireless would certainly have been compelled sooner or later to reduce their rates in step with the American companies, or lose a large amount of business to them.

The legislation required to bring the share capital of Cable and Wireless Limited into public ownership, and compensate it for having to reduce its charges under the Bermuda Agreement, was presented to the Commons by Hugh Dalton on 19 April 1946. The Government's White Paper with the text of the Cable and Wireless Bill (Cmd. 6805), issued on 24 April, stated that the Treasury Agreement of 1929 was now inappropriate and inapplicable, and would be repealed. The operating company's rights and obligations under it, however, still remained. With Government-ownership the Treasury would appoint all the directors, not merely approve two of them. The Bill dealt only with the purchasing, not the operating of the new acquisition, nor with the formation of the Commonwealth Communications Board which would require its own Bill. Freedom from political influence was necessary, commented *The Times* on 25 April, and the company should be kept independent of the Post Office.

There were two alternatives for public ownership, said the Chancellor of the Exchequer, moving the second reading of the Bill on 21 May, either direct operation by the Post Office or by a public board. Existing contracts of service would not be disturbed. Oliver Lyttelton for the Opposition thought it astonishing that the Government had not settled how the company would operate before acquiring the shares. The Tory tactic was now to attack Reith's efforts. His scheme, said Lyttelton, was 'a slavish surrender to immediate expediency, agreement at the expense of future peace and harmony in the working of Cable and Wireless'. But since the Commonwealth Governments had accepted the scheme, the Opposition would not vote against the Bill, which passed its second reading without a division.

This might have been taken as a hint to Electra House to stop spending shareholders' money on further opposition, but Wilshaw immediately had a petition deposited which was heard by a Commons Select Committee on 4, 5 and 6 July, to which he gave evidence. This time he said their opposition was in no way affected by politics or based on the question of 'nationalisation' against private enterprise. They feared it might mean the breaking up of the great British Commonwealth system of communications. On that ground they would have opposed the Bill whichever Government was in power. All the foreign concessions would be lost. His main difficulties had been the attitude of Australia, the impotence of the Advisory Committee and the dominance of its chairman.

On 24 June, Wilshaw chaired the last annual general meeting of Cable and Wireless before its shares were acquired by the Government. Richard Luff, the accountant, who had joined the Eastern Telegraph Company in 1920, became Chief Accountant in 1930 and Treasurer in 1941, was appointed a third joint managing director with Jocelyn Denison-Pender and Wilshaw. Shareholders agreed the purchase by English Electric of the whole share capital of Marconi's Wireless Telegraph Company for £3,750,000, following discussions which had been going on for two years.

In the Committee stage Dalton declined to agree to compensation for the directors, but hinted that some of them might be invited to join the new board (which, in the event, they were not). The third reading was carried again without a division on 11 July; the Lords passed it without a division on 25 July. Again Wilshaw decided to deposit petitions. The Lords Select Committee which heard them, were told on 29 July that under the present arrangement, the company owned the cables end to end. This made it much simpler to operate them. Why change things?

The answer was that in a complex operation like that of Cable and Wireless, technical efficiency was not the only criterion; there had to be an awareness of the sensitivities of an era in imperial history in which the watchword was self determination. Arrangements made before 1939 were no longer consistent with the post-war realities of Commonwealth relationships. As Lord Pakenham, moving the second reading in the Lords, observed, if the Commonwealth were Britain, one country only, it would certainly be easier to run the cables under one ownership, 'but when you have a number of separate nations, very proud of their individual sovereignty, you have to pay attention that each one regards itself as entitled to its own say in management'. It was a concept which the chairman would never have been able to understand. It was tantamount to treason. What mattered was the company's 'prestige' which was not served by playing servant to the Mother Country's offspring. He could never forgive Australia for daring to defy its London master. He told the Lords Select Committee on 31 July he had in his possession a letter from Sir Ernest Fisk, chairman of Amalgamated Wireless Australasia, asking for £62,000 in recognition of the good bargain which Fisk had pulled off for Cable and Wireless against the Australian Government.

'If this statement were in fact made', claimed Fisk in a press statement issued on 1 August, 'it is a travesty of the facts, and is one, moreover, which quite clearly implies either that I was seeking a bribe of £62,000 to betray my own government, or that I was asking for £62,000 as a reward for having already done so. There is not a vestige of foundation for either suggestion. If in fact Sir Edward Wilshaw made the statement attributed to him, I challenge him to repeat it in a place where he will not be protected by what I have been advised was an occasion of absolute privilege.' The incident related to the time when Fisk was in London helping Mr Bruce, the Australian High Commissioner, to persuade Cable and Wireless to accept the new company they were forming to replace Amalgamated Wireless.

Wilshaw was little interested in the wider issues of Commonwealth policy and impatient when told of the need to tailor the company's activities in conformity with them by the representatives of the Commonwealth sitting as the Advisory Committee chaired by a man who, as was the case with Sir Campbell Stuart, had the direct ear of the Treasury and the Imperial Defence Committee to whose deliberations he was not party. 'Instead of being an advisory committee', he told the Lords Select Committee, 'it became an inquisition, and so far from being helpful the chairman of it has done by intrigue, by misrepresentation, untold harm to the company and to the chief representatives of it, which have been very detrimental to the prestige of the company.'

But on the last day of the hearing the puff was almost out of him. It seemed inevitable, he said, that the Bill would pass, but he hoped with some amendments. The company had only opposed it because they thought it wrong and might bring disaster upon British communications. But if it was consummated, the company would do its best to avert such disaster and apply their energies to making the new conditions workable. 'For myself, I can only say that my life's work will be severed with a measure of sadness; but if the friendships I have formed in so many parts of the world, both within the Empire and foreign countries . . . can be of any service to the Government or the new undertaking in an honorary capacity, I shall consider it my privilege in the cause of British communications to place myself at their disposal.'

The Cable and Wireless Act received the royal assent on 1 November 1946, and on 19 December it was announced in the House of Commons that it would take effect from 1 January 1947, the Treasury's appointed day for the transfer of shares. The thirteen directors* would be replaced by a smaller Court of five of the Government's choosing. 'Getting rid of Wilshaw' had been finally accomplished, though he remained as Governor of the holding company now an investment trust, and insisted on continuing to occupy the flat at Arundel House. He was 67.

Sir Edward wrote an emotional Farewell Message for the staff magazine *Zodiac.*

> In the first hours of 1870 the bells of the City of London rang in the new year in which the pioneer cable companies were to begin building up a world-wide telegraph system. The willingness of thousands of small investors to hazard their savings and the enterprise of two generations of staffs have given us the British overseas telegraph system – the system we now operate. As those same bells ring out the old year, they will ring out also the era of private enterprise in British overseas telegraphy. Then they will ring in the era of State capitalism. In that solemn moment of midnight you and I will part company after fifty-three years' association. Some of you will regret the passing of the old order and some look eagerly forward to the new. But we can all take pride in what has been accomplished. Your work and mine in those years has carried on the beginnings of the pioneer, welded them into an unified whole. It has created a system which has stood the test of two world wars and, by progressive lowering of rates, given the world the cheapest system of international telecommunications it has yet known. Our service has never stood higher in fame and honour. At such a moment the State has thought fit to take control.
>
> By every constitutional measure open to me I opposed the Government's designs because I believed that they would bring disaster on the service we have created. I retract nothing. Now having finished the fight, I wish the State well with its new responsibility and earnestly hope that my misgivings may be proved to have been groundless.

---

*Lord Inverforth, president; Sir Edward Wilshaw, chairman and joint managing director with Hon Jocelyn Denison-Pender, and Richard Luff, Edward G Brooke (in Australia), Lord Courtauld-Thomson, Lieut-Colonel Ivor Fraser, Albert H Ginman (in Canada), Admiral Henry W Grant, Brigadier Henry J Lenton (in South Africa), Hon George Peel, Sir Harry Twyford, Sir John Wardlaw-Milne.

12

Towards Diversification

1947–1969

After a glass of sherry at the club, that first day of 1947, the new, Government-appointed directors of Cable and Wireless Limited hailed a taxi and drove, all five of them, to Electra House where they were met by the company secretary, Frank Lansbury, and the public relations officer, Harold Wilson.

The new chairman, 64-year-old Sir Stanley Angwin, was no stranger to the company and its affairs. Forty years in the Post Office Engineering Department, primarily concerned with radio, and Engineer-in-Chief since 1939, he was well known at Electra House and had accompanied Lord Reith on the 1945 tour of the Dominions which produced the Canberra Proposals. But his was a part-time, non-executive appointment. The chief executive of the Government-owned Cable and Wireless was the managing director, a post given to another Post Office engineer, John Innes, who had specialised in telephone systems. He had become Assistant Engineer-in-Cief, and when in 1940 he was appointed Director of Telecommunications in place of F W Phillips he successfully won the confidence of Sir Edward Wilshaw whose relations with the Post Office immediately improved. Latterly John Innes had been seconded to what became the Ministry of Fuel and Power where in 1947 he was Controller of the Coal Control Division.

The three other directors were a soldier, an accountant and a trade union official. Fifty-one-year-old Major-General Leslie Nicholls had spent his life in the Royal Corps of Signals, been General Eisenhower's Chief Signal Officer in North Africa, at Allied Force Headquarters, and Supreme Headquarters Allied Expeditionary Force, and finally served as Chief Signal Officer to Field Marshal Montgomery. Andrew Black was with accountants

Thomson McLintock who had been financial advisers to the 1928 Conference and been closely associated with the old Advisory Committee. Charles Gallie was General Secretary of the Railway Clerks Association and a member of the General Council of the Trades Union Congress. His experience in industrial relations was of the utmost value when it came to handling, with Edward Mockett, the complicated staff problems which arose from the Government's decision, not announced until 27 February, to integrate all the company's services in the UK with those of the Post Office whose scale of pay on the whole was lower.

'Cable and Wireless Limited will remain in being as a Government-owned company' the Chancellor of the Exchequer told the Commons, 'and will continue to own the assets and to operate the telecommunication services outside the United Kingdom, apart from those to be owned and operated by the respective national bodies in the Dominions, India and Southern Rhodesia. Staff of the United Kingdom system working overseas will be employees of the company. All assets in the UK will be transferred to the Post Office and the services integrated with those of the Post Office. The company as thus reconstructed will work in close association with the Post Office; the Post Office will become the UK national body for the purpose of the arrangements agreed with the Partner Governments at the Commonwealth Telecommunications Conference of 1945. Further legislation will be necessary to put the new arrangements into effect, and it is not expected that they will be in force before 1949. In the meantime the company will continue to operate as at present under the board of directors recently appointed by the Government.' It was run like any other limited liability company in Britain, having its own fixed capital and reserves, keeping its profits, receiving no operating subsidy, paying all its dividends to its shareholders who happened to be the Government who received a second contribution by way of the taxes paid on its earnings. Targets were laid down for profits 'taking one year with another', so that if there was 6 per cent one year but only 5 per cent the next, then 7 per cent was the target for the following year – though this became impracticable with galloping inflation. The company held an annual general meeting attended by representatives of the Treasury solicitor, and to have enough people attending it one share was given to Robert Harvey of the Post Office, another to Donald McMillan of the Post Office, and others. These meetings were private; the Press were not invited. But the Report to Shareholders had to lie on the floor of the House of Commons for three weeks, though no member ever took any notice and never asked a Parliamentary Question which he was entitled to do.

In Australia the position was different. There the company which the Australian Government created to take over the telecommunication assets of Cable and Wireless and Amalgamated Wireless (Australasia) had a less flexible relationship with the Australian Treasury to whom they merely had to 'hand back any surplus' and thus were never able to accumulate funds with which, as a management decision, they could take part in a scheme involving big capital investment, without getting authority from the Government. Australia's 'national body', the Overseas Telecommunications Commission (Australia), was put under the ministerial direction of the Australian Postmaster-General.

The Government of India's Overseas Communications Service, their counterpart to OTCA, also took the operation of their external telecommunications into public ownership at once. At the partitioning of India and Pakistan when independence was granted on 15 August 1947, Cable and Wireless offered to continue to operate the cable service from Karachi (now in West Pakistan), where for a time there was a training school, and to install radio equipment in Pakistan on the understanding that the Dominion should be free to take over the business should she so wish. It was a tragic time for many of the company's Hindu staff in Karachi who overnight found themselves in predominantly Moslem Pakistan. Though the changeover was, of course, known, the extent of the 'religious' violence was unexpected, and some staff had to leave their homes and their possessions and flee at once either east into India or west up the Arabian Gulf to Bahrain where the company, of course, had had a large and important station for so long. Many of the Asian staff who went to Bahrain at that time were still working there in the 1970s. The Ruler of Bahrain granted the company a new twenty year concession in 1947. The company was also entrusted with running the island's internal telephone system (as it did at Kuwait), and in 1949 opened an overseas radio-telephone service.

With security a matter of the unhappy wartime past, radio telephony now came into its own. A service was introduced to Barbados in 1946, and soon after speech communication by radio was possible between London, Malta and Athens, from Colombo to Sydney, from Hong Kong to Manila, from Nairobi to Cape Town.

But basically the company's operation was still Telegraphy. Compared with what it was to become thirty years later, the service was very straightforward: overseas cablegrams and photograms by line telegraph and HF radio, with the occasional speech link by radio-telephone for commercial and strategic purposes between certain points of the Commonwealth. Running it in 1947 were 12,500 men and women (7,000 in overseas branches, 4,900 in the UK, and 550 on cable ships) of 57 races working through 160 offices in 70 countries. The 'Via Imperial' cable system spread eastward through the Mediterranean to Egypt, and Aden, and thence to the east coast of Africa, India, Singapore, Hong Kong, Australia and New Zealand; westward across the Atlantic through Canada under the Pacific to Australia and New Zealand – the old Pacific Cable; and southward to Ascension Island and thence to South America and Southern Africa. Both north and south Atlantic branches linked in a circle through the West Indies. The radio circuits radiated from London by beam to Canada, South Africa, Australia and India, and by HF to New York, Rio de Janeiro, Buenos Aires, Santiago, Shanghai, Cairo, Nairobi, Salisbury, Athens and many other European capitals. Secondary circuits linked the Crown Colonies with London.

The operation which the new Court of Directors inherited was extensive but comparatively simple. It did not require management of any great sophistication; but it did require, and now for the first time received, a Plan.

A survey was made of the measures needed to be taken to put the system in full working order after six years of war during which cable repairs and renewals had had to be deferred. Five cables cut by the enemy were immediately restored, among them the Hong Kong-Manila and London-Singapore cables. A Ten Year Development Plan was drawn up, and the new directors set off to the far corners of the earth to gauge the requirements at first hand – John Innes and N C Chapling, traffic manager, to South America to contact the Western Telegraph's operations; General Nicholls the stations on the trunk route to the Far East and the stations of the West Indies where he was joined by J A Smale the eminent authority on radio communications who became Engineer-in-Chief on 1 February 1948 on the retirement of G H Entwisle. John Smale, aged 52 in 1947, had set up the radio relay station on Barbados for the beam circuit between London and Melbourne, and more recently the London-Barbados radio photo-telegram service, a technique with which he had been associated from the earliest days. Later in his tour he visited the other relay stations at Nairobi and Colombo, the development of which was his work.

Conditions of service and salaries at the beginning of 1947 were basically still those of 1939. As a varying degree of inflation throughout the world reflected unequal measures of hardship, it was decided to introduce entirely new salaries and allowances to replace these which had applied up to December 1946. The principle of joint negotiating machinery was adopted; committees were formed with the Union of Shop, Distributive and Allied Workers and the Association of Scientific Workers. The aim was tapered consolidation of bonus into salary, though not all bonuses were consolidated in the UK. For Foreign Service staff (F.1) a number of houses were bought and furnished by the company locally. House allowances were revised and a scheme introduced to give financial help (up to £180 a year) to managers and others with the burden of having to educate their children in boarding schools in Britain. The cost of staff to the company rose not only in these ways, but also because very many more were now being employed – in 1938 total staff numbered only 8,725.

With the signing by the Partner Governments on 11 May 1948 of the Commonwealth Telegraphs Agreement (presented to Parliament as a White Paper, Cmd. 7582, in February 1949), they formally agreed to adopt the recommendations of the 1945 Conference, to set up public corporations or nominate an existing department as their 'national body' and pass local legislation to carry out the agreement. They also agreed to set up the promised Commonwealth Telecommunications Board whose meetings would normally be in London 'but from time to time as may be found convenient meetings shall also be held in the territories of the other Partner Governments or elsewhere'. Every question was to be decided by a majority of votes. It would set up a Central Fund for the receipt of the net revenues of the national bodies. It would appoint a chief executive officer as Secretary-General at £3,500 a year, whose appointment could be terminated at the joint request of the Partner Governments. It held its first meeting on 10 November 1949 with Lord Reith as its chairman. The Commonwealth Communications Council was dissolved on that day. Reith resigned at the end of 1950 on becoming chairman of the British Colonial Development Corporation.

Britain's local legislation to give effect to the provisions of the Commonwealth Telegraphs Agreement was the Commonwealth Telegraphs Act of 1949. This stated that on 1 April 1950 the estates and interests of Cable and Wireless in land in Britain (including wireless stations), with exceptions, would be vested in the Postmaster-General, leaving the company's cable system and overseas interests in its ownership. From 1 April the bulk of the company's staff in the UK would be transferred to the Post Office and operate the services formerly conducted by Cable and Wireless. The Treasury advanced the PMG £4 million from the Consolidated Fund to enable him to pay for the transfer of the company's land and property. The Tory Opposition succeeded in making the Government agree to pay compensation 'to those who suffered loss of employment or diminution of emoluments or pension rights'. The measure received the royal assent on 31 May 1949.*

Three months earlier the compensation arbitration tribunal chaired by Lord Uthwatt which sat from 12 January to 4 February, after the company had rejected the offers made by the Treasury at the informal talks, announced its valuation of the Cable and Wireless issued share capital of thirty million £1 shares. It put the price the Government was going to have to pay stockholders at £35,250,000. But since the Treasury already held £2,600,000 worth of shares the amount payable was £32,195,000†. Cable and Wireless Holding (the private shareholders who received the compensation money) decided to distribute almost the whole amount of this compensation award. The holding company's Preference stock was repaid in full. Ordinary stockholders received £275 nominal for each £100 of stock. The Preference stock of the Eastern Telegraph Company was repaid at a premium of 50 per cent, and outstanding debenture stocks of both Eastern Telegraph and Eastern Extension were redeemed at a premium of 3 per cent.

The controversial Bermuda Agreement between the USA and the Commonwealth was now revised. 'The existence of both radio and cables is essential in the general interest of world telecommunications as a whole' stated the new text. 'Provisions of direct radio telecommunication circuits should therefore have regard to existing channels of communication.' But it was recognised that a direct radio telegraph circuit might be deemed necessary for political reasons. The Penny a Word press rate within the Commonwealth was extended to traffic between the Commonwealth and other countries. The 1s 6d a word rate between the US and the Commonwealh was raised to 2s; code rates from 1s to 1s 4d.

The big international airways corporations by now had their own private communications systems, and Cable and Wireless provided many of the facilities without operating them. But none of the smaller airlines in Europe could afford to do this, and in 1949 formed

---

* The New Zealand Post Office and the South African Post Office took their overseas telecommunications services into public ownership in 1948; the Southern Rhodesia Posts and Telegraphs Department in 1949; the Canadian Overseas Telecommunication Corporation was formed in March 1950 to acquire the assets both of Cable and Wireless and the Canadian Marconi Company; the Overseas Telecommunications Commission (Australia) came into being on 1 April 1950.

† The amount was satisfied by the issue on 1 March 1949 of £15·7 million 3 per cent Savings Bonds 1969–70 and £15·8 million 3 per cent Savings Bonds 1965–75, both of which were standing at around 102.

the Société Internationale de Télécommunications Aéronautiques (SITA), a co-operative, non-profit-making organisation, to share communications, which many junior managers saw as usurping the functions of Cable and Wireless who for a period would not recognise it. For Electra House 'the market' was still mainly one in which concessions were to be secured from foreign governments; but it soon realised the virtue of changing its stance. The demand by those who ran specialised world-wide services – air travel, shipping, banking, commodity and stockbroking – for tailor-made, private communications networks became the trend, by which time the juniors of 1949 had grown into senior managers of new thought and approach who recognised the business opportunity which the new market offered and moved to take full advantage of it.

The Colonial Wireless Sets installed at the beginning of the war became the basis of a Commonwealth radio-telephone system introduced in March 1949 at the arbitrary charge of £1 a minute – very much less than the only other radio-telephone service of the day across the Atlantic which cost four or five times as much. The simple, two-page 'Colset Licences' granted by governors to the company to operate any form of communication within their colonial territory meant the service could open without further formalities. The BBC relied on these Cable and Wireless radio-telephone circuits for transmitting the live 'voicecasts' of overseas correspondents in programmes like Radio Newsreel. Such circuits also made possible the loyal greetings from round the Commonwealth which preceded the Queen's annual Christmas broadcast; the Bells of Bethlehem were heard in homes all over Britain via Jerusalem and the Cable and Wireless mobile 'Blue Train' station in Amman.

Assisting in the operation of radio-telephony and Broadcast Transmissions in the early 1950s was J H 'Tug' Wilson, who joined Imperial and International Communications as a uniformed 'Via Imperial' messenger of 14 in 1933, one of 600 delivering telegrams to offices all over the City. Rejected by the Post Office because he was half an inch under their minimum regulation height of 4ft 6ins, he spent three years, when not pacing the streets, sitting on the benches with a hundred other boys in the Sending Out Department at Tower Chambers watching the black rats climb the pneumatic tubes, or waiting with five fellow messengers in the tiny room behind the public counter of the telegraph office at The Baltic Exchange in St Mary Axe. One afternoon a week the company sent him to a school in Kingsway where among other things he learnt to type. When he was 17 he took the school's exam and came top, which brought him the offer of a position as Clerical Assistant – others who had passed

well could become Indoor Messengers. Tug Wilson accepted the job
and began a career with Cable and Wireless which took him to the
most senior position of any member of the staff who had begun as a
messenger: Group General Manager, International Operations.

One who was born to his position in the company, John Cuthbert,
the first Lord Pender, grandson of the founder of The Eastern
Telegraph Company, died on 4 December 1949 aged 67. He had
been the first chairman of the new Cable and Wireless until 1936.
His wife Irene, daughter of Sir Ernest de la Rue, had died in 1943.
His son Jocelyn, now 40, succeeded to the peerage.

It was business as usual on 1 April 1950 for the 3,500 operating staff
at London Station, Electra House, Victoria Embankment who that
morning had become employees of the London Telecommunications
Region of the Post Office, which now ran the terminal of Britain's
overseas radio circuits and submarine cables. One person who did
not report for duty at Electra House that morning was John Innes,
the managing director, who retired on 31 March, and was succeeded
by General Nicholls. For the staff of the wireless stations at Ongar,
Brentwood, Somerton, Dorchester, Grimsby and the rest, it was a
question of returning to the department from which, with so much
controversy, they had been taken in 1929. The London Telecommunications Region of the Post Office, already busy, soon found they
were unable to handle this extra work, and on 1 October 1952 an
External Telecommunications Executive (ETE) was created to take
over all ex-Cable and Wireless functions in the UK and many of
those of the Post Office's own Overseas Telegraph Department.

Cable and Wireless Holding, now an investment trust, continued
to use its suite of offices at Electra House, and the administrative
headquarters of Cable and Wireless Limited remained there too.
Apart from this the company retained the Porthcurno cable station,
the Plymouth Cable Depot and the Exiles Club at Twickenham.

The national bodies of the Partner Governments took over the
company's stations in their countries from the same date, though the
whole operation was not completed until June. Under the new
scheme, they were regarded as partners and common users of the
Commonwealth Radio and Cable System, enjoying the use of it
regardless of the actual ownership of the various sections of it. The
expenses of running the system under the first Wayleave Scheme,
which came into force in 1950, were borne by the 'national bodies' of
each Commonwealth country in proportion to the revenue from the
traffic which they put into the system. Hence the heaviest users of

the system paid proportionately the heaviest share of the costs. Those who used it less contributed less, even though the cost of maintaining the system was the same whether it was used much or little. Thus the small user, who would be a small and comparatively poor country, was subsidised by the bigger and richer countries. So far from disintegrating, the Commonwealth system gained new cohesion derived from satisfying each community's national pride, and from loosening of Mother Country domination. So far from bringing disaster on the service, the Government's plan had given it new life. Sir Edward's misgivings had indeed proved groundless.

But, after so much had been taken away, what had Cable and Wireless left? It remained the largest part of the Commonwealth overseas telegraph system, and indeed the largest single undertaking engaged in international telegraphy in the world, with the only integrated radio and cable system outside Europe. In Britain it owned nothing operational except the shore ends of the submarine telegraph cables at Porthcurno – the half mile from the beach to the terminal cable station, which remained in company hands, and once more became a training centre too.* On 9 June 1950, Sir Stanley Angwin opened the Cable and Wireless Engineering School with W E (Dan) Cleaver, who had joined the service in 1913, as manager. It provided the last nine months of an eighteen months' course in cable and wireless engineering which began at the London section of the school at Electra House. The old quarters, which had seen such hilarity in the 1920s and had lain neglected for so long, were refurbished and accommodation provided for fifty. The theatre, the tennis courts, the cricket field sprang to life again; Mess life, of the kind they could expect on posting overseas, once again became the daily routine. It was a hundred years since submarine telegraphy began and on 28 August at Porthcurno they celebrated the centenary of the laying of the Dover-Calais cable.†

Porthcurno was the terminal of eleven telegraph cables which connected Britain with a network of some 155,000 nautical miles (186,000 statute miles) of submarine cable linking the units of the Commonwealth (both self-governing Dominions and Crown Colonies) and those foreign countries in which the company held concessions. These were its 'ocean assets', the actual lengths of copper and insulation in the sea, which under the new arrangement

---

*At about this time, too, a training school was opened in Sierra Leone for West Africans from four colonies; and in 1954 one in Hong Kong.

†At a Cable Centenary Exhibition in London a message was sent from the Science Museum which travelled 33,871 miles, was 'tape relayed' four or five times and received back in London in 53·6 seconds.

the company still owned, maintained and developed. The old Pacific Cable was still the longest in the world – its longest section from Vancouver to Fanning Island being 3,460 miles. With few exceptions the company still operated the overseas cable and wireless services of the Crown Colonies, and all the external cable services of the foreign countries in which it held concessions, notably in South America (the Western Telegraph operations), the Philippines and Indonesia. It also continued to provide radio relay services on the main Commonwealth routes. The company owned and operated 132 telegraph stations of which sixty-two were in British colonial territories and seventy in foreign territories. From these it operated 105 radio telegraph circuits; sixty-six radio-telephone and ten radio-phototelegraphy circuits; twenty-two ship-to-shore radio services; a broadcasting service in Nairobi; the technical side only of the broadcasting service in Hong Kong; and five inland telephone systems. In 1938 the system carried 231 million words; some 705 million at the peak of the war and now (in 1950) 650 million.

To maintain and develop the cables under the sea it had a fleet of eight cable ships kept at strategic points round the world. Some 1,800 miles of cable was renewed in the Red Sea on the main route to India and Far East in 1951 alone. The following year the Porthcurno-Newfoundland-Halifax telegraph which fell into disuse during the war was reinstated, and some 1,400 miles of new Telcothene-insulated cable inserted to restore it to working order. The new Harbour Grace cable of 1952 was the last major ocean telegraph cable to be laid by Cable and Wireless. The post-war 'quality assurance' drive, together with 'Telcothene' insulation, meant the fault rate on new lengths of cable was now kept to a very low level. But perhaps the time was now passed when telegraph cable needed to be durable? To many of the younger generation at Electra House such activity in aid of an obsolescent medium seemed an enormous waste of time and money, a sign that the phasing out of cable-orientated thinking was being taken at much too leisurely a pace. Radio, as it was now called, was still playing a mainly back-up role to cable. Multi-channel telegraph relays were installed in the radio relay stations at Barbados, Colombo, Nairobi, and Aden, together with electronic signal regenerators. The total staff employed overseas in all this work was some 8,200. In Britain the company employed 600.

In 1950 the company became responsible for many additional services in Hong Kong, including the engineering for Radio Hong Kong, the installation and maintenance of all telecommunications equipment associated with Kai Tak airport and the meteorological point-to-point services at the Observatory, as well as the sound equipment in the City Hall.

In competition with the rest of the world's overseas telegraph operators, the company made a special drive to provide the world's Press with the most efficient means of disseminating news rapidly at the lowest possible cost. When the Soviet-dominated North Korea invaded the southern zone on 25 June 1950 and precipitated the Korean War, and the forces of sixteen member countries of the United Nations under General MacArthur went to the aid of South Korea, this hitherto obscure part of the world became overnight news. The company immediately dispatched a field radio telegraph unit initially under N W Barnes, and then A S Pudner, as manager-engineer, to handle troops' and war correspondents' messages. The Cable and Wireless men wore battledress with the Telcom shoulder flash and blue berets of World War 2. They were attached to British Commonwealth Brigade Headquarters. Six tons of equipment were flown to Hong Kong which acted as the reception point for the Korean station, and relayed its messages to the UK. Three similar mobile radio units were established in Barbados, Singapore and Britain to provide a service at short notice wherever it might suddenly be required.

The staff of the Telcom unit in Korea were all volunteers, two operators for instance coming from Bermuda. Possible service in a war area was not held out to would-be entrants to Cable and Wireless as the kind of excitement they could depend upon. Life with the company was stimulating enough to attract the right type of young man without that. As R F C Thomae wrote (anonymously) in the recruiting booklet *A Career Abroad* issued in 1951, 'the company's work overseas provides an ideal occupation for the man who combines a keen interest in electrical engineering with a spirit of adventure and a readiness to serve anywhere in the world'. The young technician could be expected to be posted to any branch between Buenos Aires and Vigo, wrote Thomas. He must therefore be a versatile individual, prepared to take life as he found it, and adapt himself to the varying climates and conditions of service likely to be experienced during his career. There would be tennis and sailing, golf and squash. At some branches there would be riding, at practically all of them fishing – deep sea fishing at Cocos, Rodrigues and Ascension. The freedom of the open air life at tropical branches was typified by the practical day-time dress of shorts and tennis shirt. In the smaller communities the staff and their families played an important part by joining in with other residents and helping organise social evenings and dances.

In the 'island life' which each member of the staff is liable to experience from time to time in his career, a well-developed sense of tolerance and balance is essential, if staff living in isolated conditions are to lead to a happy life. Because of the careful selection of our personnel, we are fortunate in possessing a good record of contented community life at our remote branches.

In retrospect it was perhaps a somewhat rosy picture of life in an overseas Cable and Wireless station in the 1950s when air conditioning had yet to give instrument rooms the degree of comfort they enjoy today. And compared with 'commercial' pay, the salaries were not over-generous.

After passing an interview, a two-day education test and medical test (good eyesight essential), he received 55s a week for the first six months of his training, and 60s for the next twelve months, plus 15s kit allowance. After six months' 'consolidation' at a branch like Gibraltar he would go on a first overseas tour of four years – from then on they would be three. He contributed 5 per cent of his salary to a pension fund, and as a member of the 'mobile staff' would expect to retire at 55 – unless of course he had aspirations to join the Court of Directors in London, who tended to serve somewhat longer. The first all-Government-appointed directors were outside men, but executives from the company's own staff began joining the directorate soon after. N C Chapling, who became managing director when General Nicholls took over as chairman on Sir Stanley Angwin being appointed chairman of the Commonwealth Telecommunications Board on 1 April 1951, began as a junior in the head office of Western Telegraph in 1918 at the age of 15 on 12s 6d a week, and by 1947 was traffic manager.*

There were many who thought that Sir Stanley Angwin had been sent to Electra House from the Post Office to conduct the final rites over a body which within months would surely die, because of its inability to generate sufficient business to keep it in being. When it was all over they reckoned he would pick up any remains which still seemed to have a sign of life, and absorb them unobtrusively into the activities at St Martins-le-Grand. In the 1950s the Middle East and Far East did not exist as major telecommunications markets on the scale they were to become twenty years later. The rump of the truncated company, it seemed, would not have sufficient nourishment

---

* F.1 men who became directors included the Australian 'Tufty' Baker of the foreign service in Hong Kong who was elected to the Court in 1960; and W H Davies, son of G E Davies, Eastern Telegraph superintendent at Vigo, who joined the company as an operator in 1933, and was made a director in 1968; A S Pudner, the Engineer-in-Chief who joined the Court in 1969; and R W Cannon in 1973.

to keep it healthy, and given time it would inevitably wither away.

That such prognostications were proved wrong was due to a small group of people within the company who were determined to see that Cable and Wireless not only survived but acquired a fresh persona which would give it a new lease of life, animated by forces which the Old Guard had never been able to tap, and operating in fields outside those labelled 'traditional'.

The company was just holding its own financially but, though proficient in its own 'traditional' specialties, was technically behind in the *breadth* of its knowledge. As the story so far has shown, by the nature of the pre-war management, and because Eastern Telegraph not the Marconi Company became the dominating force in the 1929 merger, the company's acquaintance with radio and long distance telephony techniques had never equalled its grasp of ocean telegraphy. Under the guiding hand of J A Smale, supported in his thinking by executives like Frank Lansbury, company secretary, and H G Thomas, chief accountant, the company now made up for lost time. In this transitional period, when the company with the fine war record was preoccupied with recovering from the reverses which that war had inflicted, and in danger of merely reimposing the *status quo ante,* the Engineer-in-Chief and his group were able to ensure that adequate emphasis was placed on radio as opposed to cable, and that voice communication which, from a combination of Government insistence and Electra House indifference, had been eschewed before the war, was now put in the front rank of future planning. J A Smale enabled the company to hold its position at a time of uncertainty and improve its technical standing in the eyes of those who would have welcomed an opportunity to dismiss Cable and Wireless as experts in submarine telegraphy and little else. One example of a technical innovation from Cable and Wireless at this time was the mechanical error detecting and correcting equipment devised by its engineers and installed on the London-Sydney radio circuit. Another was the new signalling system over radio known as Double Current Cable Code which offered greater reliability than Morse and allowed for integration with Cable. It lasted until it was superseded by Five Unit Code in the mid-1950s. The automatic relaying service was also expanded; in October 1954 a relay service was opened for the New York to Bahrain circuit through Ascension Island.

The year 1954 was the company's Silver Jubilee, and to mark the occasion Cable and Wireless issued a booklet *World-Wide Communication* with a foreword by the chairman Leslie Nicholls, in which he stated that in the twenty-five years since the merger of 1929 the system had changed to one in which messages were sent by cable or wireless, or by composite circuits comprising cable and wireless links. 'The cable engineers and the wireless engineers of the past have been replaced by a new generation of telecommunication engineers conversant with the strong points of each method.'

As an aid to good communication the company operated a world Ionosphere Prediction Service to help engineers overcome difficulties of the reflecting ionosphere changing according to the time of day, the season of the year and conditions of the surface of the sun. The use of relay stations enabled messages to be sent on alternative routes where disturbances were less.

On the administration front, in 1954 the Treasury appointed one of its officials, Henry Eggers, on a ten year contract to strengthen top management (though not initially a director). The following year he was appointed 'managing director jointly' with Norman Chapling – an arrangement by which neither was required to consult the other before making a decision, as would have been the case if they had been joint managing directors.

The two of them, with chairman Leslie Nicholls, fellow directors and the modestly-sized head office staff, moved from Electra House to the new headquarters in Theobalds Road, WC, in November 1955. The building, which was named Mercury House after the classical Messenger of the Gods, was officially opened by Lord Reith in December. Cable and Wireless did not occupy all of it; two floors were let to other tenants. The investment trust, Cable and Wireless Holding, remained at Victoria Embankment, along with the Post Office Overseas Telecommunications Executive, and became the guardians of the old boardroom with its portraits of past chairmen and of the chapel, now restored after the damage it received in the war.

Within months, war threatened to disrupt the whole of the company's system in the Middle East. When, four weeks after his election as president of Egypt in June 1956, Colonel Nasser outlawed the Suez Canal Company and seized its property (which John Pender had helped to buy), British families in the Canal Zone, including those of the Cable and Wireless staffs at Suez, Alexandria and Cairo, were flown back to England. At the end of October, Israeli troops invaded the Sinai Peninsula and an Anglo-French force bombed the Egyptian airfields, but the Cairo office held firm, and the Alexandria staff got away by sea. The three operators and manager G A Kirby at Suez closed their station down, and suspended traffic on the Suez-Bombay cable which had opened at the same time as the canal in 1869. Kirby's men were told to stay at home, but on 5 November the Governor, fearing the mob might raid their flats, had them moved to safer quarters, along with fifteen other British expatriates. They had to leave everything behind except a few personal belongings and a blanket, and boarded a motor coach which took them to a primitive camp. During the drive, as one of them, V G Pullen, recollected in 1978, they hid their heads in their jackets so that the excited crowds in the streets listening to the BBC reports of the invasion from loudspeakers hung on lamp posts, did not recognise them as 'the enemy' whose planes were strafing their airfields. At the camp their main diet was sweet halva.

The Suez War heralded the end of the company's business in Egypt, for the Government issued a directive to cut the Alexandria cables.

The installation, maintenance and operation of an internal tele-phone system by land line was simple and straightforward, using techniques shared by engineers all over the world. In the 1950s inter-continental telephony was conducted mainly by radio. Cable and Wireless provided certain radio-telephone services between London and Accra, Athens, Barbados, Bermuda (the first), Cyprus, Hong Kong and Malta. But ever since its introduction in 1927 engineers had tended to regard radio-telephony as uncommercial due to its dependence on atmospheric conditions and its notorious insecurity. The search for a reliable means of carrying the high frequencies required for voice transmission under the sea could be said to have begun in 1891 when Siemens Brothers made the first Anglo-French telephone cable. Only the economic depression stopped the laying of a transatlantic telephone cable the 1,800 nautical miles between Ireland and Newfoundland in 1928. The invention by Imperial Chemical Industries of 'polythene' in 1933 gave the search for an inter-continental submarine telephone cable a new boost, but the war put an end to further experimentation.

It was not until 1956 that the British Post Office, the American Telephone and Telegraph Company (ATT) and the Canadian Overseas Telecommunications Corporation (COTC) jointly com-pleted the first Trans-Atlantic telephone cable, TAT-1. It spanned some 1,950 miles and provided initially thirty-six high grade telephone circuits between Oban in Scotland and Clarenville in Newfoundland, and sixty between Newfoundland and the mainland of Canada. Ninety per cent of it was manufactured by Submarine Cables Limited, a merger of the two British companies Siemens and Telcon, at Ocean Works, Erith, which they built for the purpose. It had conventional helical armour wires for protection. Flexible amplifier/repeaters, small enough to pass through the laying gear of the cable ship, were wound into the armour as bulges in the cable. But they were too small to house the components required to enable signals to pass in both directions, and a separate cable had to be laid each way, and required 4,200 miles of cable. This was highly uneconomical. The construction cost of £12½ million was some fifty times higher per circuit-mile than a land system providing 960 circuits.

There was no doubt, however, that TAT-1 was an outstanding success both technically and commercially. ATT embarked on four other similar schemes including a TAT-2 which in 1959 gave voice communication between Britanny and Newfoundland. Since one voice channel of this co-axial cable could act as twenty-five telegraph channels it heralded a complete change of scale. 'Bandwidth' emerged as a factor which was to play an increasingly important role as the years went by. It was the first trans-oceanic cable telephone system to be integrated with national and international systems, and was described as 'the dawn of a new era in global communication'.

TAT-1 was not ideal, but it worked, and it was the first of its kind. Four years before it opened for service, the British Post Office had launched a joint research project with Cable and Wireless and British submarine cable and repeater manufacturers at their Dollis Hill laboratories, to devise a new trans-oceanic telephone cable which disposed of all the objections encountered hitherto.

Their answer was a 'lightweight' cable without external steel wires – heavy armour was not needed in the depths – which maintained its strength by means of a high tensile steel wire threaded through its centre, round which was wrapped a copper tape with an outer conductor of light aluminium. The co-axial interior had an external plastic jacket, initially polyethylene. It achieved a new balance, not twisting or kinking when pulled.

For laying in the ocean depths, the lightweight cable required a newly developed deep-sea valve-amplifier/pressure housing (a repeater) to be inserted every twenty-six miles along its entire length. By this means the number of telephone channels the cable could carry for simultaneous transmission was sixty. The repeater was designed with reliability as the most important factor, as each repeater was required to operate on the seabed for twenty years without requiring attention.

The formula devised by Reg Halsey, the Post Office's Director of Research, and his joint team at Dollis Hill in 1952 (simultaneously, it should be said, with the Bell Laboratories of America) was in

Tˢᵇ¹ᵃˡH·C·

THE ISLANDER

10p   **ASCENSION ISLAND**   20¢

ISSUE NUMBER 351   Friday, 14 April 1978   ISSUE NUMBER 351

LATE WERE WE- I WONDER WHY?

The frustrations of our "Islander" readers prevailed again last week when the "Islander" failed to be distributed on Thursday evening. This is probably an understatement of the whole situation as several "Islander" staff were not only extremely frustrated but sorely disappointed with the lack of adequate support - again for the second week. My sincere personal thanks to those who put forth extra hours last week allowing the "Islander" to be published.

In evaluating the situation, I feel there are a lot of people willing to help but not aware of our needs. At present, the "Islander" staff is approximately 50 percent manned with active volunteers. If we are to have a newspaper <u>next</u> <u>week</u>, the following positions/jobs need to be filled. Would you please seriously evaluate your schedules to see if you can't squeeze out some time to help with the "Islander". The more volunteers we have, the less work for everyone.

    a. Four fairly proficient typists are needed.

    b. Two feature editors are needed, each sharing the workload.

    c. A social editor for the Georgetown Saints Club is needed.

    d. A subscription secretary to handle the mailing of "Islanders is needed.

    e. A distribution secretary and assistant to handle distributing the the paper on the island are needed.

    f. Four more printers are needed to form a team where 2 or 3 are scheduled weekly to actually print. A production manager is needed and will be selected from a qualified volunteer printer.

    g. Six more staplers are needed to add to the stapling team. If six can be scheduled for an edition, the job would only take about 30 minutes. Contact Lorraine Manning G221/G438.

I would like to see more St.Helenians involved as a large portion of our population are St.Helenians. Most jobs on the "Islander" can be done with a bit of training which is readily available and some without any training - there's a place for everyone.

Cont. on Page 5

113–15 George Warwick, Regional Director, Hong Kong and General Manager, Hong Kong (*left*) in 1979 ran the largest branch in the Cable and Wireless Group, with a staff of 2,300 97 per cent of whom were locally engaged. (*Below*) the Hasler computerised Telex exchange, the first stored program-controlled equipment developed specifically for use in the international and national Telex services. By August 1977, three computerised Telex exchanges were operating with a capacity of 10,500 subscribers. (*Opposite top*) Programmers at the console of the computer at the Message Switching Centre.

116 There was no more isolated a posting than one of the two islands off the west coast of Africa, Ascension (*see* 112) and St Helena (*pictured above*).

117 On the walls of the old Eastern Telegraph Court Room (boardroom) at Electra House, Victoria Embankment still used by the investment trust which was once Cable and Wireless (Holding), hang the portraits of the Pender family. (*From left to right*) John Jocelyn, 2nd Lord Pender (1907–65); John Cuthbert, 1st Lord Pender (1882–1949); Sir John Pender (1815–96), the second of two copies of the second version of the portrait by Sir Hubert von Herkomer (*see Appendix B*); Sir John Denison-Pender (1855–1929).

118–19 Straight aerials as well as round ones at Mount Misery, Barbados (*above*) and Cape d'Aguilar, Hong Kong (*below*).

120 John Bird, in 1979 Managing Director, Communications Systems and Services, Cable and Wireless.

121 Richard Cannon, in 1979 Managing Director, Public Telecommunication, Cable and Wireless.

122 Peter McCunn, who was appointed Group Chief Executive of Cable and Wireless with the title Deputy Executive Chairman, following the resignation of the Managing Director A. A. Willett on 1 June 1977, is seen here with Her Majesty the Queen and HRH Prince Philip, Duke of Edinburgh, at the reception held at St James's Palace in London on 2 May 1979 to mark the company's golden jubilee. Presenting a bouquet to Her Majesty is Mr McCunn's granddaughter Emily.

every way as significant to the development of long distance communication as the inventions of Cooke and Wheatstone in 1837, Bell's in 1897, Marconi's transatlantic breakthrough in 1901 and his short wave achievement of twenty years later. Presumably the Post Office, whose engineers had led this enormously successful piece of research, were in no doubt that they would be entrusted with applying it to the next major project for a trans-oceanic telephone cable which came along.*

This, it was decided, should be from Britain to Canada, where the chief executive of the Canadian Overseas Telecommunications Corporation (COTC) was Douglas Bowie, who had begun his career as a Cable and Wireless operator serving all over the world, and had become manager of their station at Halifax.

The decision to build CANTAT, as the Canadian Trans-Atlantic Telephone was called, came just at the time when Leslie Nicholls was succeeded as chairman of Cable and Wireless by the distinguished civil servant who had been Ernest Bevin's dynamic manpower chief in the war, and was now Permanent Secretary to the Ministry of Labour, Sir Godfrey Ince (who also had a reputation as a poet and a tennis player). He at once lent his great authority and drive to the effort being made to win for the company the participation in CANTAT which would at last give it recognition in the field of overseas telephony. Under his leadership and guidance it was decided that Cable and Wireless should *own* half the cable from territorial waters off Oban to the middle of the Atlantic, and COTC from there to Corner Brook in Newfoundland, and up the St Lawrence into Canada. Ownership involved being the contracting party without having any traffic interest. It meant specifying the design, placing the order, purchasing it, taking delivery and laying it, but not operating it, which at the UK end was the role of the Post Office who paid the owner for the use of it, and took half the revenue.

So it was a joint Post Office/Cable and Wireless delegation which went to Canada to establish how the project would be handled and apportion the various responsibilities.

For those who regarded Cable and Wireless as down and nearly out – particularly perhaps at St Martins-le-Grand – this ending of a time-honoured principle must have come as a shock. It opened a door which was to lead to activities far beyond merely overseas telephony, and gave the company the technical confidence, other

---

*In 1953 Submarine Cables Limited laid a telephone cable, 197 miles in length, for the British Army of Occupation between England and Germany. It was the longest in the world then existing. The following year they laid a 307 mile telephone cable from Scotland to Norway.

than in its traditional field, which it had hitherto lacked. A A Willett who joined Frank Lansbury's staff in the Company Secretary's office in 1948 considered this entry into long-distance, 'secure', intercontinental voice telecommunication as the biggest break in modern Cable and Wireless history. He was the first secretary of the Commonwealth Cable Construction groups.

Cable and Wireless and COTC, with the aid of the British Post Office, planned CANTAT in Ottawa throughout 1957 and, though it was still on the drawing board, decided to use Reg Halsey's British lightweight cable and British repeaters. CANTAT was to be a single cable, able to take communication simultaneously in both directions – originally sixty circuits along its 2,100 mile route. With two-thirds the attentuation of the TAT-1 cable, it needed less repeaters and achieved 70 per cent greater circuit capacity. A Joint Transmission Team was formed to work on the laying ship and the shore terminals to ensure that the transmission objectives were achieved during laying.

The revival of Cables, which as telegraphs were becoming obsolete, as a result of the co-axial voice transmission development, brought a new demand for the services of cable ships to lay, maintain and repair them. Over the next five years four new cable ships were to be built, designed specially to handle coaxial cables, one for the Post Office the *Alert,* and three for Cable and Wireless, the *Retriever, Mercury* and *Cable Enterprise.* In 1958 an international Cable Damage Committee was formed to afford protection to cables from damage by trawlers. It was composed of a consortium of cable companies and administrations from Britain, USA, Canada, France, West Germany, Italy, Denmark, Norway and Sweden.

In 1958 a Commonwealth Communications Conference in London evolved a plan for a Commonwealth Round-the-World Telephone System on the lines of CANTAT, still being planned and to include it. It would be owned jointly by all the Commonwealth countries concerned. This would give alternative routings in the event of failure. It would cross the Pacific from Vancouver to New Zealand and Australia, thence to Malaya, Ceylon, India and Pakistan, to South Africa* and back to the UK via West Africa and Gibraltar. The 30,000 nautical mile project was estimated to cost £88 million. Britain agreed to participate as a partner with the Commonwealth countries, and Cable and Wireless were made

---

*The Union of South Africa declared itself a republic outside the Commonwealth on 31 May 1961, and withdrew from the project, knocking a serious hole in it, so that the original concept of circling the world through Commonwealth territories had to be abandoned.

responsible on behalf of Britain for financing, laying and maintaining the UK share of the project, while the British Post Office designed and engineered the cable. As many functions were interrelated, there was a close working partnership between company and Post Office.

At the Pacific Cable Conference in Sydney in 1959 the governments of Britain, Canada, New Zealand and Australia studied the technical and financial aspects of the second leg of the Round-the-World Telephone System, a Pacific section from Vancouver to Sydney (COMPAC) for introduction early in 1964. The first leg, CANTAT, was opened on 19 December 1961 with Her Majesty Queen Elizabeth II in London speaking to John Diefenbaker, in Ottawa with the words 'Are you there, Mr Prime Minister?', and describing the project as a skilful and highly imaginative enterprise, 'a splendid example of co-operation between Commonwealth countries'. At a press conference in Lancaster House the same day Norman Chapling, managing director, told reporters the cable was the most advanced, efficient and economical telecommunications link in the world, and had now been handed over to the Post Office for operation in the public service.*

CANTAT had eighty telephone circuits finally, the quantity which all Commonwealth 'experts' reckoned would satisfy demand for the next twenty years Their forecast was out by a factor of 100. With the increase of quality over HF Radio and no increase in price, traffic doubled in the first two months and by the end of the year had trebled. In fact the eighty circuits were able to cope with the traffic for only three years. From that time on all telecommunications traffic was to double every three years.

At a conference in Kuala Lumpur in 1961 a South East Asia section of the Commonwealth Round-the-World Telephone System was planned to give Hong Kong, Sabah and Singapore voice communication with Australia – SEACOM costing £23·6 million of which Cable and Wireless were to contribute 46 per cent.

As a Crown Colony, Hong Kong, unlike the self-governing, sovereign nations of the Commonwealth, had never had to establish its own overseas communications company. In November 1948 Cable and Wireless had made a new agreement with the Hong Kong government to provide all the colony's external services and link

---

*On 9 December 1961 the Post Office cable ship *Alert* sailed from Southampton to lay a new eighty circuit system between Bermuda and New York. This cable was owned and *operated* jointly by Cable and Wireless Limited and the American Telephone and Telegraph Company (ATT). Thus with the opening of this system in 1962, Cable and Wireless Limited had at last arrived in the business of providing and operating international voice communications over a submarine cable.

them to the national network of the Hong Kong Telephone Company. The first public telephone booths were introduced at Union House and the Peninsula Hotel in 1949, and for the first time the public had access to the overseas radio-telephone for which a new receiving station was opened on Mount Butler in 1950. The first leased telegraph circuits came into operation in 1954. As, from the mid-1960s onwards, more and more land was reclaimed at the foot of the hills on Hong Kong Island, skyscrapers were built which changed the old skyline out of all recognition. The colony rapidly acquired new importance as a commercial, industrial, financial and tourist centre, and its build-up as the company's key telecommunications station in the Far East reflected this phenomenal growth. By the same token, the existence of an efficient and swift telecommunications system linked to the entire outside world itself attracted businessmen, industrialists and bankers to Hong Kong where the first on-line data transmission system in British banking history was soon to be installed in the colony's branch of The Chartered Bank.

'Data' – as in Electronic Data Processing – was the term applied to written information – words, figures, graphs, formulae – instantaneously transmitted by line or radio and received not, as on a teleprinter (Telex) by which most telegrams and cablegrams were now sent, on paper, but thrown electronically on to the back of a small glass screen (similar to a television set) known as a Visual Display Unit (VDU), or bigger, wall-mounted apparatus. The decade saw a Data Communications revolution with new codes, higher speeds, faster response times, greater accuracy, lower costs, and its wide adoption for airline, hotel and theatre bookings, stock exchange and bank information systems, commercial intelligence agencies and the like.

Data was a welcome newcomer to the expanding range of telegraph traffic. But in 1961, it seemed, consideration was to be given to a startling new advance in the *means* of radio transmission. The British Post Office discussed it with the US authorities in Washington in February of that year.

Was it possible, they asked each other, to throw into space, and to maintain in orbit round the earth, man-made satellites carrying equipment which would receive, amplify and re-transmit radio signals sent to it from earth? They thought it worth trying to find out, and HMG and the US Government agreed to hold a series of tests. 'Telstar' the first man-made telecommunications satellite designed by Bell Telephone Laboratories, and weighing 170 pounds was launched from Cape Canaveral in America at dawn on 10 July 1962. It hurtled in an elliptical orbit around the earth at speeds up to

18,000 miles an hour, 3,000 miles away at the furthest point and 600 miles at the nearest. The British Post Office built a giant reflector at Goonhilly Downs in Cornwall, 85 feet in diameter and weighing 870 tons, to follow Telstar. The system was capable of handling more signals than all the telephone cable and radio channels which then linked America with the rest of the world. The experiment was successful. It introduced no new means of communication. The 'wireless' of Marconi was still used for transmission, but transmitted waves at super high frequency (s.h.f.) were used. Unlike the lower frequency waves which either follow the curvature of the earth or are reflected back to earth by the ionosphere, the super high frequencies pierce the ionosphere and are normally lost in space. However, the satellite, an orbiting radio repeater, was placed so as to intercept these signals and re-radiate them to aerials and associated receivers back on earth. The techniques had been made possible by man at last mastering, originally for military purposes, the mechanics of launching rockets into space, which had so far belonged to the realm of science fiction and the dreams of visionaries like Jules Verne and H G Wells. In February 1963 in the United States the Communications Satellite Corporation (COMSAT) was formed, but it was some years before the satellite technique was developed to the stage of it taking its place alongside the traditional methods.

In the meantime, on 2 December 1963, the second link in the Commonwealth Round-the-World Telephone System, COMPAC, was opened by HM The Queen with an inaugural call of 14,000 miles from London to Ottawa, Sydney, Wellington and Suva (Fiji). It was described by Sir John Macpherson, a distinguished Commonwealth Governor, now chairman of Cable and Wireless, as 'the world's largest single telecommunications project'. On the platform with him was Sir Alec Douglas-Home, the Prime Minister. The Suva-Auckland section had been opened the previous year, but on 3 December 1963 a second COMPAC opening ceremony was held in the Grand Pacific Hotel, Suva, to switch the telephone cable through to Vancouver from where voice transmissions could be sent across Canada by microwave radio* to the east coast and the terminal of

---

*A 'terrestrial' method of transmitting radio signals horizontally across the surface of the earth at super high frequency used in radio communication via satellite. The link must be direct, without obstruction, between transmitting and receiving aerials which usually take the form of 'dishes' mounted vertically on towers. These dishes are small in comparison with the dishes of a satellite earth station or a tropospheric scatter station (see later). Because communication by microwave radio is restricted to two stations within line of (radio) sight of each other, signals are sent over long distances, as across Canada, by a chain of such stations.

the CANTAT telephone cable which took them to Britain and Europe. Shortly after the COMPAC telephone cable was opened, all telegraph traffic was diverted through it, and the sixty-year-old Pacific Cable, to the building of which Electra House had put up so much opposition, was switched off and put into retirement. It was never worth trying to recover it and it has been used for many years now by the University of Newcastle upon Tyne for a scientific project concerning the changes which take place in the earth's magnetism. This is carried out by measuring the earth current flowing in the cable, which still remains intact over a long distance. The station on Fanning Island, the scene of so much drama, was closed.

The days of the electric telegraph submarine cable, without the capability for telephony, were numbered, but at the same time submarine cable as such was holding its own against the ether as the medium of transmission in a way few would have predicted earlier. But then co-axial cables were carrying, not the electrical charges of the old telegraph lines, but electro-magnetic, radio waves at radio frequencies. Moreover with co-axial cable there was no limitation of frequency, no question of over-crowding the ether with channels. Once the best frequency for voice and music was determined, *every* cable could carry that. The waves would not leak from one cable to another. Five co-axial cables lying side by side could all be carrying signals on the same frequency without interfering with each other.

'Twenty-five years ago,' Reg Halsey, who had been a part-time director of Cable and Wireless since 1959, told the Institution of Electrical Engineers in October 1961, 'it appeared that cables must be unchallenged for communication overland and radio communication would be unchallenged across the oceans. Five years ago it seemed that just the reverse was true. Micro-wave radio relay systems providing thousands of telephone circuits or their equivalent, were coming into increasing use on land, and world-wide ocean cable networks were being planned. Now, in 1961, we see signs of a possible further reversal, namely waveguides, a form of cable capable of providing tens of thousands of circuits overland, and radio relay via earth satellites across the oceans.'

Telecommunication techniques were changing rapidly and so was what used to be called the British Empire. In July 1963 the Commonwealth Telegraphs Agreement of 1948 was brought up-to-date to accommodate the 'national bodies' of the Federation of Rhodesia and Nyasaland, the Federation of Nigeria, Ghana (the Gold Coast), the Federation of Malaya, and Cyprus. From 1

January 1964 the company's operations in Nairobi, Mombasa and Dar-es-Salaam were transferred to East African External Telecommunications Company Limited, of which Cable and Wireless had 40 per cent of the capital, the other 60 per cent being held by East African Posts and Telegraphs on behalf of Kenya, Tanzania and Uganda.

The growth of Hong Kong continued apace. Between 1963 and 1964 the number of telephone calls received at the new Central 36-position telephone exchange increased by 40 per cent. By 1964 Hong Kong was the largest branch in the Cable and Wireless system, with a staff of 1,000. Its Signals Centre for airline traffic, on the lines of that opened in Singapore in 1955, was handling 250,000 messages a month. A contract was placed for common control automatic exchange equipment in 1964, similar to that used in the COMPAC 'gateway' exchange at Sydney. It would provide direct dialled connections to Malaysia, Australia, North America and Europe, and constitute an important step towards achieving world-wide subscriber dialling. Total cost £750,000.

In 1964 too the representatives of eighteen nations met in Washington to discuss the final details of a global commercial communications satellite system costing £71 million, of which the United States would pay 61 per cent and Britain 8.4 per cent (£6 million). The USSR declined to join. The ground (earth) stations would be owned by the countries in which they were located. Signature of interim agreements in August 1964 brought the International Telecommunications Satellite Consortium (INTELSAT) into being. On 6th April 1965 the consortium launched from Cape Kennedy the world's first commercial communications satellite, INTELSAT I ('Early Bird') with a capacity of 240 two-way voice channels or one two-way television channel. Weighing 85 pounds, it was positioned 23,000 miles up, off the east coast of Brazil. When it became operational on 28 June, it established the first continuous satellite link between the United States and Europe, including Britain's earth station run by the Post Office at Goonhilly. President Johnson in Washington spoke to British Prime Minister Harold Wilson in London. 'I want to emphasise,' said Tony Benn, the Postmaster-General, at the opening ceremony, 'that a satellite communication system will be complementary to existing and future submarine cable systems. . . . We have now passed out of the purely scientific experimental stage and completed the first step in setting up an eventual world-wide system of entirely reliable and high quality telecommunication links by radio which will be unaffected by the hazards and vagaries of existing short wave radio systems.'

Telephone traffic across the Atlantic in 1965 was growing at a rate of 16 per cent a year.

The International Telecommunications Satellite Consortium operated under these arrangements until permanent agreements were concluded in 1971 and came into force in 1973. Under these permanent agreements INTELSAT became the International Telecommunications Satellite Organisation with a Board of Governors which made all the decisions concerning the design, construction, operation and maintenance of the system. The Governor for the UK usually had two alternates, one of which was provided by Cable and Wireless under the financial agreement between the company and the Post Office, by which the company participated in the UK investment share in INTELSAT in the proportion which utilisation by Cable and Wireless earth stations in British territories bore to the total UK utilisation.

Cable and Wireless became directly involved when in 1965 it contracted to provide and operate a satellite earth station on Ascension Island to give support communications for the US Apollo space project to land a man by rocket on the moon. A 42 ft diameter, steerable, parabolic dish aerial system was designed to communicate via satellite with the American earth station at Andover, Maine, as part of a communication link between the NASA tracking station on Ascension and the Goddard Space-Flight Centre. Both the INTELSAT/NASA negotiations and the Ascension Island project were the overall responsibility of A A Willett, with Dave Wilkinson on the engineering front and Harry Madigan as Project Installation Engineer. Built by the Marconi company and costing £1 million, the station at Devil's Ashpit was handed over to Cable and Wireless on 21 September 1966. It was officially opened on 18 February 1967 by Henry Eggers who, when Norman Chapling had retired in 1965 on reaching his 65th birthday, had become sole managing director, with Edgar Howitt who had joined the company in 1924 as deputy managing director.

For the Apollo project Ascension Island was one of three remote earth stations which, with three tracking ships, helped NASA monitor the historic journey to the moon.* But the contract with NASA was for ten years, and the station had spare capacity which was used for commercial channels to the US and elsewhere. The property of Cable and Wireless, it was the UK's first overseas civil satellite communications facility.

---

* The company's HF radio station on Bermuda played an important part in the link, between a NASA tracking ship and the Goddard Space-Flight Centre, in the critical moments after blast-off.

The experience the company's engineers gained from the earth station at Ascension led to the building of two more, at Hong Kong and Bahrain, each with the capability of transmitting and receiving more than 300 telephone channels and costing £2½ million, to work to INTELSAT III satellites launched in synchronous (stationary) orbits above the Pacific and Indian Oceans. An Atlantic satellite soon followed.

It must have been all very bewildering for Sir Edward Wilshaw who, on reaching the age of 84, in 1964, finally decided it was time to retire as Governor of Cable and Wireless Holding, and became President. He never made the Mansion House but, with considerable skill, made the company, now shorn of its telecommunications shares, into one of the most profitable investment trusts in the kingdom. He died, aged 88, in 1968. He left £114,650 (£65,350 net).

Because of the increasing number of entities within the Commonwealth partnership it became apparent to the partner governments that the old arrangements for 'imperial' communications had become outdated. As a result of decisions taken at Commonwealth Government conferences in 1965 and 1966 a new 'Commonwealth Telecommunications Organisation' (CTO) was set up with a Commonwealth Telecommunications Council (CTC) meeting once a year. They handled the operational and business relationships of the relevant Commonwealth telecommunications entities. A permanent administrative office, the Commonwealth Telecommunications Bureau, was established to administer the decisions and agreements of Council. It operated with a staff of Commonwealth experts under a General Secretary located at 28 Pall Mall, London. Through their collective activities the now many national bodies in the Commonwealth partnership were less isolated and better informed about world telecommunications than any other association of free nations. Cable and Wireless was not represented separately on the Council but its interests were covered by the Representative of Britain who was a nominee of the Post Office, Britain's national body, and by the Representative for British Overseas Territories and Associated States.

The Postmaster-General had been right to welcome a technique which was unaffected by the vagaries of 'ordinary' short waves – the commonly called 'HF Radio'. However, other techniques were also available to link distant communities. One of these techniques was tropospheric scatter radio transmission. Tropospheric scatter links operated in the Ultra High Frequency (UHF)* band, and made use

See page 346 for footnote.

of the phenomenon that the transmitted beam of radio waves would be scattered by inhomogeneities in the troposphere, allowing a small part of the radiated signal to be received at locations far beyond the horizon. High transmitter power and large aerials were required, usually at each end of such a link, but costs were lower than those of constructing a satellite earth station. The system could provide a capacity of at least sixty channels over distances of several hundred miles. It met the needs of the West Indies exactly.

In 1965 Cable and Wireless (West Indies) Limited, in co-operation with the American Telephone and Telegraph Company (ATT), announced the Eastern Caribbean Expansion Project using tropospheric scatter systems inter-connecting Barbados, St Lucia, Antigua, Tortola and thence to St Thomas and the United States. The existing system between Barbados and Trinidad would be expanded. The new network would carry additional telephone, Telex, leased services and telegraph traffic, and have automatic dialling. An 80 circuit co-axial cable would be laid from Tortola to Bermuda. On return from a tour of the Caribbean in 1968 Donald McMillan, an ex-director of the External Telecommunications Executive who had become chairman, said the unexpectedly rapid rate of growth in the area meant the company planned an even greater expansion. A A Willett, who in January 1973 took over as managing director from Edgar Howitt who had been with the group for forty-nine years, was later to announce a £15 million expansion of facilities in the West Indies to give a system ten times the capacity, and to include a 160 circuit co-axial cable between Jamaica and the Cayman Islands.

This was a measure of the company's confidence in its future in an area from which, after the Cuban Crisis of 1962, American telecommunications companies had made a bid to oust the British

---

*The repetition rate of a wave is referred to as its frequency and the unit for one repetition in a second is the hertz (Hz). The spectrum normally used for radio communication lies between 3 kHz to 300,000 MHz and this is divided conventionally as follows:

3 kHz to	30 kHz Very Low Frequency (VLF)
30 kHz to	300 kHz Low Frequency (LF)
300 kHz to	3000 kHz Medium Frequency (MF)
3 MHz to	30 MHz High Frequency (HF)
30 MHz to	300 MHz Very High Frequency (VHF)
300 MHz to	3000 MHz Ultra High Frequency (UHF)
3000 MHz to	30,000 MHz Super High Frequency (SHF)
30,000 MHz to	300,000 MHz Extremely High Frequency (EHF)

Radio waves of frequency below a nominal 30 MHz are reflected back to earth by the ionosphere depending on the angle of incidence while those above 30 MHz pass out into space.

influence. Continental Telephone Corporation, the second largest internal telephone company in the USA after ATT (Bell), had for some time run the telephone system of Port of Spain (Trinidad), Grenada, Barbados, and Jamaica. In 1966 Continental tried to persuade the Trinidad Government, who were part owners of the internal telephones, to allow them to take over external communications from Cable and Wireless whose concession, which did not involve partnership with the government, had expired some years earlier, and was now continuing on a year to year basis. Cable and Wireless put in a counter-offer which in 1969 the Trinidad Government accepted, at the same time renewing Continental's internal contract. Cable and Wireless were no longer to run the service on their own but through a joint company in which the Trinidad Government had a 51 per cent interest and Cable and Wireless 49 per cent. Trinidad and Tobago External Telecommunications Company (TEXTEL) began operating on 1 January 1970. The Government provided a chairman and three directors, and the company the other three directors. The general manager was a Cable and Wireless man, Douglas Buck, who was not a director, however.

By now Cable and Wireless had had wide experience in handling this type of situation. Going into partnership, establishing good relations with ministers, exercising patience, showing understanding, making concessions to national pride, was positive policy. It bore fruit in the joint companies established in Africa in 1964, East African External Communications Limited and Sierra Leone External Telegraph Limited. Where a government took over 100 per cent, with no Cable and Wireless participation, as in Nigeria in 1966 when the Nigerian External Telecommunications Company (NET) took over when the joint company established in 1963 was acquired by the Nigerian Government, an African was general manager with a Cable and Wireless man (Douglas Buck again) as his Assistant. Naturally, when joint, or wholly Government-owned companies were created, existing staff were retained. Cable and Wireless always had a clause in the transfer agreement that these should be treated no worse than under their regime. Under a ten or twenty year concession the company provided all the capital. The territory's needs were determined by normal market research methods; the necessary equipment was bought and installed, and it remained the company's property. Some kind of 'licence fee' was generally paid, or a royalty which could be a percentage of profits or of local collections. The government then entrusted the running of the service entirely to the company who took all the revenue and with it hoped to pay off the capital invested during the period of the concession. Not all joint company considerations were the same – and the partnership was not always with a government. In July 1967 the company joined with its American competitor to form Cable and Wireless/Western Union International Inc to acquire the assets of Cable and Wireless (West Indies) in Puerto Rico, St Thomas and St Croix in the American Virgin Islands.

The main build-up was taking place entirely with the company's own resources in the total-concession areas of the Middle and Far East, where Bahrain and Hong Kong were now the company's largest branches. The company had been the first to introduce Telex to Bahrain in 1963 and it had spread to the rest of the Gulf; in 1966 it opened tropospheric scatter links between Bahrain, Doha (Qatar) and Dubai. The same year it inaugurated a tropospheric scatter circuit between Hong Kong and Taiwan (Formosa).

The big projects planned earlier in the decade now came into operation one by one. The full SEACOM, the third section of the 23,000 mile Commonwealth Round-the-World, wide-band co-axial submarine cable, was opened on 30 March 1967. The first section of this South East Asia Commonwealth Cable, from Singapore to Jesselton in North Borneo and Hong Kong, had been opened on 1 March 1965. The second section took the cable from Hong Kong to Guam, Madang (New Guinea) and Cairns in Australia, from where there was a connection to the COMPAC cable in Sydney by microwave radio and co-axial land line. SEACOM had 80 circuits between Singapore and Guam, and 160 circuits between Guam and Cairns. Queen Elizabeth II opened SEACOM with an address heard at all stations on the cable, each of which had their own ceremony.*

The Bahrain earth station was opened on 14 July 1969 by the Ruler of Bahrain who made a telephone call to Prince Philip in London. A call to Washington on 24 September 1969 opened the earth station on Stanley Peninsula in Hong Kong working to the Pacific satellite – when it was announced that a second 'Big Dish'

---

*The cable ship *Mercury* was built specially for laying the Commonwealth co-axial cable. In the period of laying CANTAT in the North Atlantic only one large laying ship, the Post Office *Monarch* was available. For COMPAC and SEACOM, where very long steaming distances were required to and from the cable and repeater factories in the UK, it had been foreseen that two large cable laying ships would be required to speed up the actual installation period in order that these systems could become operational in the shortest time possible. Even with two large ships to share the work it took five years to complete the actual cable laying work for COMPAC and SEACOM.

reflector and station would be built on the same site to work to the Indian Ocean satellite.

What Sir Godfrey Ince in the previous decade had foreseen as the Golden Sixties had certainly fulfilled their promise. In the two financial years 1965-67 alone the company invested some £18 million in new plant and equipment, all from its own funds without resort to the money market, and 95 per cent of it spent with British manufacturers. In being self-financing Cable and Wireless were unique. The company had made big profits in the war and in those pre-union days had kept their wage bill low. They ended the war with £12 million in gilt-edged investments. Though Government-owned they had no access to the finance facilities open to 'national-ised' industries in Britain, no access at good rates to the National Loan Fund, no subsidy of any kind – not that they wanted one. But having all their operations overseas everything they did remained at political risk. This had always been the case, but it was very much more so now.

At the end of the 1960s management considered that the concession market could not but shrink, and the assumption was made that by the end of the century it would probably have disappeared altogether. Though contracting geographically, concessional business was, however, becoming increasingly profitable. It included the Leased Circuits (Private Systems) business, which was the knitting together of customer requirements and fitting them into segments of the existing public network, hiring exclusive time twenty-four hours a day for 365 days for five or ten year periods, and supplying the terminal equipment.

The first private systems had been, as noted, within the field of airline operation. Because, in those early days, they did not conform with the world's operators' views on how telecommunications should be run, they were outlawed. There was a feeling in pre-Wideband days – and not only at Cable and Wireless – that it was intolerable that groups engaged in a common activity, such as civil aviation, coffee growing, oil, shipping, banking, should pool their messages to secure reduced rates. Telex had been similarly resisted because of the damage it would do to traditional telegraphic traffic; and the leasing of complete circuits was resisted because it would reduce the Telex traffic. The static network of the day, whatever it was, must not be upset; the customer must employ what he found and make the most of it. The existence of SITA was a fact, but for years, as seen, OTCA, Cable and Wireless and others refused to recognise it and

charged SITA's separate airline members full public rates. But when
Don Michell persuaded the management at Mercury House to
change their minds, annual leased circuit revenue quickly mounted
to £1,800,000.

In 1971 the bankers took a leaf out of the airline operators' book,
and sixty-eight of them commissioned a study of the possibility of a
private communications network to carry increasing international
payments traffic. Two years later 239 banks combined to launch the
Society for Worldwide Interbank Financial Telecommunications
(SWIFT) with HQ in Brussels.*

Telecommunications operators such as Cable and Wireless had
already realised they could no longer control the *whole* of the market,
but had to let the market control, or rather guide, them. They could
not dictate what a customer could, and could not, do. They could
justifiably insist that he did not overload the system, but in future
their only form of regulation was to be price – have what you want,
but the more exotic the more expensive. It was a tide which the new
management at Mercury House had no wish to turn back, and well
before the 1970s Leased Circuits had become 'traditional' practice.

There was a future for the company in playing out its traditional role until the final curtain, however far away that might be.
In 1969 Cable and Wireless would have been the choice of most
governments which still sought to have their Public Telecommunications (PT) run for them by an outside agency. Whatever damage
had been done to the company's standing by its association with
colonial imperialism had worn off. The fact that Cable and Wireless
had no home base, and no attachment to any particular manufacturer, was an advantage. In studying its specialised market it had
acquired a reputation for comprehending the developing world and
being able to work in it. It had perfected the art of establishing the
right relationship with foreign governments and being members of
overseas communities – a legacy of Eastern Telegraph days. When it
came to obtaining *and retaining* concessions, no matter the wording on
the document, these were the factors which counted. Along with the
more obvious ones of high technical efficiency and the credibility
which it had won by always being able to see a complicated
operation through to the end, they constituted an enviable asset
which made the temptation to keep to what was familiar and safe a
very great one.

But, as Peter McCunn who joined the Traffic Department in 1947

---

*When the first stage became operational in May 1977, 270 banks, including
several in North America, were taking part, but at the last moment the British
banks opted out.

and became a director in 1969, observed ten years later, in that way the company might have finished up in a blaze of residual values and topped up the pension funds, but there was no good reason for allowing it to 'finish up' at all. Thus, in 1969 the Court of Directors decided to make indefinite survival their objective. In making an almost metaphysical decision to keep alive a body, a corporate structure, which had acquired so real but indefinable a character, they were thinking of the staff – overseas, in Britain, and in the marine service – who were the successors of the hundred-year-old tradition traced in these pages.

To survive, Cable and Wireless would have to acquire an alternative activity while at the same time maintaining and developing the traditional business. For some time the company would ride two horses, making as good a pace as circumstances allowed on Old Faithful and training up a Young Hopeful in the expectation that it would prove a winner.

13

Quiet Efficiency
1970–1979

The final decade of the first fifty years in the life of the operating company created in 1929 was coloured by the decision of the Court of Directors to deploy the expertise inherent in the company from now on in a very much wider market than hitherto – on any basis where there was money to be made – with growing emphasis on 'Projects'.

Henry Eggers retired on 31 October 1969 after fourteen years as managing director and was succeeded by Edgar Howitt. On 1 November a new organisational structure of Head Office came into operation to enable management to pursue the new two-pronged objective, adopting in part a report written by PA Management Consultants Limited recommending, among other things, a computer based management information system allied to strict budget control.* A Market Forecasting and Research Department was set up in 1970 under Andrew Boa.

The year in which these foundations were laid for a new tradition bade farewell to what John Packer, today's recorder of Porthcurno history, called 'the age of polished mahogany and lacquered brass'. On 31 December 1970, Porthcurno ended its life as a cable station, and The Tunnel in the cliff was closed. 'As the heavy flywheel of the No-Break slowly came to rest, and power was removed from drives and ancillary circuits of the London-Carcavelos-Lisbon TG1 cable,

---

*The first public intimation of the existence of the diversification policy was the sentence in the company's annual report for the year ending 31 March 1970 under the heading 'Reorganisation of Head Office': 'The new structure will provide an organisation which, in addition to administering the Group's telecommunications operations, will provide the capability to explore and develop new avenues of business allied to telecommunications.'

a hundred-year-old link in the chain of world communications in West Cornwall was broken. The "whispering galleries" lapsed into silence, disturbed only by the water dripping from the vast granite caverns.' For Richard Angrove, who wrote those words, it was indeed the end of an era. Along with Sid Chappel, Bill Harvey and Arthur Thomas, who also witnessed the final act of closure, he had spent his whole working life at PK. However, he continued to serve the Engineering College where the new generation of Cable and Wireless staff learnt the new techniques which had supplanted the old ocean telegraphy now finally laid to rest.*

Helping PK in this task were the training colleges in Bahrain, Barbados and Hong Kong. A £4 million investment programme for Barbados included a new Cable and Wireless Telecommunications College at Wildey to replace the Dover Centre school. It was opened by William Demas, Secretary General of the Caribbean Free Trade Area, in October 1972. A month later another project in the programme was opened by HRH Princess Margaret, the EC$12 million Barbados Earth Station, working to the Atlantic satellite, capable of more than 1,000 telephone channels and giving direct links to Jamaica, Britain, USA and Canada.

The need for highly trained engineers in the West Indies was increased by the decision, taken in 1970, to upgrade the existing tropospheric scatter system, which provided communication between the islands, with one capable of giving the extra voice circuits which, it was reckoned, would be required in 1977. The method chosen was 'line-of-sight' SHF microwave radio, for which the hilly nature of the islands, except Barbados, was an asset – though the hills were not always on the right side. It was impossible to link Barbados and St Lucia by a line-of-sight link; the mast on Mount Misery at the former was only 1,000 ft above sea level and it needed to be double that. So the existing tropospheric scatter service was upgraded from 80 voice circuits to 240 with access to the 800 mile East Caribbean Microwave System at St Lucia with its 960 simultaneous telephone circuits reaching north to Tortola in the British Virgin Islands and south to Trinidad. The chain was going to be the longest series of island-hopping microwave links in the world. Ron Hobbah was the Regional Engineer in charge. There was some doubt whether microwave radio could be made to work over the long sea distances of the Caribbean but the data base which Cable and Wireless engineers in London had established through

---

* The first half of the course ceased to be held in London just before Head Office moved from Electra House to Mercury House in 1955, and the whole of the eighteen months basic course now took place in Cornwall.

tropospheric scatter enabled them to design a system to meet criteria more exacting than any previously encountered. It was an engineering achievement which added greatly to the company's technical capability in the eyes of the rest of the world, and was a useful 'sales tool' when it came to selling systems in the future. It took six years to plan and complete, and involved placing masts for the dish antennae, able to withstand hurricane, earthquake and lightning, on peaks inaccessible to all but cargo-carrying helicopters, donkeys and engineer Dave Parker. Lifting cement blocks to the top of Saba was perhaps the most hazardous undertaking, but it was the accident of this little island's position that made possible the long last link between Antigua and Tortola, which became the Caribbean Operations Centre under Richard Pilcher and was the terminal of the 1,600 kilometre cable to Bermuda. A hovering helicopter lowered the blocks to the site from the sky, but the project team had to cut a stairway of a thousand steps in the undergrowth and climb it every day to build the vital repeater station in the clouds. Such stations were necessarily unmanned, though monitored, and their batteries were continually charged by the mains and standby diesel engines lower down the mountainside which switched on automatically when the local supply cut out. One day the power may come from solar panels or windmills.

For an area with a different order of economy (compared, for instance, with Hong Kong) the EC$14 million investment by Cable and Wireless (West Indies) and its associate Trinidad and Tobago External Telecommunications Limited (Textel), was considerable. The return was not reckoned to reach the company average. Though difficult to justify commercially, there were political considerations; and, in any event as always, the presence of a modern telecommunications system was itself expected to attract trade and commerce, and so improve the economy. Trinidad had struck oil and was developing fast under the influence of a stable political regime. Textel's earth satellite station opened for service in 1971. In Jamaica the need to expand the system which the company ran under Government licence became apparent in the early 1960s. The company's application for a long-term extension of its licence was granted, but the Government reserved the right to terminate it, and purchase the company's assets, whenever it wished. Plans were well advanced to build an earth satellite station and then, in November 1970 the Jamaican Government exercised this option and had a limited liability company formed, Jamaica International Telecommunications Limited (Jamintel) in which it held 51 per cent, and Cable and Wireless (West Indies) 49 per cent, of the J$15 million share capital.

The Jamaican Government had the option of increasing its share to 75 per cent at any time during the life of the agreement and, on expiry in 1988 of purchasing the entire holdings. The staff at Jamintel of some 360 technicians were eventually all Jamaican, most of whom received eight months' development training at the Cable and Wireless college at Barbados.

With self-determination in all things, including telecommunications, the guiding principle of the developing world, joint ventures such as these, and the partnerships formed on the other side of the globe with the Rulers of Sharjah and Ras Al Khaimah, were to be an important part of the pattern of the company's Public Telecommunications business from now on. But there were to be exceptions. Contrary to the trend, in August 1970 the Government of the Yemen Arab Republic entrusted the company wholly with running the country's international telecommunication services. A Cable and Wireless branch was opened in the capital, Sana'a, on 1 April 1971.

The revenue from shared operations was naturally lower than that from wholly owned enterprises, and pre-tax profits suffered in consequence. The year 1970/1 saw them reduced by £2½ million from £7½ million to £5 million. In the summer of 1971 the directors reviewed the probable future level of capital expenditure in relation to available reserves, and decided to go to the Money Market for a medium term loan – the first in the company's history. There was no question, of course, of tapping the Public Purse. Such was the company's record of profitability and stability there was considerable competition for the privilege of being the leaders. Its borrowing capability was never in doubt. A City Syndicate – a consortium of banks – was formed for the purpose, and with it the company negotiated a credit facility of up to £19 million to be drawn before April 1974 and repayment five years later. In the event only £6 million was ever used. Group profit jumped to £9 million in 1972/73 and to £14·7 million in 1973/74.

In 1971, on the basis of the Post Office handling all UK external communications, the company sold to the Corporation its half share in CANTAT. Ownership had established Cable and Wireless as the UK arm in the inter-continental co-axial submarine cable field. Through it the company was now fully launched as a major participator in Britain's development of Wideband telecommunication systems, and there was no purpose in holding on to the launching pad. The financial agreement reached with the Post Office over CANTAT was part of an overall deal involving a number of exchanges of rights and ownerships. The Post Office had operated CANTAT from the start; now they owned it as well, and Cable and

Wireless were relieved of the cost and responsibility of maintaining it. The company still retained the UK share of the capital investment in both the COMPAC and SEACOM systems.

Though the administrative headquarters of the group was in London, its largest operational centre was in Hong Kong which had been the biggest branch for more than a decade and was now bigger and more important than ever. The first fifteen storeys of the fine New Mercury House, designed by Fred Hammond, chief architect, and members of the Cable and Wireless Architectural Services Group in association with David Ho of Hong Kong, was opened in 1972. The Message Switching Centre was first computerised in 1969 and the installation in May 1972 of the first fully computerised Hasler Telex Exchange introduced automatic connections (subscriber dialling) and internal Telex services. The company now operated two Big Dish earth stations both at Stanley; one working to the Pacific satellite, opened in 1969, and the other to the Indian Ocean satellite, opened in 1971. By the end of 1972 they were handling a million outgoing telephone calls annually, and traffic generally was growing at a rate of 20 per cent a year.

Hong Kong was thus the natural choice for the establishment of the first company, as opposed to the department in London, created exclusively for Private Systems Business. In October 1972 'Cable and Wireless Systems Limited' was formed with registered office in Hong Kong, to provide specialised communications systems, services and products in non-franchise areas of the Far East. John Rippengal was appointed chief executive. Within a year it had secured two major contracts, an 18-station Microwave System for the China Light and Power company in Hong Kong and an HF Radio System for BS Brunei State Telecommunications. It followed these with projects which included a Microwave System for the Electricity Generating Authority of Thailand, a satellite earth station for the remote Central Pacific island republic of Nauru, Speech Processing and Control Equipment for Hong Kong International Airport, and long range ground to air HF Radio for Gulf Air in the Arabian Gulf.

The slowly developing diversification exercise was not without setbacks. The aim had been a £50 million turnover with accumulated launching losses having been recovered by 1980. By the middle of the decade the latter amounted to £6·7 million – accumulated overall group profit in the same period was some £55 million. The early days of diversification necessarily involved nursing infant enterprises which might or might not grow into money-makers. Half of the £6·7 million loss came from the liquidation of a small manufacturing company which Cable and Wireless acquired in Hong Kong in 1972 called Coltronics Limited. With so much activity taking place in Hong Kong and South East Asia, it was hoped that this would complement in the Far East the function of the Development and Production Unit at Smale House, London, which for several years had been developing and designing equipment – the 'one-off engineering special' – to meet the diverse requirements of concessional and non-concessional business all over the world. At first there was spare capacity at the Coltronics factory, and the company undertook to assemble cheap pocket electronic calculators. Encouraged by the profitability of the initial, modestly-sized contract the general manager, without authority, accepted an order for 960,000 calculators and contracted to buy £10 million worth of components. It soon became obvious that the order could only be completed at considerable loss but it proved impossible to cancel the components. Costly litigation and settlements out of court ensued, and Coltronics Limited was wound up.

Following an account of the affair, accompanied by accusations of mismanagement, in a journal called *Social Audit,* based on photocopies of confidential company documents given to its editor by a chartered accountant in Cable and Wireless employ, the national press added their comment; there was a government (Departmental) investigation and a statement in the House of Commons which imputed no blame on Cable and Wireless. However, the company itself agreed with hindsight that some disclosure in the previous year's accounts would have been desirable. There was never any quesion of a 'cover up'. The world at large considered the loss the result of taking a commercial risk involving a possibility of failure of the kind any competitive enterprise must always be prepared to take.

The age of electronic calculators was far removed from the days in which Cable and Wireless had its beginnings – in fact by precisely one hundred years. On 6 June 1972 a reception was held in London on board the headquarters ship *Wellington,* moored on the Thames at Temple Stairs, to celebrate the formation of John Pender's Eastern Telegraph Company incorporated on 6 June 1872.

In spite of political, economic, military and technical crises, and the doom-laden prophesies of Sir Edward Wilshaw, few could deny in 1972 that the Century of Service* constituted a success story. As many of the company's customers were wont to remark, if Cable and Wireless had not existed, Britain would have had to invent it. And

---

* The late K C Baglehole, the company's archivist, produced a book of that name in 1969 to cover, in brief, the events of 1868–1968.

contrary to the forebodings in 1950 the truncated company of twenty years later was extremely healthy.

Cable and Wireless contributed to, rather than took from, the national exchequer, so the Treasury's concern was merely that it should maintain its record of profitability. Five nominee shareholders – three from the Ministry and two from the Treasury – attended the annual general meeting to conduct the formal business required by English company law. The Government did not appoint the directors for any fixed period, but *all* retired at *each* AGM and offered themselves for re-election, which gave the Government the power to remove any of them on that occasion. Part-time outside directors, like the Senior Director of External Telecommunications at the Post Office, normally held office for two to three years; but the inside, full-time executive directors had to justify their positions each year by performance. The chairman was told on appointment that in the normal course he could expect to remain in office for three years, perhaps five. As part-time, non-executive, he was not concerned with management. This was the prime responsibility of the chief executive, the managing director, who was appointed by the Court of Directors from among its members, with Shareholder approval. As the man responsible for the execution of policy and managing the Cable and Wireless group in all its ramifications, the status of 'MD/Ln' in the world of international telecommunications was second to none.

In 1972 this was Edgar Howitt, and it was during his time that, at the prompting of A A Willett and R A Rice, through H G (Peter) Lillicrap, the radio and television adviser to the Post Office, who had succeeded Donald McMillan as chairman in February, the company agreed with the Government eight areas of the telecommunications market in which it could justifiably diversify, and be left to do so as currently constituted.*

---

*In their report on Cable and Wireless written in 1975, the Select Committee on Nationalised Industries, commenting on the growth of non-concessional business stated: 'When requirements of this nature had arisen in places where the Cable and Wireless held concessions, it was natural for the company to be called on to meet them and it had therefore become involved in such activities, though on a limited scale, over a number of years. More recently the group had seen a need to develop this sector of business outside concessional areas, since it was becoming apparent that this was not the kind of optional fringe activity but something that had to be developed for the company's long term survival. Accordingly at the end of 1972 an approach was made to the Government who agreed, early in 1974, that the company might go ahead provided that certain activities were not undertaken in the United Kingdom; and more generally that the company should continue to seek to avoid promoting foreign sources of supply when there was a British equivalent. In this connection it is perhaps important to stress that non-concessional business is an activity which has only been going at maximum effort and with full Government approval for little more than two years.'

Such diversification required the development of new marketing skills. Persuading a government that the company was capable of handling their telecommunications and obtaining a concession was done by its top people. Once the concession was obtained, a Public Telecommunications service needed no promotion at a lower level. The service sold itself on quality. So people had to be found inside the company, and outside, who could supply the missing expertise. Some £1 million was already being earned round the fringes of the company's concessional operations, and those involved in this were brought back to Head Office to form the nucleus of a commercially minded Private Systems Business department. The marketing operation was divided into two. To ensure that the company acquired its due share of whatever concessional business was still being offered, this activity was concentrated in an International Operations Division; and non-concessional business in a Communications Systems and Services Division (CSS) – successor to the existing PSB – to head which John Bird, deputy managing director of Ultra Radio, was recruited by A A Willett. Roving representatives were dispatched to developing countries to see what opportunities existed. The company's main offering was now the small satellite earth station (Standard B) which, although it required more power from the satellite than the larger Standard A station, and therefore meant a higher fee to INTELSAT, cost only £500,000–£1,000,000 (about the same as a comprehensive HF radio installation), compared with the £5 million or so of the larger Standard A station.

Over the coming six years C&W were involved with Standard B earth stations in Mauritius, Seychelles, Sana'a, The Gambia, Maldives, Malawi, Bermuda, Belize, Gibraltar, Tonga, The Solomon Islands, New Hebrides – all but one 'concessional'.

The need for diversification was underlined within a year when in 1973 the Brazilian Government decided not to renew the PT concession which they had first granted the Western Telegraph Company for sixty years in 1873 and then for another forty years in 1933. With the departure of Peter Fontaine, the last Cable and Wireless Regional General Manager, South America, ended a hundred year association representing a rich vein of original entrepreneurial investment – in personalities and money. All foreign operators left Brazil between 1971 and 1973 but it was likely to be some time before well-entrenched British procedures were totally abandoned and Brazilians stopped calling a telegram a 'Western'. But on the whole the company's new marketing thrust aimed at alternative business had the effect of stimulating new franchises.

In 1973 also China invited Cable and Wireless to lay a

300 channel telephone cable from Hong Kong to Canton – a fitting climax to the ten years which Brian Suart had spent as general manager in Hong Kong, after being Engineer-in-Chief, Mount Butler Receiving Station, from 1954 to 1957. In July 1974 another joint company was formed with Philippine interests called Eastern Telecommunications Philippine Incorporated, and this company took over the assets in that region of the old Eastern Extension Australasia and China Telegraph Company. The general manager was M V Bane. The profits began rolling in, and a major problem of the 1970s was how to share them equitably. The new Commonwealth Telecommunications Financial Arrangements, in which twenty-six of the twenty-eight partner governments took part, were introduced on 1 April 1973 in an attempt to take account of the new wideband facilities, submarine telephone cables and satellite communications. It replaced the old Wayleave Scheme which had been operative since 1950.

There was no question of tackling the non-franchise market worldwide. A third of the market was seen to lie in the United States and another third in Western Europe. Japan accounted for 15 per cent and the whole of the rest of the world for another 15 per cent. The European market was growing more quickly than the US market, and by the 1980s was expected to represent a larger proportion than USA. It was tempting to start across the Atlantic where English was spoken and business and government methods, taxation and customs were similar to those in Britain. But management thought it would be easier to control a new business which was nearer home. Distance, plus the establishment of the European Economic Community which was expected to offer opportunities for the exploitation of British knowledge, decided the company to mount its first CSS exercise across the Channel, though already a procurement agency had been established in New York. The latter was Cable and Wireless Technical Services Inc, which, apart from purchasing equipment, acted as a listening post for any development likely to offer CSS a platform for future action in the part of the world which for so long had set the pace in telecommunications technology.

In 1973 Cable and Wireless formed Eurotech SA as a small shell company for market research purposes as the first stage in its European CSS venture. The plan was to create new companies throughout the European Common Market (not to acquire existing firms) as providers of telecommunications systems and equipment which ideally had been designed and produced within the Group. It was seen as a totally controlled, trans-national operation with a wholly-owned holding company, Eurotech BV in Holland, which offered tax advantages. Eurotech BV then set up its own wholly-owned trading subsidiaries which in a few years had spread to Belgium, France, Germany, Italy (two), Holland and Denmark. There were no local shareholders.

HMG publicly backed the new policy. 'I believe the company has a great deal to offer,' stated Eric Varley, Secretary of State for Industry in his written answer to a Parliamentary Question, 'and that it is important that it should be free to compete with the private sector on an equal footing.'* With the support of HMG and the Post Office Corporation, the company joined International Aeradio Limited in 1974 in forming Energy Communications Limited to provide an integrated off-shore communications network in the British oil exploration sector of the North Sea.

Cable and Wireless had been on the Order of Reference of the Select Committee on Nationalised Industries since 1969, but no opportunity occurred to deal with it until November 1975. Following the publication of their journal of 31 August 1974 with the piece on Coltronics, *Social Audit* wrote to the chairman of the Select Committee on 2 September making a formal request that the company's affairs should be examined. But at the session of 3 December 1975 the chairman, Sir Donald Kaberry, stated; 'To make it absolutely clear, this committee had decided to investigate the accounts and reports of Cable and Wireless as a nationalised undertaking before, and not in consequence of, the reports that were published in the *[Social Audit]* journal.' Before taking evidence from the company's chairman and managing director, and witnesses from the Treasury, Department of Industry, Foreign Office and *Social Audit,* the members of Sub-Committee C visited the company's operations in the Caribbean, Hong Kong and Bahrain.

The outstanding impression conveyed by Cable and Wireless, stated the Committee in the conclusion of their report (No. 472) published on 22 July 1976, was the quiet efficiency with which it went about its business of providing world-wide telecommunications services.

> So much so, that it can almost be said to be taken for granted in the areas in which it operates. Its origins go back more than 100 years (the Cable Depot at Singapore was celebrating its centenary when the Sub-Committee visited it last October) and during this time it has built up a tradition and reputation in which Britain can take pride.

---

* In 1974 the Ministry of Posts and Telecommunications was abolished and its staff absorbed in the Department of Trade and Industry which became the sponsoring department for Cable and Wireless.

The enterprise is well-founded and profitable, and earns for the country considerable amounts of foreign currency. Its expertise, backed by the British Government, will continue to be in demand for the foreseeable future. The efficiency of its communications services and the high standard of facilities that it provides, serve both to attract and to promote commercial activity in centres such as Hong Kong and the Arabian Gulf. It has made a material contribution to growth in developing countries in the provision and operation of sophisticated equipment and in training local staff in technical and management skills. It maintains a very presentable image at all its various installations around the world.

Inevitably it reflects some of the corresponding weaknesses of its strengths – too great a dependence on traditional loyalties and methods of working; the remnants of a traditional pay and career structure (now being replaced by job evaluation); difficulties of local staff relations in certain areas, combined possibly with a degree of paternalism; a tendency towards over-centralisation. However, these are all aspects of change to which the Company is adapting itself; there is every reason to suppose that it will be just as successful in doing so as it has been in meeting the challenge of new technologies.

As to the future, the impression gained during the Sub-Committee's visits to installations overseas was that the public international telecommunications services provided by Cable and Wireless are of such apparent excellence that many (if not most) administrations would be reluctant to run the risk of disrupting them by dispensing prematurely with the high standard of efficiency and expertise offered by the Company. For this reason, we hope and believe that the phasing out of the Company's concessional business will be a slower rather than an accelerating process, and that the goodwill, which is so noticeable a feature of the Company's external relationships, will lead to an increasing number of joint enterprise undertakings in host countries rather than expropriation. Nevertheless, there will inevitably be a progressive reduction in conventional business which, if the Company's skill and enterprise are to survive, as we believe they should, must be replaced by profitable new business in the non-concessionary field. lthough Cable and Wireless has been involved in this sector on a limited scale for some time, it is only within the last two years that it has been able to go ahead at maximum effort and with full Government approval. The market has been carefully researched and the Company is confident of its ability to succeed; we see no reason why it should be hamstrung in its desire to extend its activities in this direction.

The extent to which a wholly-owned Government organisation such as Cable and Wireless should venture into keenly competitive market trading, presents certain problems; but there is no attempt on the part of the Treasury or the sponsoring Department to inhibit the Company from going ahead on these lines. Having seen something of the progress that has already been made we consider that the Group should be given every encouragement to promote manufacture and marketing through subsidiary companies of products and services which are supplementary or complementary to its main objectives and activities as a world-wide communications organisation – though we agree with the Treasury that it could be unwise to stray beyond the field of telecommunications, in its broadest sense, into unfamiliar territory. There are other commercial enterprises, which are better fitted by hard experience in international trading, to venture into the manufacture and marketing of products which carry unusually high attendant risks. Not that every project on which Cable and Wireless embarks can be expected to be profitable; risks have to be taken and, with them, the risk of failure. But the Company has to be judged on its overall performance. Having examined it on that basis, we believe that Cable and Wireless deserves the continued encouragement and approbation of the Government and of Parliament.

One way to have encouraged Cable and Wireless would have been by giving more generous remuneration to those whose stewardship had brought the results they so fully approved. The Government was guided in this respect by Lord Boyle's Review Body on Top Salaries whose first report had been issued in 1971. In 1974 it turned its attention to the salaries of 'the chairmen and directors of public bodies of a commercial character empowered by statute to engage in substantial trading operations', senior civil servants, judges and senior officers of the armed forces.

All those members of the public service, stated the Review Body in its Review No. 6 of December 1974 (para 7), were responsible in different degrees for the efficiency and effective deployment of very considerable resources of national wealth and manpower. It was of the highest importance that such public bodies

> should be able to attract and retain men and women of outstanding talent and ability; it would be very damaging to the national interest if they were to become deprived of their share of that talent through a system of rewards that came to be regarded as plainly inadequate and inequitable.

On the specific case of the nationalised industries they observed:

> We believe that direct comparison with the private sector must be the overriding principle for assessing pay levels for the top posts in the nationalised industries, though we do not go so far as to suggest that it would be appropriate to reflect the highest levels outside. But we see no realistic alternative . . . to an approach which recognises that

management responsibilities in these industries are closely comparable to those in the private sector, and must be remunerated
accordingly.

They recommended that the salaries of directors of the Cable and
Wireless Category (the fifth) should be in the scale £11,500–£15,000,
and that of the chairman £22,500. The chief executive/managing
director should be paid £15,500–£19,000.

The Cable and Wireless directors' report presented to The
Shareholder at the annual general meeting of 28 October 1976
contained the following paragraph:

> The chairman and other non-executive directors feel bound to
> record their concern at the inadequate remuneration of the executive
> directors who have received no significant increase in pay since 1972
> despite the recommendation by the Top Salaries Review Body for
> substantial increases in its report to the Government at the end of
> 1974. While implementing the corresponding recommendations relat-
> ing to the Civil Service and other elements of the public sector, the
> Government decided to defer consideration of the recommended
> increases for members of the public boards until they had received the
> report of the Royal Commission on the Distribution of Income and
> Wealth. Although that report was published on 29th January 1976,
> no further pronouncement by the Government has yet followed. One
> of the consequences is that directors' salaries have been very
> substantially overtaken by those of the company's senior officials
> whose pay is aligned to that of the Civil Service. Apart from the
> manifest injustice to the individuals, there is great danger for the
> company's continued efficiency and success in a situation in which
> any vacancies on the Court itself cannot be filled either from within
> the company (since it is unreasonable to expect that senior officials
> will be willing to accept a substantial reduction in pay) or from
> outside (because the executive directors' salaries at their present level
> are so widely out of line with the market rate).

Those key men in the front line who ran the group's operations
across the world in its various regions, were outside this particular
fight. They were not members of the central Court of Directors, so
their salaries were fixed by the company not the Government,
adjusted to the appropriate local cost of living – in Bahrain, for
instance, it is the second highest in the world, next only to Tokyo.
Responsible for areas in which Cable and Wireless business was
concentrated – the Arab World, the Philippines, Hong Kong, North
America, Europe, the Far East – their careers had mostly begun with
the two and a half years' basic company training as telecommuni-

cation engineers as those of all F.1 men had done. They had won their top positions in the field by becoming in addition managers, administrators, educators, debt collectors, diplomats, generators of new business, maintainers of the good public image, and genial hosts to 'visiting firemen' of every description. While never losing sight of the first requirement, which was to make the hardware work – a matter of continual wonder to laymen and races who had never managed to develop a technological class – the success of such people, and indeed of all branch managers in the field, was equally due to their understanding of what made *humans* work. In all they did, regional directors, regional chief executives and branch managers who organised groups of office workers and technicians of different cultural backgrounds, had the backing of their own families, particularly their wives who willy nilly becme an integral part of every overseas Cable and Wireless scene. Many wives found it merely a continuation of the life they had known as children, their parents having had careers with Western Telegraph, Eastern Telegraph, Imperial and International, Cable and Wireless, and perhaps themselves been employed as a secretary or assistant in the company.

David Izard took the course at PK at the end of the century, made the usual tours of overseas branches, got drafted into cable ships in World War 1, and then back on the station circuit in the 1920s and 1930s. His wife and family followed him round, and at Bermuda his daughter Patricia became friends with the latest young Assistant Engineer out from England on his first appointment, George Warwick. Patricia made a second tour of the world as Mrs George Warwick, and then made her home in Hong Kong where her husband, who had served the company thirty-five years, was General Manager and later Regional Director.

Although it was the company's policy to train men of the country in which they operated to take over the duties of expatriates, the 20–30 per cent growth in business each year made it necessary to plan new techniques calling for new skills. As a result the number of expatriates remained fairly constant, though the proportion of local staff necessarily increased. At the end of the decade, Hong Kong branch for instance had more than 2,000 Chinese and only 60 British. A major activity of Middle East, Far East and Caribbean branches was training suitable youngsters as telecommunication technicians to take over the middle jobs below that of 'engineer'. But human communication had to come first, and where instructor and student had no common tongue, one or the other had to learn a new language. In the more sophisticated Gulf states most had received a grounding of English at school and many had then been to college in the United States or Britain; but in a town like Sana'a, capital of the (North) Yemen Arab Republic, where the Cable and Wireless contingent were the first non-diplomatic English ever to operate there, the Yemenis (apart from those who had come up from Aden) spoke only Arabic which had remained static since Mahomet had used it in the seventh century. But in Hong Kong mastery of Chinese was not considered necessary.* All instruction was given in English, and the Cantonese-speaking inhabitants of Hong Kong who hoped for employment at New Mercury House, or any of the radio stations in the Colony, had to learn technical English. For those with higher ambitions the branch ran introductory management courses and gave customer contact training.

Graduates in electrical engineering were the hard core of the young Chinese whom Jim Bairstow, Controller of Training, had to make available each year to allow for expansion and promotion, and to take the places of the fifteen or so technicians who left the branch annually, either by retirement or wastage. Some seventy to ninety went through his school every year, but finding suitable candidates was difficult. For those selected had not merely to be capable of absorbing the knowledge they were given in the original course but to show a potential for developing. A youth trained on HF Radio had to have the capacity in his mid-thirties of graduating to transistors, integrated circuits and high speed data. To enable them to do this 'In-service Courses' had been devised to upgrade their knowledge. Repairing faults had been simplified with the coming of the 'card' technique; but for every technician who was able to change a card only a few acquired the greater skill required to mend it. But more and more would now match the achievement of Fung Hak Ming who became the first Chinese Engineer-in-Charge of Stanley Earth Station, manned by three expatriates and forty-two Chinese, with its two Standard A dishes in which, in spite of the hundred mile an hour winds – or perhaps because they afforded protection from them – unknown birds had built their nests and refused to be shifted.

Aptitude to learn telecommunications technology was not always accompanied by an ability to learn a language. In Bahrain those

---

*Like Arabic, Chinese had to approximate modern scientific terminology with portmanteau words made up from traditional ideographs. The Chinese characters for 'Cable and Wireless Limited' spelt out 'Great Eastern Electric Message Bureau' – in other words still the Chinese equivalent of The Eastern Telegraph Company with the prefix 'Great' added, presumably to be upsides with The Great Northern. No one as yet has managed to contrive a set of Chinese characters to denote 'Cable and Wireless'; but the chosen Arabic hieroglyphs mean 'The Lightning and No Wire Company'. With no word for 'electricity' Arabic speakers have to rely on the ancient word 'lightning'.

who showed signs of having both were taken through a City and Guilds syllabus comprising a year of orientation, a year at the Camborne Technical College in Cornwall, and a third year back in Bahrain. In 1975 the year at Camborne was phased out and students from Bahrain, Dubai and Sana'a took a two year course conducted at Bahrain Training College by a staff of one Greek, two Indian and twelve British instructors under Officer-in-Charge Peter Turton. With sixty students attending at the end of the 1970s, it was soon hoped to have the full complement of ninety which the new buildings allowed for.

It was reckoned that those who made the grade would need another three years on the job before they could be regarded as fully qualified members of the 1,400 strong staff at Bahrain – of which in 1978 only 125 were F.1. Its wide-ranging telecommunications system had made Bahrain a commercial and financial centre second to none in an area where oil revenues had brought other states very much greater wealth. The telecommunications build-up which began in 1947 as seen, and had been accelerated following the concession of 1968 which was going to expire in July 1982, had given the island a new role.* The pearl fishing industry on which its economy once depended was no more; the first oil well to be found in the Gulf was all but spent; cheap natural gas had given birth to cheap aluminium smelting; Saudi Arabian oil was refined and ships of all nations repaired. But none of these activities justified the frenzied hotel and office building on the reclaimed land at Manamah. It was the availability of instant, cheap telephone, Telex, high speed data, facsimile and television communication which had attracted the money-brokers, the off-shore banking units, the off-shore traders, the international airline operators and news agencies like Reuters with their Monitor Project, the shipping companies and stockbrokers who were giving Bahrain its new prosperity.

Apart from the impressive technological achievement which must always take precedence, as was the case in Hong Kong and the areas controlled by Norman Brooks, Regional Chief Executive in the Caribbean, and Maurice Bane, Regional Director, Philippines, a Cable and Wireless overseas exercise was a triumph of human co-ordination, of successful orchestration of a polyglot staff which at Bahrain was British, Asian and Arab. Basil Leighton, Regional Director, Arab World, and General Manager, Bahrain, had long

---

*An Emiri decree giving a concession of this sort took precedence over all other legislation; and by it the Emir had the right to 60,000 'words' a year free of charge, though in Wideband days, when traffic was measured in minutes, this had become somewhat out of date.

learnt to live with two weekends, Saturday/Sunday for the Christians (applicable to many of the 'mobile' Anglo-Indians who had come from the old communications centre of the Arabian Gulf, Karachi, at the time of the division between India and Pakistan, and were Catholic not Hindu), and Thursday/Friday for the Moslems; seeing that the annual staff party was not a *Christmas* celebration; ensuring no one felt insulted by having a telephone line laid along the top of the purda wall of his Wimpey house in the new town of Isa; taking care to make his telephoning to Mercury House within the four and half hours of each week in which his working day overlapped that in London; keeping himself informed of local political developments and never allowing himself to become involved in them. But making the circuitry work was his prime responsibility – and it was certainly impressive. Through any of the nine automatic exchanges of the Bahrain Telephones internal network run by Cable and Wireless, under the direction of Alec Sherman, anyone could dial in from outside Bahrain and be connected via the satellite station direct to London. Those who only wrote Arabic could confidently telegraph their business associates abroad in the knowledge that the Message Switching Computer in Bahrain would switch telegrams written in Arabic script.

The United Arab Emirates announced their intention of taking control of all the local subsidiaries of both Cable and Wireless and International Aeradio, and on 1 September 1976 an Emirates Telecommunications Corporation (Emirtel) came into being in which Cable and Wireless and International Aeradio together had a 40 per cent holding, and the UAE 60 per cent. They took over the Standard A earth stations at Dubai, and the one 120 miles away at Abu Dhabi. There were others close by at Doha (Qatar), Oman, Ras al Khaimah, and Bahrain.

The company's predecessor, the Eastern Telegraph Company, had built its business round Britain's expanding second empire, but no tradition existed to justify its presence in the first empire which George III had lost well before the electric telegraph had ever been invented. The company's first foothold in the USA, as seen, had been the procurement company; and for five years, as well as buying equipment, Cable and Wireless Technical Services Inc. had successfully kept a watch on developments, as planned, and reported back to headquarters. With the formal establishment of the Communications Systems and Services Division, John Bird made the New York company in addition one which could sell other manufacturers'

123–4 In the 1960s the main build-up by Cable and Wireless was taking place in the total-concession areas of the Middle and Far East. The company had introduced Telex to Bahrain in 1963 and it had spread to the rest of the Gulf; in 1966 it opened tropospheric scatter radio links between Bahrain, Doha (Qatar) and Dubai. (*Above*) a telegraph operator at Bahrain; (*below*) a radio-telephone ship-to-shore operator at Bahrain coastal station.

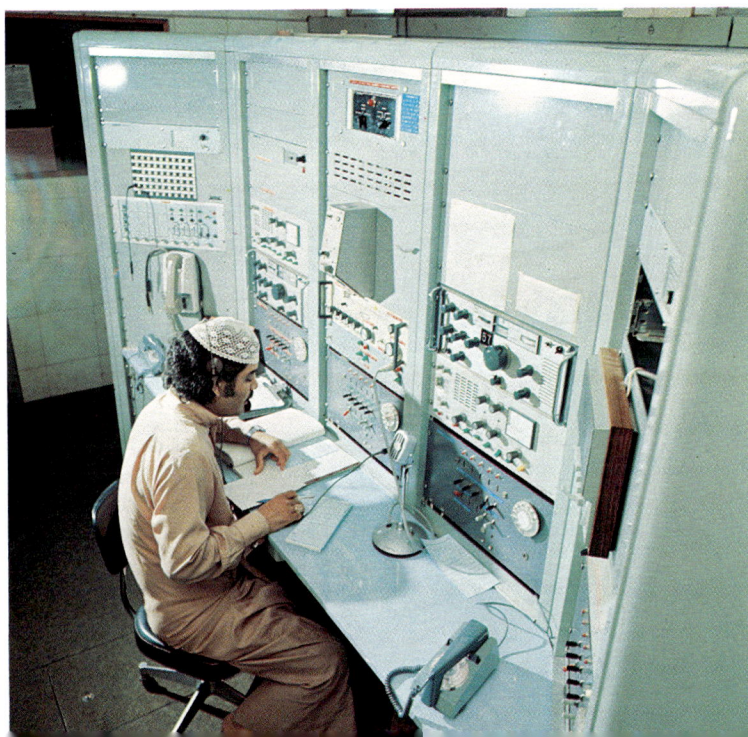

125–8 Installation and maintenance of private communications systems for airline companies and civil aviation authorities at international airports had become big business for the Cable and Wireless Group by the 1970s. (*Above left*) air traffic control in the Seychelles; (*above right*) flight information centre at Kai Tak airport, Hong Kong; (*below left*) airport systems switching centre in Barbados; (*below right*) typical reservations check desk.

129–32 The overseas image: New Mercury House, Hong Kong (*above left*); the Wildey complex, Barbados (*above right*). The design of earth satellite station aerials has changed. The dish for the earth station built in West Africa in 1978 (*below left*). A Standard B dish operating in the Indian Ocean is seen (*below right*). In the future even smaller dishes are expected to work to satellite systems directly from roof tops in towns and remote rural areas.

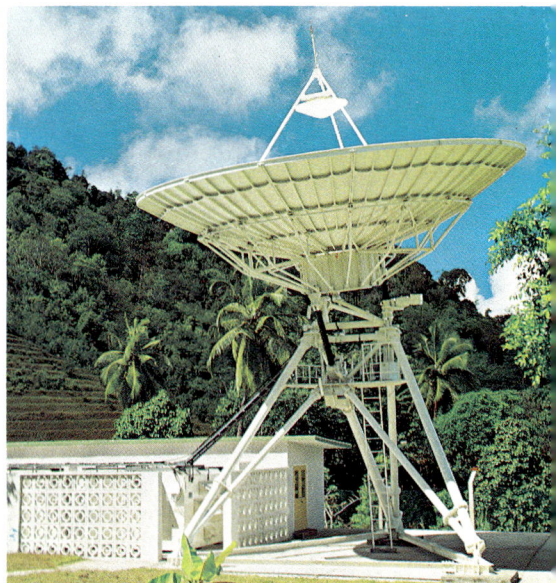

133–36 On 31 December 1970 Porthcurno, Cornwall (*above*, seen from the air) ended its 100-year life as a cable station but continued as the Engineering College which, together with training schools in Bahrain, Barbados and Hong Kong, supplied the operators and engineers to run the group's operations on the China coast, in the Persian Gulf, in the Caribbean and in the small but vital links in such places as the Seychelles (*below*) and Ascension Island (*right*).

137–8 F.1. (overseas) staff at home on leave receive a pleasant reminder of the England they left behind in the gardens of the Cable and Wireless Exiles Club at Orleans Park, Twickenham (*above*) which the Eastern Telegraph Company acquired as a recreational centre in 1920. Though the group is administered from its London HQ, its largest branch is Hong Kong where an earth station working to the Pacific satellite was opened in 1969, and a second, working to the Indian Ocean satellite, in 1971 (*below*) the site at Stanley.

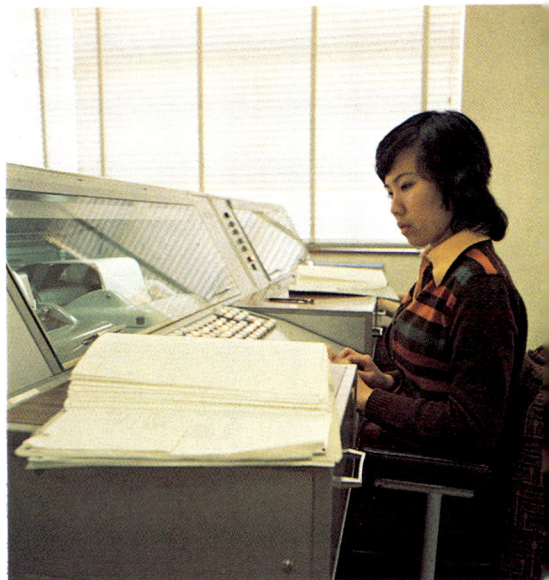

139–41 The Message Switching Centre at Hong Kong (*above left*) was first computerised in 1969, and a fully computerised Telex exchange was installed in 1972. But the system still needed the human touch. In 1979 Hong Kong branch of Cable and Wireless had a mainly Chinese staff of 2,300 – one of them is seen operating a computer (*above right*). The latest technology is also applied to the installations in the West Indies. (*Below*) locally-trained girls at a cordless telephone switchboard on Antigua.

142–5 It might have been expected that the services of cable ships would have been dispensed with in the era of radio-by-satellite; but they were in greater demand than ever with the introduction of the submarine coaxial telephone cable which could provide telegraph (Telex), facsimile, high-speed data and television channels as well as voice communication. (*Above right*) c.s. *Cable Venture*, acquired, converted and put into service in 1977; (*above left*) loading cable into her tank. (*Below left*) the ship's band strikes up amid the repeaters and buoys in the hold of c.s. *Mercury*; (*below right*) a cable's final destination – landing the shore end.

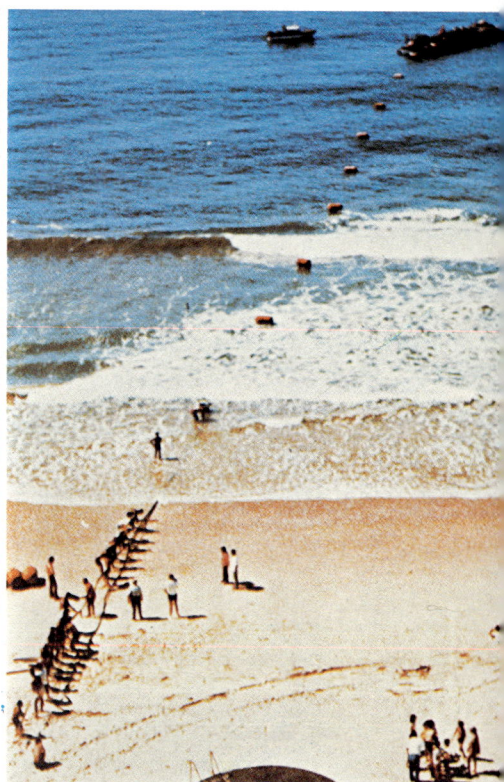

products, and to a smaller extent message switches and similar equipment designed at Smale House. It also offered consultancy. An eye – and an ear – could be kept open too for any major project in North America, Central or South America (which had notoriously long gestation periods) where the company's expertise and experience could be brought to bear. An outpost in New York with these wider objectives justified a local presence with management of the highest calibre. In 1975 Ronald Hobbah was sent out to New York to head the CSS exercise in the US which John Bird now regarded as one of the division's main thrusts.

That first year Hobbah began by putting 'seed capital' in a telephone management company called TDX Systems Inc. He found that the achievements of Cable and Wireless in other parts of the world were better known than he had thought. Its reputation for providing a reliable service in so many different circumstances and locations, the quality of staff and the *breadth* of their knowledge, had made a lasting impression on Americans used to the high-powered sales talk and assurances of the conventional 'rep'. For of course Ron Hobbah, like most Cable and Wireless business generators, was an engineer, and would never make promises which he knew the company could never keep. His experience in telecommunications was second to none – for four years engineer in charge of Cape D'Aguilar radio station in Hong Kong; Regional Engineer for the East Caribbean Microwave Chain; Supervising Engineer (Project Manager) of the COMPAC co-axial terminal at Suva; and before these, spells at Cocos and Ascension and the usual F.1 tour.

While in charge of the COMPAC operation at Suva, Fiji, in the early 1960s Ron Hobbah came to know at first hand of the staunch dependability and professionalism of the other arm of Cable and Wireless overseas, the Marine Service. He came to appreciate the skill of navigating a heavy cable ship through a narrow passage blasted for it through a coral reef, as the *Retriever* had to do to land the shore ends with their heavy armoured sheathing of the line and earth cables to Auckland. On that occasion Jock Black had taken his bagpipes and piped them in as they floated on buoys towards the trench leading to the terminal where the Supervising Engineer personally connected all four lines (the two for Hawaii were laid at the same time), praying they were labelled correctly.*

---

*Whilst the Commonwealth co-axial cables were being laid in the period 1961–67 the American Telephone and Telegraph Co. (ATT) were similarly occupied with laying two 128 circuit co-axial cables across the Atlantic and also a trans-Pacific cable of the same capacity between San Francisco and Tokyo. All the laying work was carried out by the American cable layer *Long Lines*. All these systems used valves in the repeater amplifier.

With the completion of the Commonwealth co-axial cable system the two laying ships *Monarch* and *Mercury* were next engaged on laying a 360 circuit co-axial cable between Cape Town and Lisbon for the South Atlantic Cable Company. At the end of this work in 1969 it became apparent that there were now insufficient new cable systems being planned to keep two large cable laying ships fully occupied. Cable and Wireless then bought the *Monarch* from the Post Office and after an extensive refit renamed the ship *Sentinel* and based her at Bermuda as a repair ship for operations in the North Atlantic and Caribbean area.

The cable system from Cape Town to Lisbon was in fact the last major system to use valves in the repeater amplifier. From 1969 onwards transistors were used in the repeater amplifier as they offered the attractive prospect of a wider bandwidth for co-axial cable systems and thereby a larger number of telephone circuits. Throughout the early 1970s a large number of 480 circuit systems (i.e. CANBER), 725 circuit systems (TAT 5 and Florida – St Thomas), 845 circuit systems (Trans Pacific No. 2), 1,380 circuit systems (i.e. CANTAT No. 2), were installed. In this work the Cable and Wireless laying ship *Mercury* was kept fully occupied laying the British co-axial systems. On CANTAT No. 2 and more recently the Las Palmas – Venezuela system, again a 1,380 circuit system, *Mercury* carried out a total of nine loading and laying operations which was more than she carried out on the COMPAC and SEACOM. In 1976 ATT laid another transatlantic system (TAT 6) between USA and France with a capacity of 4,000 telephone circuits. The latest British systems to be laid are capable of carrying 5,520 telephone circuits.

Repairing the world's co-axial cables was now a co-operative activity under the Atlantic Cable Maintenance Agreement and other agreements covering the rest of the world. The role of the company's seven cable ships changed in the 1970s to that of trouble shooters ready to sail at a moment's notice to restore communication for a cable owner who paid the company a fee under a long term maintenance agreement. To cover the Pacific, for instance, the company had three cable ships, the *Retriever* based at Fiji, the *Recorder* at Singapore, and the *Edward Wilshaw* at Madang, New Guinea.

The captain of the *Cable Enterprise,* the Cable and Wireless cable ship built in 1964 and based at Vigo in northwest Spain, would receive a telephone call on board ship from ATT in New York reporting a break and, having confirmed with London, would take his ship into the Atlantic, find and repair the break. The ship had both propeller and rudder at the bow. When operating in shallow water the captain could command the ship from the 'Bow Control' console on deck. On the job the complex of eighty (sixty Spanish and twenty English) worked twenty-four hours a day in continuous shifts. The crew numbered sixty of which half were deck hands, and the rest were the cable technicians. In the hold were Flat Fish grapnels for holding and cutting a cable in sand or mud, a Gifford chain section for use when the seabed was rocky, and a host of other devices. Alan Pentney, the company's Marine Superintendent, reckoned that a new repair ship for the 1980s would cost in the region of £12 million. It would need a speed of 16 knots and its hull would have to be strengthened against ice for working in the Arctic, and fully air-conditioned for the hot and humid climates of the tropics.

When Peter Lillicrap retired as part-time, non-executive chairman in October 1976 he was succeeded by Edward Short who had just completed a distinguished political career in which he had served as a Cabinet Minister in several of Harold Wilson's administrations, been Government Chief Whip, Postmaster-General and Secretary for Education and Science. In April 1972 he succeeded Roy Jenkins as Deputy Leader of the Labour Party, and in March 1974, at the age of 61 became Lord President of the Council and Leader of the House of Commons. He had resigned from the Government in April 1976 and announced he would not again contest the Newcastle Central constituency which he had represented since 1951. He was given a life peerage and took the title of Lord Glenamara.

The salary of the three-day-a-week chairman was £9,080 with which he was well content, but he sympathised with his managing director earning £12,828, (half the Boyle recommendation) and his four executive directors with £10,330, who had stated their intention of declining to stand for re-election to the Court of Directors in protest at the low level of remuneration, and at the failure of the Government to act on the Boyle Report which had been shelved for two years. A National Economic Development Corporation report had criticised the Government's policy, and some hundred board members of nationalised industries had indicated their willingness to join a state industries group pledged to obtain Government agreement to implement the scales recommended by the Top Salaries Review Body. As seen, the Court of Cable and Wireless had publicly expressed its concern in its 1976 report. But HMG were determined not to break their Prices and Income Code. In November 1976 Lord Glenamara told the Press he had been discussing the Cable and Wireless situation with the Government. He had been amazed at the number of hours which his directors worked. 'If these sort of experts are to be kept in this country, particularly in the public sector, then the pay problem must be solved. They must be adequately rewarded. Over the past few years their differentials have gone to pot. Some senior people earn £3,000 more than their bosses.'

It was no quibble to insist that Cable and Wireless was not a nationalised industry but a Government-owned commercial company (like Rolls-Royce) which took nothing from the Exchequer but, because of the high quality of its management, contributed large sums to it by way of dividends and taxes. It had unfortunately become bracketed with loss-making nationalised industries which were a burden on the taxpayer. To claim separate treatment as a 'special case' only compounded the contention that the company was One of Them – only Special. It was not just the Labour Government which jibbed at acting on the Boyle Report; Edward Heath had deliberately delayed it. Both administrations were concerned at the effect implementation would have on organised labour fretting under the constraints on wages imposed by a Pay Policy designed to curb inflation. To the managing director of Cable and Wireless, A A Willett, it seemed that HMG were a little concerned whether the directors continued in office or not. As a result, he saw the company's management structure being even further distorted, damaging its ability to recruit and retain the expert employees it needed to maintain its highly successful record of recent years. He resigned on 1 June 1977, and he took the post of bursar at St Antony's College in the University of Oxford. He was 53. His departure at the peak of a career which had only just reached its climax and still had much to contribute in maintaining and increasing the momentum for which he had been so largely responsible, was greatly deplored. In his message in the 1977 report, Lord Glenamara commented it was quite unrealistic to assume that either Cable and Wireless or any other publicly-owned enterprise would be able to maintain its high grade management if the Government continued to be unwilling to pay salaries which were comparable to those in the private sector.

> The Government's problem in fighting inflation is considerable and well understood, but unless the situation in the board rooms of publicly-owned companies is faced with courage and resolution the public sector will sooner or later be deprived of adequate management. A company such as ours with its operation spread over the

globe and operating in the most advanced technologies must continue to have management of the highest calibre.

As a temporary measure, the corporate function of managing director was taken over by Peter McCunn and John Bird as joint chief executives. The latter had been made an executive director from 1 June. Subsequently Peter McCunn was made group chief executive with the title Executive Deputy Chairman; John Bird, Managing Director, Communications Systems and Services; and Richard Cannon, Managing Director, Public Telecommunications. This permitted some salary increases which had previously been unacceptable to HMG. Ernest Potter joined the company from Cammell Laird as Director of Finance and Corporate Planning, without a seat on the Court. Arthur Cooke remained as company secretary.

The new corporate structure at the top reflected the two-pronged nature of the programme to 1999. In December came another joint company with a local administration, though the Fiji Government had only 10 per cent interest in Fiji International Telecommunications Limited (Fintel), with an option to increase it to 51 per cent in the following five years. A satellite earth station had been opened earlier in the year at Suva. In May 1977 the Guyana Government, celebrating its eleventh year of independence from Britain, announced its intention of taking over the assets of Cable and Wireless for £222,000. But that summer saw the opening on the other side of the world of the largest capacity telecommunications cable system in South East Asia which the company had constructed in conjunction with Kokusai Denshin Denwa of Japan and Eastern Telecommunications Philippines Inc, in which Cable and Wireless had a 40 per cent interest. OLUHO, as it was called, linked Okinawa, Luzon and Hong Kong, cost £28 million and provided 1,840 telephone circuits between Hong Kong and the Philippines, and 1,600 between Okinawa and Luzon. The company agreed to take a 20 per cent interest in the Philippines-Singapore cable planned by the Association of South East Asia nations which opened in mid-1978, and invest up to £16 million to secure 6 per cent ownership of the proposed IOCOM cable connecting India with Malaysia. Altogether four new concessions were won in 1977 and in February 1978 the New Hebrides Government invited Cable and Wireless and France Cableset Radio to run their overseas telecommunications for fifteen years, through a joint company formed for the purpose, Hebritel, with a Standard B earth station.

It looked as if maybe 'traditional' business, which in 1977/78 still accounted for 84 per cent of the company's activities, was not to disappear as quickly as expected; though on 31 March 1978 ended a hundred year association with aden when the Peoples Democratic Republic of the Yemen took over the once key station at the southern entrance to the Arabian Gulf which has played so prominent a part in this story. Happily the billiard table in the old Eastern Telegraph Mess, on which so many F1 bachelors had demonstrated their dexterity, was carefully packed away and shipped north to Sana'a to join the earthy nine hole golf course and the squash court to provide recreation for another generation of Exiles.

How they would fare at Aden and other sites where regimes had decided to 'go it alone' remained to be seen; but wherever Cable and Wireless had handed over to other administrations, it had always done its best to see that the nationals who had worked for the company were trained to the highest standards, which ensured that the new organisations were able to carry on efficiently and maintain the momentum left behind. But of course, as technology became more abstruse with computerisation, fully electronic exchanges and the rest, it was that amount more difficult to assimilate. Every year the feat of preventing the system from going dead for long periods at a time and losing the confidence of its users, became more and more exacting.

Cable and Wireless became fully stretched wrestling with the kind of problem presented to it by the Danish proprietors of the Maersk Line who wanted a high speed data network between the Far East, Copenhagen and the United States handled through computers, so that a user at any point on it could receive an answer in six seconds. The company suggested three circuits, Copenhagen-Hong Kong, Hong Kong-Tokyo, and Copenhagen-New York, and won an order to design and organise the system, provide the terminal equipment and 'facilities management' – running the entire private network. It was extremely complicated and it was a year before all was running smoothly. At Hong Kong the company provided data transmission facilities at all speeds, ranging from 300 'bits' (binary digits) a second to 9,600. The volume of traffic through New Mercury House operations centre was increasing 17 per cent a year, with 1,400 new Telex subscribers a year. A second computerised Telex exchange had been installed in 1974 and a third in August 1977 which increased the capacity to 10,500 subscribers. Early in 1978 it was decided to install a £4½ million automatic telephone exchange, and a second operations centre was planned for the mainland of Kowloon.

The six year growth in CSS activities on which the long term future of the company depended, carried out through forty-seven centres separate from Cable and Wireless branch offices, was slow but just about on target and revenue for 1977/78 improved by 63 per cent over the year before, from £18 million to £29 million. Total investment had increased fourfold to £12½ million. The division strictly fulfilled the requirement to provide British products where available and only use foreign equipment when this was not possible. Its range of activities covered electronic Hotel Systems (Musak-type background music, in-house movies, conference voting equipment, morning calls, paging, fire alarms), Mobile Radio Networks, Marine RT and Telex, Audio and Video Systems, Personnel Location, Security Systems. But the major 'alternative business' was turnkey projects. In March 1978 the company prepared to embark on the biggest contract in its history when the Ministry of Defence on behalf of the British Government signed a contract with the Saudi Arabian Government to supply and install a complete telecommunications network for the 35,000 strong Saudi Arabian National Guard (SANG). Cable and Wireless were designated the prime contractor. Worth at least £200 million, this military contract was won in the face of fierce foreign competition, particularly American. The chairman claimed its execution over five years 'could mean the beginning of a new era for Cable and Wireless'. The final period would involve training the National Guard to operate the system.

Eurotech's turnover had reached more than £3 million in four years, and was beginning to show a profit. Don Jones took over from John Parsons as head, under John Bird, of Eurotech's operations both in Britain and Europe, which were amalgamated. Cable and Wireless UK Services Limited, with its own head office in London and six branches in the provinces, offered a range of services including the provision of equipment and the design, installation and maintenance of private systems anywhere in the UK – a direct result of the removal of statutory limitations to the company's activities at home.

Ron Hobbah's office in New York ceased to be a Cable and Wireless Group purchasing exercise the day he was asked by Pan American Airways to install equipment in the Panam Building and at Kennedy Airport. A second company, Cable and Wireless (North America) Ltd was created. The Panam contract led to slow penetration of other airlines and banks, but it was obvious that the company would take a long time to grow into a large domestic organisation in the US from grass roots. Acceptance in the competitive American market would be more likely to come through the acquisition of existing companies, of which three became subsidiaries in 1977/8. A controlling interest was acquired in TDX Systems Inc of Washington in which capital had already been invested, as seen, and a six-year-old New York company, Incotel, was purchased which specialised in computerised telecommunication services and systems, and became C & W Incotel Ltd. But the most important acquisition was the Carterfone Communications Corporation, bought for $16,250,000. This admitted the company into the profitable, multi-billion dollar 'inter-connect' business which Tom Carter's famous lawsuit challenging ATT's sixty-year-old monopoly made legal. The victory revoked a Federal Communications prohibition on connecting 'foreign attachments' to the end of telephone company lines such as Carter's device linking mobile radio units on oil rigs.

In 1978 Cable and Wireless (North America) Limited, with its three operating companies, was consolidating its CSS business and, with the object of expanding it still further, established a separate office in Connecticut. At the same time there were representatives in New York and San Francisco.

The revenue of the Cable and Wireless Group with its forty subsidiaries and associates operating in seventy countries and employing 10,650, had more than trebled in five years – £56,000,000 in 1974, £71,000,000 in 1975, £104,500,000 in 1976, £157,500,000 in 1977, £177,000,000 in 1978. In 1974 the Group paid HM Treasury £1,809,000 in dividend and £7,025,000 in tax, a total of £8,834,000. In 1977 it gave The Shareholder £6,600,000 in dividend and £32,634,000 in tax, a total of £39,234,000. There was a recession in 1977 and high inflation, so £18,000,000, as opposed to £13,000,000 in 1976, was set aside to provide for the replacement of assets at increased prices. It was the year the East Caribbean Microwave Chain came fully into operation which meant no further capital expenditure and the promise of a mounting contribution to overall revenue. Revenue increased in 1977/78 by £19,600,000 but costs by even more – £23,300,000. So the profit before tax and after supplementary depreciation fell from £45,500,000 to £35,800,000, but even so was more than three times the £10,023,000 of 1974.

The return on capital employed rose from 11·6 per cent in 1974 to 40·7 per cent in 1977; and on 24 March 1977 reserves were capitalised, thus doubling the issued capital to £60,000,000. In the 1978 Accounts the 1977 return on capital was shown as 31·8 per cent, with 21·1 per cent for 1978 and capital employed at £182,300,000. Reserves stood at £65,000,000.

The most significant statistic, however, was that the Group was now responsible for 2 per cent of the country's *net* invisible earnings.

Before all else, young men entering the service of Cable and Wireless in 1979 had to learn to live with computers. The Computer Revolution had come to stay. The union of computers and telecommunications was perhaps its most impressive manifestation to date. The ability to store, retrieve and switch information was superhuman. A 2,000 line, computer-controlled Telex exchange examined the nature of an incoming call, made the necessary switching to service it, automatically activated the caller's teleprinter to spell out the message that the distant correspondent was engaged; logged the call when it was eventually made, and recorded the charge on the caller's account. When 'asked', it told the supervisor what lines were in order, which were out of order, and how many were waiting for calls. All this without human intervention. But if it encountered a problem that was beyond its electronic brain it automatically handed itself over to the live operator.

By 1979 the volume of traffic, and the decisions to be made arising from complicated routing permutations, had become more than human hands and brain could contend with, and switching systems both for telephone and message systems made extensive use of computer control.

With digital replacing analogue techniques, allowing for fully electronic telephone exchanges taking advantage of computer technology, the telegraphic arts were paradoxically returning. In 1979 the greatest demand was still for voice channels, but the market for the transmission of high speed data received on the screen of a visual display unit was fast catching up, along with that for Facsimile. The Computer Revolution had made possible Data Banks which could only be 'accessed' over telecommunications links, Electronic Funds Transfer which did away with sending cheques through the post, and Electronic Voting which (when it comes) will do away with having to walk to the polling booth and personally dropping a voting paper into a ballot box.

As the report of the Post Office Review Committee (the Carter Committee) published in July 1977 stated,

> The high speed transmission and routing of information will in effect eliminate distance. These new services and many others will have a profound effect on the organisation and conduct of business and could have an equally great effect on the lives of private citizens. Although many of these developments are in progress now, it is likely to be many years before they take the form of common services available for widespread use. A great deal of trial and market assessment will be required in order to prove their value, but they have one major factor in their favour, namely that they all depend on microelectronics, and this technology is becoming cheaper and cheaper.

Within two years of that statement the Microelectronics Revolution, with its 'miracle' silicon micro-chips, was threatening to change the nature of the Computer Revolution. Cable and Wireless were naturally concerned with all technological development, since it determined the style of service it could offer its customers. But group policy was to buy research in the product; as a service industry it did not see product research and development as falling within its province. It made a point of patronising a British supplier whenever his equipment was demonstrably superior, and backed by an efficient spare parts service. Cable and Wireless engineers specified the quality they required, which might not be 'the best' but the most reliable in a given environment. They looked for high performance in terms of MTBF (Mean Time Between Failures) and MTTR (Mean Time To Repair). What can you guarantee? was the question asked of every potential supplier by the Cable and Wireless Quality Assurance Department. Without itself undertaking manufacture, it was vital that the group's engineers should be able to convince customers in the Arabian Gulf, in Malaysia, in the Caribbean, that the systems they were recommending attained the quality on which (rather than price) Cable and Wireless had for so long staked its reputation.

The group insisted that a customer should worry not so much about the cost of buying a telecommunications system as the continuing cost of owning one – the expense of maintaining it over its required lifespan. And it was never afraid to warn a customer of the waste of ordering a system designed to last longer than he needed. The group's central product assessment effort was accommodated in Smale House, named after former Engineer-in-Chief, J A

Smale CBE, AFC, where ideas generated throughout the Group were developed and experimental models of equipment produced for evaluation in the working environment. It had many successes to its credit, particularly in telegraph channelling and error-correcting apparatus.

The group's own effort was mainly geared, however, to help the sale of complete 'turnkey' systems from concept and tendering to installation, handover and training. In 1979 the market potential for this was regarded as enormous. As deputy-chairman Peter McCunn saw it, in the next decades the rest of the world must join the family whose relations, by leave of modern telecommunications techniques, had become so intimate. There was a shortage of money and resources, and ever-present political problems, but Africa, South America and Arabia could not afford to stay out in the cold for very much longer. Once the Third World had caught up with the public communications system, it would then seek to develop systems by making private and discrete use of a segment of that network, the implementation of which would give enterprises like Cable and Wireless their opportunity.

Throughout North America there were highly developed, commercially operated, custom-built private systems – the packet switched Telenet in the USA and Datapac in Canada (competing with the circuit switched Infoswitch). Because of inroads being made into US legislation on Freedom of Individual grounds, the demand for turnkey private systems in the US was about to become too large to be met solely by the US telecommunications industry. It seemed probable too that in the light of the report of the National Committee on Computer Networks published in October 1978, the 1980s would see a change of attitude to the establishment of private systems in Britain.

The Court of Directors' stewardship in the fifty years since the opening on 30 September 1929 of the public service which resulted from the shotgun wedding of cable and wireless interests, cannot have been other than gratifying to a Treasury to whose coffers the group had made such welcome contributions both in profit and tax. It had been a commercial success story of which, thanks to the efforts of an imaginative and highly effective Public Relations Department headed since 1971 by Joseph Tobin, no one could be unaware. Its consistent performance owed much to the tenacity with which the company insisted on keeping the dual character given it by being the offspring of two hitherto irreconcilable parents.

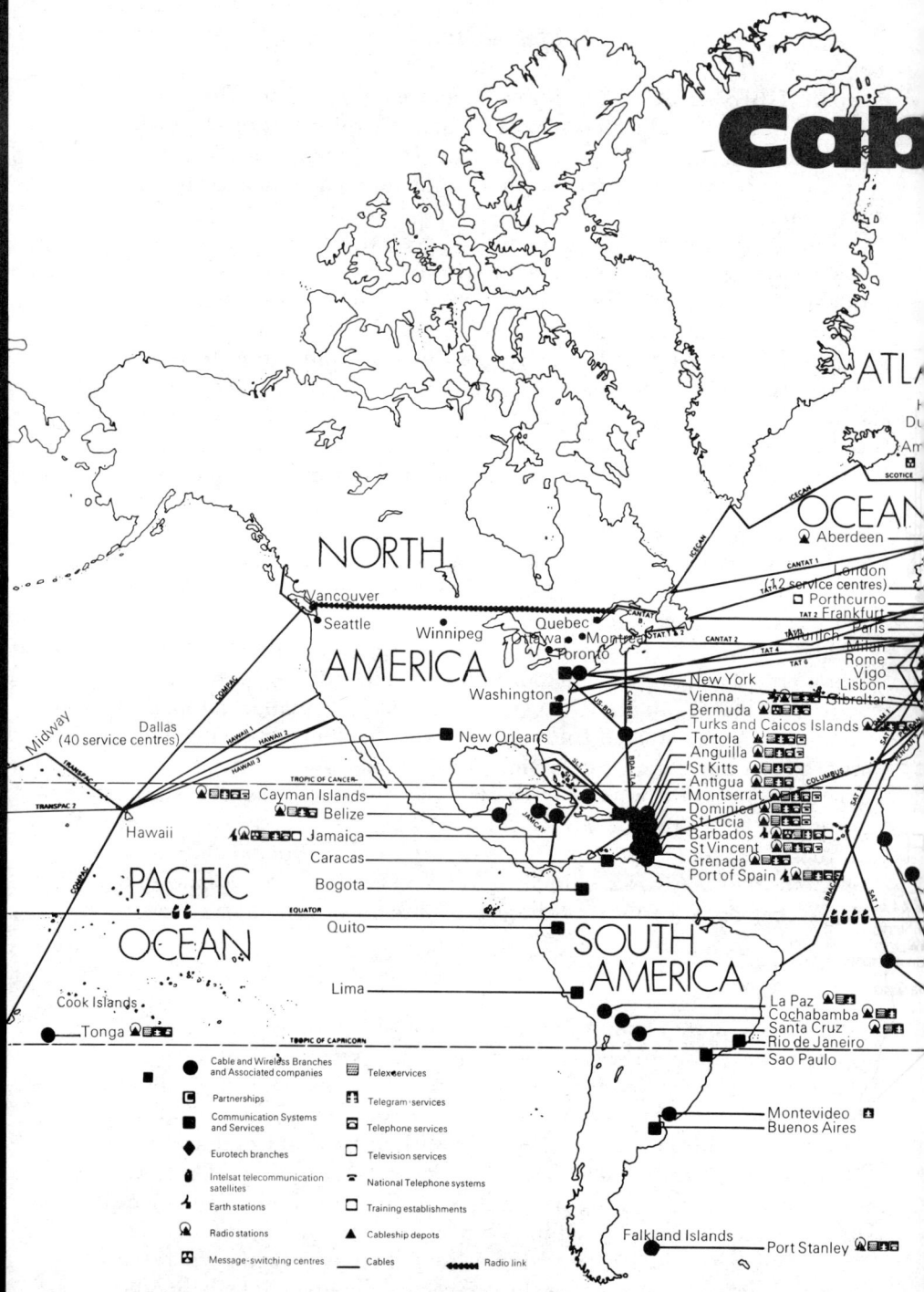

Cab

ATLANTIC

OCEAN

NORTH

Vancouver
● Seattle
Winnipeg
Quebec
Ottawa ● Montreal
Toronto

AMERICA

Washington

Midway
Dallas
(40 service centres)
New Orleans

Hawaii 1
Hawaii 2
Hawaii 3

TRANSPAC
TRANSPAC 2

Hawaii

COMPAC

TROPIC OF CANCER

PACIFIC

OCEAN

EQUATOR

Cook Islands

● Tonga ▦▤▥▨

TROPIC OF CAPRICORN

Quito

Lima

SOUTH

AMERICA

○ Aberdeen
London
(12 service centres)
□ Porthcurno
TAT 1 Frankfurt
TAT 3 Munich Paris
Milan
TAT 4 Rome
TAT 6 Vigo
Lisbon
Vienna
Gibraltar
New York

Bermuda ▲▦▤▥▨
Turks and Caicos Islands
Tortola ○▦▤▥▨
Anguilla ▲▦▤▥▨
St Kitts ○▦▤▥▨
Antigua ▲▦▤▥▨
Montserrat ○▦▤▥▨
COLUMBUS
Dominica ▲▦▤▥▨
St Lucia ○▦▤▥▨
Barbados ▲▦▤▥▨
St Vincent ○▦▤▥▨
Grenada ▲▦▤▥▨
Port of Spain ▲▦▤▥▨

Cayman Islands ▲▦▤▥▨
Belize ▲▦▤▥▨
Jamaica ▲▦▤▥▨
Caracas
Bogota

La Paz ○▦▤
Cochabamba ○▦▤
Santa Cruz ○▦▤
Rio de Janeiro
Sao Paulo

Montevideo ▤
Buenos Aires

Falkland Islands ● Port Stanley ○▦▤▥

1979

Legend

■ ●	Cable and Wireless Branches and Associated companies	▦	Telex services
▣	Partnerships	▤	Telegram services
■	Communication Systems and Services	▥	Telephone services
◆	Eurotech branches	□	Television services
▮	Intelsat telecommunication satellites	☎	National Telephone systems
⚓	Earth stations	□	Training establishments
○	Radio stations	▲	Cableship depots
▨	Message-switching centres	—	Cables
		◄◄◄	Radio link

e and Wireless

A STEP AHEAD OF TODAY'S COMMUNICATIONS

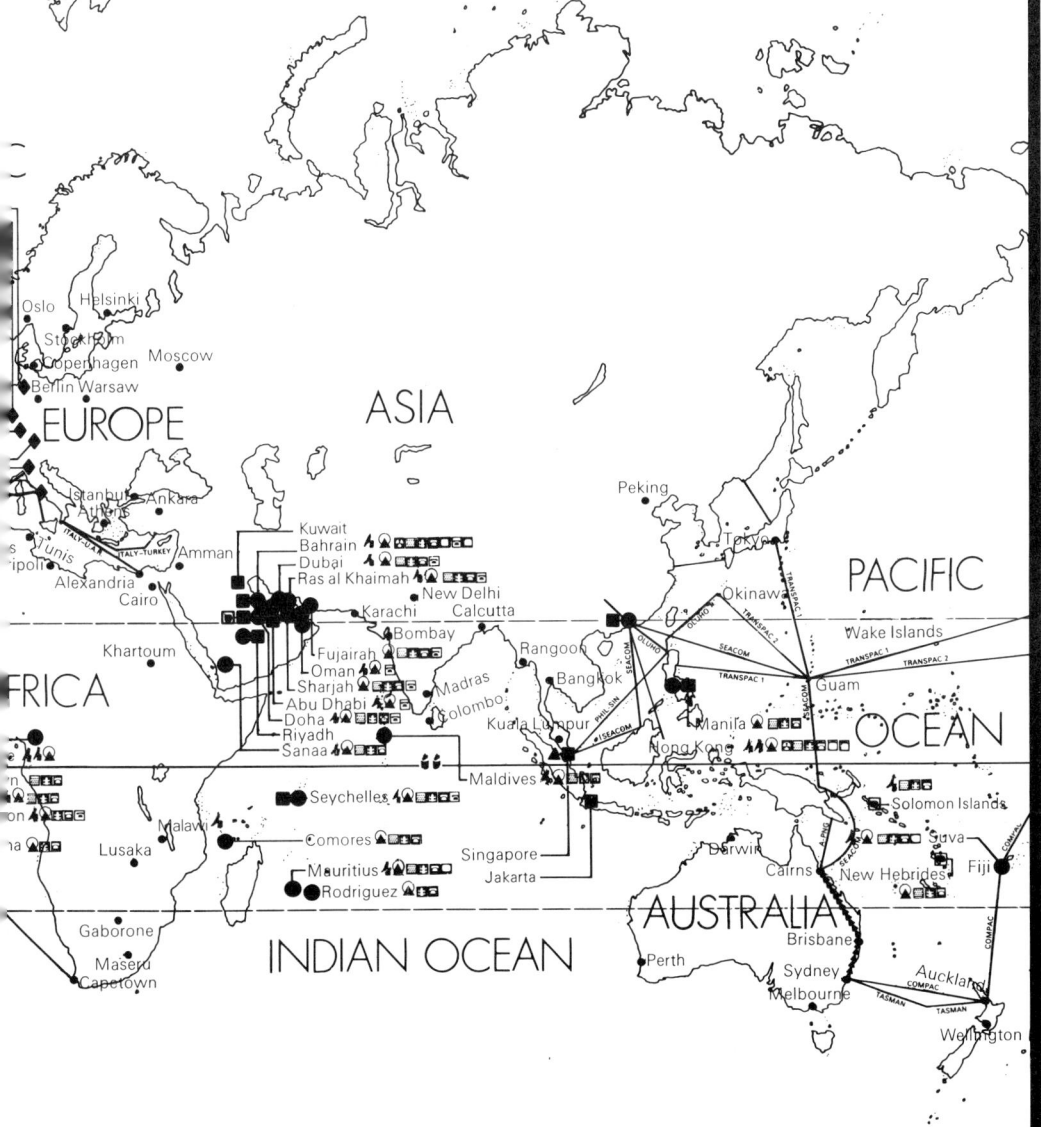

EUROPE

Oslo　Helsinki
Stockholm
Copenhagen　Moscow
Berlin Warsaw

ASIA

PACIFIC

Istanbul　Ankara
Athens
ITALY-U.K.　ITALY-TURKEY
Tunis
Tipoli
Alexandria
Cairo

Kuwait
Bahrain
Dubai
Ras al Khaimah

Peking

Tokyo
Okinawa

Wake Islands

New Delhi
Calcutta
Karachi

Bombay

Rangoon

TRANSPAC 1　TRANSPAC 2

Khartoum

Fujairah
Oman
Sharjah
Abu Dhabi
Doha
Riyadh
Sanaa

Madras
Colombo

Bangkok

SEACOM

Guam

FRICA

Kuala Lumpur

Manila

OCEAN

Maldives

Hong Kong

Seychelles

Solomon Islands

Malawi

Comores

Singapore
Jakarta

Suva

Lusaka

Mauritius
Rodriguez

Darwin

Cairns

New Hebrides

Fiji

Gaborone
Maseru
Capetown

INDIAN OCEAN

AUSTRALIA

Perth

Brisbane

Sydney
Melbourne

Auckland

COMPAC

Wellington

TASMAN

Such a stand inevitably led it into pursuing a course which contrasted with that of the United States which, as the acknowledged leader in the field, no one could afford to ignore. The Penders promoted Cable to the detriment of Beam Radio for reasons of finance not utility, acknowledging Marconi's telegraph without wires merely as a convenient, cost-effective cable interruption device. After 1929 their successors had to sit by and see economic reasons similarly motivating the United States, with their big investment in INTELSAT, to bolster Radio at the expense of Submarine Cable which Mercury House saw as being economically advantageous on high volume, short routes with a maximum length of two to three thousand miles (such as Bermuda to New York), while regarding radio-by-satellite as the more economic medium for long haul routes, whether the traffic was dense or light.

So far as quality of transmission was concerned – distortion and background interference – there was nothing in 1979 to choose between the two; engineering standards were the same. However, the user of a radio-by-satellite link had to adapt his conversational manner to the inter-loop delay of a quarter of a second which became a more noticeable half second with a double hop. The effect of the delay made it somewhat more difficult to interrupt the talker, but it was observed that conversation therefore became more polite.

It was expected in the next decade that radio-by-satellite would be operating on different frequency bands and with different methods of modulation, and that the satellites themselves would become more sophisticated. Nineteen-seventy-six witnessed the launching of the MARISAT system by which COMSAT provided high quality, direct radio communication by satellite to receivers with small dish aerials installed on the decks of ships at sea and oil rigs. In the years that followed plans were laid to extend the use of satellite communication to small communities. Such a system would bring isolated outposts, entertainment and mental stimulation, education and political information by television and sound broadcasting, as well as telephone and Telex links. It would give them a sense of belonging to Time and sharing a World with the rest of the human race, which *ekstasis* and estrangement so easily removed. When perfected, this was the kind of operation in which Cable and Wireless, it seemed to many, was admirably suited to participate.

In 1979 the distance at which the use of wires was preferable was on the point of shrinking. With a physical guided connection the advantages were seen to pile up as the distance grew less, and the volume between discrete points rose. The early 1980s were likely to witness the development of shallow water optical fibre systems –

and overland* too – though full distance, trans-oceanic systems for major routes were expected to take more time to perfect. The manufacturers were experimenting in making submarine optical fibre cable systems, but they had yet to design a suitable repeater, for which, however, they had established the principles. A submarine cable consisting of a bundle of insulated optical fibre strands was likely to be no slimmer than a co-axial cable. It had to be strong enough to lay and recover. It would be powered by the same means as a copper wire, of course, but the cable system could be much cheaper to install as compared to a conventional co-axial cable system with the same number of telephone circuits.

With inter-continental traffic doubling every three years, the introduction of submarine optical fibre cable could not come soon enough. The transmission of high speed data by 8-unit code to interlink computers of one country with those of another in the private computer language through copper wire co-axial cable and radio was widespread, but it seemed that the galloping demand for more channels would only be fully met by augmenting radio with cables which flashed the bits round the world at the speed of light.

Little was predictable in a field in which the pace of technological change was unprecedented. But in 1979 it was safe to assume that the advantages of the two mediums would continue to leap-frog the one over the other for many years to come. But the Story So Far indicated that the Cable and Wireless Group would with difficulty abandon its policy of safety in numbers, of providing optimum telecommunications systems by wireless *and* cable, if only because the two were never interrupted for the same reason.

The euphemistic use of the word 'interruption' to describe the long periods of silence to which the old telegraph cables were so frequently reduced covered the achilles heel of an enterprise which clung to the 'superior', because older, medium to which Eastern Telegraph had so emotional an attachment. Unimaginative management and unadventurous shareholders took pride in looking down on an upstart system which a growing number of users considered complementary but they saw as a threat to the capital locked in their ocean assets, the dividends it earned them, and comfy, job-for-life salaries. But once shorn of its 'national' responsibilities Cable and

---

*A line system through which electronic signals can be transmitted at frequencies higher than by satellite radio, because the insulated conductor (core) is no longer of copper but of glass, the width of a human hair, through which the impulses can zig-zag at the speed of visual (optical) light. It was hoped that each pair of fibres would carry up to 1,900 telephone circuits, though the actual distance between each repeater was not yet determined.

Wireless relished the role of freelance and exploited its potential to the full. But no swashbuckling. In the knowledge that the reputation it prized most was for dependability, the keynote was disciplined thoroughness. The flair for devising complicated communications for a demanding client like Maersk – which it had in full measure – was of little avail without the dedication and expertise required to keep the system in working order. And the reputation for reliability of that kind depended on the quality, enthusiasm and morale of the staff, 10,000 in number, whose efforts in earning £177 million in 1978, mainly from overseas activities, gained for the company the Queen's Award for Export Achievement 1979.

The highly skilled and highly adaptable men and women who had opted to run so diverse and successful an operation all over the world as that directed from Mercury House in London were a *corps d'élite* whose services were recognised by the Head of State, whose Government owned the group, by attending a reception to mark the operating company's golden jubilee.

A Prince of Wales and a Duke of Cambridge had come to the home of the Penders at Arlington Street to help celebrate previous anniversaries, as seen. But the honour bestowed by Her Majesty Queen Elizabeth II that 2 May 1979 on directors and representatives of managerial, executive and clerical grades in every branch in the UK and overseas, by inviting them to one of *her* homes, was unique in the company's history.

The life of even so healthy an organism as Cable and Wireless can allow no interruption. But to pause in order to look backward – and forward – after fifty years was as invigorating as it was refreshing, and there could have been no more inspiring place in which to do so than the ancient royal palace of St James's.

Appendix A

The Old 'PQ'
(from J E Packer's *Porthcurno Handbook*)

About halfway between Porthcurno beach and the Logan Rock there is a curious white stone pyramid close to the coastal path. 'A beacon for ships' is the obvious conclusion, but whilst this is true, the story behind its construction is more involved than first appearances suggest. Stand by the pyramid and look out to sea and you will notice rusty iron railings set into the rock at intervals, and appearances which suggest a pathway descending steeply towards the shore from the left foreground. The path ends abruptly above a sheer rock face, and here two vertical channels have been cut into the cliff and faced with masonry. Bathers on the beach have stared up at those precipitous slots and wondered; and to the observant visitor the site bears the obvious marks of some long forgotten enterprise. Sinister ideas spring to mind; wartime installations in the face of the cliffs perhaps, or a pipeline to supply fuel to gathering navies in the wake of invasion. Those old enough begin to recollect project 'Pluto'.

The men who hacked their way down the cliff, suspended by ropes from easier slopes above, did their breakneck labour in 1880 and the intent was neither sinister nor war-like. Their carefully fitted masonry has lasted the century well, and originally covered and protected a cable. The cable was brought to a little wooden hut at the top of the cliff, its distant end was in France, and it formed part of the trans-Atlantic telegraph route Cornwall-Brest-Nova Scotia. It was owned by 'La Compagnie Français du Telegraph de Paris à New York', a company whose cumbersome title was soon contracted to 'PQ Company' after the initials of one of its directors, a Monsieur Pouyer-Quertier.

The original Brest cable was linked by landline to a telegraph office in Penzance. At this point the records seem to conflict with the evidence before one's eyes. According to old cable charts and literature, a second Brest cable was laid directly into Porthcurno beach in 1918, yet there are plainly two separate channels in the cliff below the white pyramid. Certainly by 1930 diversions had taken place; both Brest cables landed at Porthcurno and the Eastern Telegraph Company had become responsible for the equipment which linked them with London.

The old black hut on the cliffs had outlived its usefulness as a cable house and took on a new lease of life as a summer holiday chalet. It survived in this role until the 1950s when the National Trust acquired the stretch of coast and had it demolished. As it had become very dilapidated and something of an eyesore this action no doubt pleased the hikers and photographers, but it raised a cry of protest from local fishermen who had been using it as a convenient leading mark for navigation. The white stone pyramid was therefore built on the exact site to serve this purpose, and hopefully to please all parties.

There can be few submarine cables which have had such an abrupt and precipitous landing, the reason for not using the beach at Porthcurno in the first instance was no doubt simply the fact that the Eastern Telegraph Company, at that time a rival concern, already had prior claim to its use.

Appendix B

Emma Pender to Hubert Herkomer

18 Arlington Street,
S.W.
April 26 [1888]

Dear Mr Herkomer,

I feel really in a great dilemma, but after very anxious consideration I am sure it is wisest to say frankly to you what is my feeling concerning my husband's portrait. Owing to illness I was unable to go to your Studio in Ebury Street, to see the original picture, but yesterday the replica arrived here. May I confess to being most sadly disappointed? It seems to me wanting in expression of the vigor and strength which Sir John's ability and great force of character mark upon his features so plainly.

To me this picture represents a much heavier, older and more common place man.

I cannot make up my mind to present it to the Telegraph Boards, as I should not like him to be thus handed down to posterity. And it appears to me to be only right to tell you of this before I express it publicly, for unfortunately I must now give my reasons in withholding the picture.

It is most painful to have to write thus. But you are too GREAT I know, to think I do wrong in candidly owning my opinion of the work. And trusting you will therefore forgive me any annoyance I may cause you, Believe me

Yours very truly
Emma Pender

Appendix C

CABLES AND WIRELESS LTD

The proposed extent and allocation of capital in the new company in 1929:

5½ per Cent. Cumulative Preference Shares of £1 each:

£
20,000,000 to Eastern and Associated Telegraph Companies
3,500,000 to Marconi's Wireless Telegraph Co., Ltd.

23,500,000

7½ per Cent. Non-Cumulative 'A' Ordinary Shares of £1 each:

£
13,200,000 to Eastern and Associated Telegraph Companies.
8,000,000 to Marconi's Wireless Telegraph Co., Ltd.

21,200,000

'B' Ordinary Shares of £1 each:

£
3,150,000, viz., 35 per cent, to Eastern and Associated
Telegraph Companies.
5,850,000, viz., 65 per cent, to Marconi's Wireless
Telegraph Co., Ltd.

9,000,000

53,700,000 Total

Votes	Cables	Marconi
Preference Shares (If dividend in arrear – 1 vote for 10 shares)	—	—
7½ per cent. 'A' Ordinary Shares – 1 vote per share	13,200,000	8,000,000
'B' Ordinary Shares – 2 votes per 3 shares	2,100,000	3,900,000
	15,300,000	11,900,000
	56·25 per cent.	43·75 per cent.

NOTE

The New Company to acquire:

(1) From the Marconi Company the whole of its Ordinary, Preference and Debenture Capital, which would be satisfied out of the above consideration.

(2) From the Cable Companies the whole of their Ordinary Share Capitals, leaving the Preference and Debenture issues undisturbed.

387

Appendix D

Allotment of full share capital of
Imperial and International Communications Ltd
in 1930

Shares were allotted as follows for the purpose of acquiring the whole or part of the undertakings of the companies mentioned and to provide the purchase price of the Pacific cable, W. Indian telegraph system and the Imperial Atlantic cables.

	In respect of the transfer of communications assets	At par in respect of cash provided for the purchase of Government undertakings	Total
Eastern Telegraph	£10,626,608	£552,947	£11,179,555
Eastern Extension	6,533,050	287,957	6,821,007
Western Telegraph	4,724,845	239,333	4,964,178
Marconi	4,225,517	186,249	4,411,766
African Direct	330,397	—	330,397
Europe and Azores	132,689	—	132,689
European and S. African	1,414,439	—	1,414,439
W. African	206,105	—	206,105
London-Platino Brazilian	193,481	—	193,481
Pacific and European	89,587	—	89,587
River Plate	31,928	—	31,928
W. Coast of America	224,868	—	224,868
	£28,733,514	£1,266,486	£30,000,000

Appendix E

The Group in 1929

Imperial and International Communications Limited,

formed with a capital that was limited at its inception to £30,000,000, acquired the communications assets of the:

Eastern Telegraph Company Limited
Western Telegraph Company Limited
Eastern Extension, Australasia and China Telegraph Company Limited
Eastern and South African Telegraph Company Limited
Europe and Azores Telegraph Company Limited
West African Telegraph Company Limited
African Direct Telegraph Company Limited
Pacific and European Telegraph Company Limited
West Coast of America Telegraph Company Limited
London Platino-Brazilian Telegraph Company Limited
River Plate Telegraph Company Limited
Société Anonyme Belge de Cábles Télégraphiques
Marconi's Wireless Telegraph Company Limited

as well as controlling interests in the:

Indian Radio Telegraph Company Limited
Marconi Radio Telegraph Company of Egypt S.A.
Wireless Telegraph Company of South Africa Limited

and interests in:

Amalgamated Wireless (Australasia) Limited
Companhia Radiotelegraphica Brazileira
Radio Suisse S.A.
Transradio Chilene Compania de Radiotelegrafia Limitada
Transradio Internacional Compania Radiotelegrafica Argentina
Trans-Oceanic Wireless Telegraph Company Limited
Russian Company of Wireless Telegraphy and Telephony

While it has since acquired the:

Direct West India Cable Company Limited
Halifax and Bermudas Cable Company Limited
Cuba Submarine Telegraph Company Limited
Indo-European Telegraph Company Limited
British East African Broadcasting Company Limited

and controlling interests in:

West India and Panama Telegraph Company Limited
Companhia Portuguesa Radio Marconi
Peruvian Telephone Company
Guayaquil Telephone Company

In all 32 Companies

In addition to these it also took over the:

Pacific Cable Board's Cables
The West Indian Cable and Wireless System worked by the Pacific Cable Board
The Imperial Atlantic Cables

and, on a lease for 25 years, the

Post Office Beam Services

(from first (1929/30) annual report of Imperial and International)

Appendix F

Directors of
Imperial and International Communications and Cable and Wireless Ltd.
from 1929 to 1979

Sir Charles Stewart Addis, K.C.M.G.
Frederick Robert Stephen Balfour, D.L., J.P.
The Rt. Hon. Baron Abertay of Tullybelton
Sir Frederick James Barthorpe, H.M.L.
The Rt. Hon. The Earl of Bessborough, C.M.G.
Sir Basil Phillott Blackett, K.C.B., K.C.S.I.
Col. The Hon. Arthur Grenville Brodrick, D.L., T.D., A.D.C.
The Rt. Hon. The Earl of Clarendon, D.L., J.P.
The Rt. Hon. The Lord Denison-Pender
Admiral Henry William Grant, C.B.
Henry Charles Hambro
The Rt. Hon. The Earl of Inchcape, G.C.S.I., G.C.M.G., K.C.I.E.
The Rt. Hon. The Lord Inverforth, P.C.
Francis Alexander Johnston
The Rt. Hon. Frederick George Kellaway, P.C., J.P.
Major Harry Lefroy, P.C., J.P.
The Marchese Guglielmo Marconi, G.C.V.O., LL.D., D.Sc.
John Francis O'Malley, F.R.C.S.
The Rt. Hon. The Earl of Midleton, P.C., K.P.
The Hon. Arthur George Villiers Peel, D.L., J.P.
Lorcan George Sherlock, LL.D.
Admiral of the Fleet, The Rt. Hon. Lord Wester Wemyss, G.C.B., C.M.G., M.V.O.,
    D.C.L., LL.D.
Sir Norman Roderick Alexander David Leslie, C.M.G., C.B.E.
The Rt. Hon. Lord Courtauld-Thomson, C.B., K.B.E.
Sir Edward Wilshaw, J.P., F.C.I.S., K.C.M.G.
Herbert Arthur White
Sir Harry Edward Augustus Twyford, K.B.E.
Edward George Brooke
Brigadier Henry John Lenton
The Rt. Hon. The Lord Reith, P.C., G.C.V.O., G.B.E.
Sir John Wardlow Milne, K.B.E., M.P.
Albert Harry Ginman
The Hon. John Jocelyn Denison-Pender, C.B.E.
Lieut. Col. Ivor Fraser
Richard Edmund Relfe Luff, C.B.E., A.C.A.
Col. Sir Arthur Stanley Angwin, K.B.E., D.S.O., M.C., T.D.
John Innes, C.B.
Maj. Gen. Sir Leslie Burtonshaw Nicholls, K.C.M.G., C.B., C.B.E., M.I.E.E.
Andrew Black, A.S.A.A.
Charles Neill Gallie
Col. Alfred Howard Read, C.B., O.B.E., T.D., D.L.
Frederick Ivor Ray
Norman Charles Chapling, C.B.E.

Kenneth Anderson, C.B., C.B.E.
William Alfred Wolverson, C.B.
Henry Howard Eggers, C.M.G., O.B.E.
Col. Donald McMillan, C.B., O.B.E., B.Sc., M.I.E.E.
Sir Godfrey Herbert Ince, G.C.B., K.B.E., B.Sc., LL.D.
John Keith Horsefield, C.B.
Reginald John Halsey, C.M.G., B.Sc., F.C.G.I., D.I.C., M.I.E.E.
John Fletcher, C.B.E.
Harold Carew Baker, O.B.E.
Sir John Stuart Macpherson, G.C.M.G., M.A., LL.D.
Edgar George Leonard Howitt, O.B.E.
Henry Ernest Matthews, O.B.E.
Archibald Anthony Willett, F.Inst.M., F.C.I.S., F.B.I.M.
Cyril James Gill
Wilfred Horace Davies, M.B.I.M.
Anthony Serle Pudner, M.B.E., C.Eng., F.I.E.E., F.I.E.R.E.
James Hodgson, M.A.
Harry George Lillicrap, C.B.E.
William Ratford Richard Haines, F.B.I.M.
John Graham Salkeld, M.A., F.C.A.
The Rt. Hon. The Lord Glenamara, C.H., LL.B.
Peter Alexander McCunn, M.B.I.M.
Richard Walter Cannon, C.Eng., F.I.E.E., F.I.E.R.E.
John Louis Warner Bird
Roderick Alexander Rice
David Berriman, M.A., F.I.B.
Archibald Johnstone Kirkwood, M.B.E.
Patrick Michael Meaney

CHRONOLOGY

1815 John Pender born in Scotland.

1837 The Year of the [Electric] Telegraph – world's first patent by Cooke and Wheatstone in England.

1840 Charles Wheatstone suggests submarine telegraph from Dover to Calais.

1842 Samuel Morse tries out submarine telegraph cable in America.

1845 Michael Faraday recommends gutta percha as submarine cable insulating medium; S W Silver & Co devise way of coating wire with it.
Jacob and John Brett register The General Oceanic Telegraph Company.

1849 John Brett forms The English Channel Submarine Telegraph Company.

1850 ECSTC lays telegraph under Dover Straits (but it breaks down after six days).

1851 London-Paris telegraph opens 13 November – world's first submarine telegraph.

1852 English & Irish Magnetic Telegraph Company formed – John Pender a director.

1853 EIMTC build London-Dublin telegraph with Port Patrick-Donaghadee sea link.

1854 John Brett's Mediterranean Telegraph Company open telegraph Genoa-Corsica; fail to lay Sardinia-Algeria cable; abandon Europe-North Africa telegraph.
Glass, Elliot, cable manufacturers, formed.

1856 The Atlantic Telegraph Company formed – John Pender a director.

1857 ATC's first attempt to lay Atlantic Cable fails.
Capt Rassloff, a Dane, plans telegraph to link North and South America.

1858 ATC lays Atlantic Cable (13 August); but it breaks down on 18 September.
The Red Sea and Telegraph to India Company fail to link Suez, Aden & Karachi.

1859 HMG appoint Committee of Enquiry n the Construction of Submarine Telegraph Cables.
Publication of Parliamentary Blue Book *The Establishment of Telegraphic Communication in the Mediterranean and with India.*

393

1861 *The Telegraphic Journal* founded.
Turkish Government open inland telegraph from Constantinople to Baghdad.

1862 Government of Bombay form Indo-European Telegraph Dept, and plan telegraph from Karachi to Baghdad.

1846 At instigation of John Pender, The Gutta Percha Company merge with Glass, Elliot to become The Telegraph Construction & Maintenance Co (Telcon).

1865 First telegraph opened between India and Europe.
River Plate Telegraph Company formed; lays telegraph between Buenos Aires and Montevideo.

1866 Anglo-American Telegraph Company (co-founder John Pender) lay Atlantic Telegraph.
International Ocean Telegraph Company open telegraph between Florida and Cuba.

1868 Anglo-Mediterranean formed; lay telegraph from Malta to Alexandria.
Siemens form Indo-European Telegraph Company.

1869 John Pender forms British-Indian Submarine; Falmouth, Gibraltar & Malta; British-India Extension; and China Submarine.
West India & Panama formed to link West Indian islands.

1870 John Pender forms British Australian.
Falmouth, Gibraltar & Malta open cable station at Porthcurno, Cornwall.
Completion of Porthcurno-Bombay cable (the Red Sea Line).
Post Office take over British inland telegraphs and those to Europe.

1872 John Pender merges companies into The Eastern Telegraph Company (6 June).
Commercial opening of All-Sea Australia to England Telegraph (15 November).
Opening of telegraph from West Indies to US and Europe.
Platino-Brazileira, Western & Brazilian, and Direct Spanish formed.

1873 John Pender forms The Globe Telegraph & Trust Company; re-groups Far East companies into The Eastern Extension, Australasia & China Telegraph Company.

1874 Black Sea Telegraph Company formed.

1876 West Coast of America Telegraph Company formed (Peru-Chile).
Eastern Extension open telegraph from Sydney to New Zealand.
John Pender seeks to merge Anglo-American with Direct United States.

1877 Conference in New South Wales discusses possibility of colonial governments themselves laying cable to North America.
John Pender buys Black Sea Telegraph Company.
Alexander Graham Bell demonstrates telephone to telegraph engineers in London.

1878 Trade Depression.
Sandford Fleming of CPR proposes Pacific Cable Canada-Australia.
Platino re-formed in UK as The London Platino Brazilian Telegraph Company.
Death of Harry Pender, John Pender's eldest son, in Naples.

1879 International Telegraph Conference in London.

1879 John Pender forms The Eastern and South African Telegraph Company with Telcon.
Death of James Clerk Maxwell who predicted existence of electro-magnetic waves.

1880 British courts declare a telephone is a telegraph in law.
E&SA complete cable to Cape Colony.

1882 Riots in Alexandria.

1883 Conference in Sydney proposes cable from New Caledonia to San Francisco.

1884 Royal Commission recommends direct cable West Indies-Halifax.

1885 Eastern Telegraph and Brazilian Submarine form The African Direct Telegraph Company, with HMG subsidy.

1887 Queen Victoria's Jubilee Conference hears Sandford Fleming's ideas for a Pacific Cable between Canada and Australia.
50th Anniversary of the Electric Telegraph; and of John Pender's 30 years in Submarine Telegraphy – he is knighted.
Heinrich Hertz confirms existence of electro-magnetic waves.

1888 Revolution in Brazil; change in position of Western & Brazilian.

1889 The Halifax & Bermudas Cable Company formed

1890 Cecil Rhodes plans overland telegraph across Africa.
Sir John Pender lays second submarine cable to Africa.
Eastern Telegraph duplicates telegraph to Australia.
Bermuda-Halifax-UK cable opened.

1891 World's first submarine telephone service opened London-Paris (1 April).

1892 Pacific & European Telegraph formed.
Cecil Rhodes forms African Transcontinental Telegraph Company.

1893 Europe & Azores Telegraph formed to operate Portugal-Brazil line.

1894 Silver Jubilee of Submarine Telegraphy to Far East; banquet and reception attended by Prince of Wales.
Colonial Conference in Ottawa agree on desirability of Pacific Cable; P & T Conference in Wellington offer to subsidise it.

1895 The Amazon Telegraph Company formed.
F G Creed invents Morse keyboard perforator – forerunner of teleprinter.

1896 Death of Sir John Pender, aged 80; Lord Tweeddale becomes chairman, and John Denison-Pender, managing director, of Eastern Telegraph group.
Guglielmo Marconi comes to England from Italy and takes out patent for wireless.
Duplex working introduced on cables.
Joseph Chamberlain appointed Colonial Secretary.
Imperial Pacific Cable Committee meet in London (report not published till 1899).

1897 Guglielmo Marconi forms The Wireless Telegraph & Signal Company.
Queen Victoria's Diamond Jubilee.

1898 Marina House, Hong Kong, opened.
Sir Sandford Fleming advocates State-controlled Cable for whole Empire.

1899 South American companies merge to form The Western Telegraph Company.
Outbreak of Boer War; Eastern Telegraph lay submarine cable to South Africa via Ascension Island and St Helena.

1899 HMG appoint Cable Landing Rights Committee.
Marconi sends wireless signals across English Channel.

1900 Boxer Rebellion in Peking.
Marconi re-names his company 'Marconi's Wireless Telegraph Company'
and takes out master patent 7777 solving problem of syntony.
Telcon receive contract to make and lay Pacific Cable.
Sir John Wolfe Barry becomes chairman of Eastern Telegraph group.

1901 Jubilee of Submarine Telegraphy.
Eastern Telegraph plan Indian Ocean Cable to link Australia with South
Africa via Mauritius, Rodriguez and Cocos.
Marconi transmits wireless signal across North Atlantic from Cornwall to
Newfoundland.
Pacific Cable Bill receives royal assent (17 August).
Six colonies form Commonwealth of Australia.

1902 Eastern Telegraph group moves head office from Winchester House, Old
Broad Street to Electra House, Moorgate.
Sir John Wolfe Barry succeeds Lord Tweeddale as chairman.
Eastern Telegraph install monitoring wireless station at Porthcurno.
Indian Ocean Telegraph (Durban-Adelaide) opens, 1 November.
Pacific Cable connected between Canada and New Zealand, 31 October;
service opened to public 8 December.
Coronation Conference (King Edward VII) discuss Sandford Fleming's ideas
for state-controlled imperial communications.

1905 Wireless Telegraphy Act allows Marconi to operate ship-to-shore service.

1906 Eastern Telegraph launch *The Zodiac* house magazine.

1907 Marconi starts limited commercial wireless service across North Atlantic.

1908 City businessmen hold Cable Reform meeting in Mansion House, London.

1909 Imperial Press Conference call for lower cable rates.
British Post Office buy Marconi's ship-to-shore wireless station.
Marconi opens full transatlantic wireless service.

1910 Marconi conceives plan to link whole British Empire with wireless.

1911 Cable Landing Rights Committee reject Marconi's Imperial Wireless
Scheme; suggest HMG operate it and Marconi build it.
Coronation Conference (King George V) discuss Marconi Plan and possible
state-owned cable between England and Canada across North Atlantic.

1912 The 'Marconi Scandal' delays building of Imperial Wireless Chain.
William Ash retires from Porthcurno after 34 years as Superintendent.

1913 Post Office place contract with Marconi for Imperial Wireless Chain.

1914 Outbreak of World War 1; German raids on Fanning Island *(Nürnberg)* and
Cocos Islands *(Emden)*. PO cancel contract for wireless chain.

1915 Training School started at Porthcurno.

1916 Marconi re-investigates possibility of short wave propagation.

1917 Sir John Denison-Pender becomes chairman of Eastern Telegraph group.

1918 World War 1 ends with defeat of Germany and her allies.

1919 Marconi's new Imperial Wireless Scheme considered by Imperial Wireless
Telegraphy Committee (Norman Committee) who reject it and produce
own scheme.

1919 Amalgamated Wireless (Australia) formed; place contract with Marconi for long wave station.
Eastern Telegraph buy Meadowbank, Twickenham, as staff recreation centre.

1920 Eastern Telegraph London Training School opened in Shepherd Walk, Hampstead.
HMG accept Norman Report; but their scheme never carried out.

1921 Wireless telephone link made between Carnarvon and Australia.

1922 50th anniversary of formation of Eastern Telegraph Company; staff reception at Meadowbank; banquet and fête in Regents Park.

1923 Marconi build long wave station for South African Government.
Sir Campbell Stuart appointed Canadian representative on Pacific Cable Board.
Imperial Economic Conference told of plan for a Pacific Cable Board West Indies System.

1924 Marconi's trials in his yacht *Elettra* confirm reliability of short wave beam wireless.
Sir Robert Donald's Committee recommend that the Post Office own and run UK terminals of new short wave, beam Imperial Wireless Chain; and HMG accept their report.
Experiments at Carnarvon in radio transmission of still pictures between UK and USA.
First stations opened in PCB West Indies System.

1925 Eastern Telegraph introduce revolutionary 'Regenerator' principles.

1926 Post Office station at Rugby establish first radio-telephone link across North Atlantic (commercial service opens 7 January 1927).

1927 Short wave, beam Imperial Chain in full operation by September.
Eastern Telegraph have discussions with HMG over wireless competition.
Sir John Denison-Pender meets Lord Inverforth of Marconi; discuss merger.

1928 Imperial Wireless and Cable Conference meet to examine situation arising from Beam Wireless competition with Cable services.
Conference told Marconi and Eastern Telegraph had decided to 'fuse'.
Conference report recommends fusion in a private company.

1929 Imperial Telegraphs Bill receives royal assent.
'Imperial and International Communications Company' and 'Cables and Wireless Ltd' formed; Treasury Agreement of 29 May.
Combined wireless and cable service opened to public 30 September.
World Economic Crisis.

1931 Statute of Westminster makes Dominions autonomous communities.
I & I C introduce D/N Scheme to reduce staff.
Investigating Committee (Greene Committee) enquire into I & IC.

1932 J C Denison-Pender becomes chairman of I & IC (later 1st Lord Pender)

1933 I & IC move administrative HQ to Electra House, Victoria Embankment.

1934 I & IC change name to 'Cable and Wireless Limited'.
Picture transmission service opens London-Melbourne.

1936 Spanish Civil War; trouble for Direct Spanish staff.
Edward Wilshaw becomes chairman and sole managing director of Cable and Wireless.

1937 Coronation Conference (King George VI) review imperial communications.
      Empire Rates Committee sit under Edward Wilshaw.
      'Colonial Wireless Scheme' conceived.
      Death of Guglielmo Marconi.

1938 Imperial Telegraphs Act embodies heads of agreement reached by empire
          governments, giving Cable and Wireless a new 'charter'.
      Munich Crisis; emergency operation centre set up at Electra House, SW.

1939 Outbreak of World War 2.

1940 The Tunnel built at Porthcurno (opened 31 May 1941).
      Expeditionary Force Messages (EFMs) introduced.

1941 Electra House, Moorgate, put out of action by German bombing.
      Japanese bomb US naval base in Hawaii, and the US enter war.
      Japanese attack and occupy Hong Kong; British Cable and Wireless staff
          interned.
      Singapore falls to Japanese; wireless relay station site switched to Barbados.

1942 Australian Telegraph Conference in Canberra discusses imperial communi-
          cations policy without representative of Cable and Wireless.

1943 Cable and Wireless staff in battle areas organised as Telcom units in uniform.

1944 Commonwealth Communications Council review communications system of
          entire Commonwealth.
      Sir John Reith becomes a director of Cable and Wireless.
      The Fisk Scheme (for imperial wireless networks); the Reith Plan (for a single
          Commonwealth public corporation); the Anzac Scheme (for national
          corporations in each Commonwealth country).
      Electra House, Victoria Embankment, hit by German flying bomb.
      Allied invasion of Europe.

1945 Lord Reith and party set out on mission to Commonwealth; draft 'Canberra
          Proposals' in Australia and secure assent of other countries.
      End of War in Europe (7 May); Labour Government win British General
          Election.
      Cable and Wireless oppose recommendations of Reith Mission Report at
          Imperial Communications Conference in London (July).
      HMG accept recommendations of Conference report; set up Commonwealth
          Telecommunications Board.
      Chancellor of Exchequer announces Cabinet decision to transfer telecom-
          munication services operated by Cable and Wireless to public ownership.
      Sir Edward Wilshaw mounts campaign to prevent transfer.
      Bermuda Conference meets to discuss USA-Commonwealth telecommunica-
          tions; Bermuda Agreement on rates signed 4 December.

1946 Cable and Wireless petition against Cable and Wireless Bill, and Select
          Committees of both Commons and Lords hear evidence.
      Cable and Wireless Act receives royal assent (1 November); Treasury
          Agreement of 1929 repealed.
      English Electric acquire Cable and Wireless (Holding)'s interest in the
          Marconi Company.

1947 Cable and Wireless Act operative from 1 January; Sir Arthur Angwin
          becomes chairman in place of Sir Edward Wilshaw who heads investment
          trust (holding company).

1949 Bermuda Agreement revised.

1950 Cable and Wireless Engineering School opens at Porthcurno.
Telcom units provide press communications in Korean War.

1951 Sir Leslie Nicholls becomes chairman of Cable and Wireless.

1955 Cable and Wireless move headquarters to Mercury House, Theobalds Road.

1956 First transatlantic telephone cable (TAT-1).
Egyptian Government sequestrate Cable and Wireless cables after Suez Campaign.
Sir Godfrey Ince becomes chairman of Cable and Wireless.

1957 COTC and Cable and Wireless plan CANTAT.

1958 Commonwealth Communications Conference plan Commonwealth Round-the-World Telephone System.

1961 Washington Conference consider possibility of radio-by-satellite.
CANTAT opened.

1962 Sir John Macpherson becomes chairman of Cable and Wireless.

1963 COMSAT formed in the US.
COMPAC opened.

1964 INTELSAT formed; first commercial communications satellite launched.

1965 Cable and Wireless build satellite earth station on Ascension Island.
Cable and Wireless (West Indies) announce Eastern Caribbean Expansion Project using tropospheric scatter systems.

1966 Commonwealth Telecommunications Organisations formed.

1967 Death of Sir Edward Wilshaw, aged 88.
SEACOM opened.
Donald McMillan becomes chairman of Cable and Wireless.

1970 Porthcurno closes as a cable station.
H G Lillicrap becomes chairman of Cable and Wireless.

1971 Cable and Wireless sell half share in CANTAT to Post Office.

1972 East Caribbean Microwave System planned.
New Mercury House, Hong Kong, opened.
Centenary of Eastern Telegraph Company.

1973 End of group's 100-year association with South America.

1975 Select Committee on Nationalised Industries examine Cable and Wireless.

1976 Edward Short (Lord Glenamara) becomes chairman of Cable and Wireless.

1977 A A Willett resigns as managing director of Cable and Wireless; Peter McCunn appointed Group Managing Director and Deputy Executive Chairman.

1979 GOLDEN JUBILEE OF CABLE AND WIRELESS.
Reception in St James's Palace attended by HM The Queen.

BIBLIOGRAPHY

PART I (to 1897)

Sir Stanley Angwin, *The Romance of Porthcurno Cable Station* (typescript) address to London Cornish Association, 12 October 1948

The Atlantic Telegraph Company, *The Atlantic Telegraph, A History of Preliminary Experimental Proceedings and a Descriptive Account of the Present State and Prospects of the Undertaking,* London: Jarrod, 1857

K C Baglehole, *A Century of Service, Cable and Wireless Ltd 1868–1968,* London: Cable and Wireless, 1969

E C Baker, *Sir William Preece, FRS, Victorian Engineer Extraordinary,* London: Hutchinson, 1973

F G C Baldwin, *The History of the Telephone in the UK,* London: Chapman & Hall, 1938

R M Black, *Electric Cables in Victorian Times,* London: Science Museum, HMSO, 1973

Brian Bowers, *Charles Wheatstone,* London: Science Museum HMSO, 1975

Charles F Briggs and Augustus Maverick, *The Story of the Telegraph,* New York: Rudd & Carleton, 1858

Charles Bright, *Submarine Telegraphs, Their History, Construction and Working,* London: Crosby Lockwood, 1898

Lady Frederick Cavendish (ed. John Bailey), *Diary* (18 August 1858), London: John Murray, 1927

William (Dan) Cleaver, *A History of Porthcurno* (typescript), 1953, Cable and Wireless Archives, London

E L (Jimmy) Cozier, *Barbados: One Hundred Years of Communications* (typescript), Barbados, 1972

A Danashery (?), *Rough Notes on taking over the cable to the Continent* for Major J V Bateman Champain, Indo-European Telegraph Department, 27 October 1868, India Office Records ref 10 R, L/PWD/7/177

Eastern Telegraph Company Limited, *Reports and Accounts and Proceedings at General Meetings* (from 6 December 1872), Cable and Wireless Archives, London

Fairplay, 18 and 25 July 1890, 'Cable Ships, Their Adventures and Misadventures'

F H C Farver, *The Associated Cable Companies in South America 1866–1922,* with Foreword by Mr Farver dated 20 April 1934 (typescript), Cable and Wireless Archives, London

Bernard S Finn, *Submarine Telegraphy, The Grand Victorian Technology,* London: Science Museum HMSO, 1973

G R M Garratt, *One Hundred Years of Submarine Cables,* London: Science Museum HMSO, 1950

Michael Angelo Garvey, *The Silent Revolution,* Edinburgh 1852

W T Glover & Co Ltd, *Glover's Centenary 1868–1968,* Manchester 1968

Lesley A Hall, *The Indo-European Telegraph Department 1865–1931,* India Office Records ref L/PWD/7

W T Henley's Telegraph Works Co Ltd, *W T Henley,* London, 1953

Dionysius Lardner, *The Electric Telegraph Popularised* (Lardner's Museum of Science no 31), London: Walter & Maberly, 1855

J E Packer, *Cornish Cable Communications* (typescript), 1974

J E Packer, *PK Notebook* (typescript), 1974

J C Parkinson, *The Ocean Telegraph to India,* Edinburgh: Wm Blackwood, 1869

W H Russell, *The Atlantic Telegraph,* London, 1865 (David & Charles reprint, 1972)

George Saward, *The Transatlantic Submarine Telegraph: A Brief Narrative of the principal incidents in the history of the Atlantic Telegraph Company compiled from authentic and official documents* (privately printed), London 1878

Willoughby Smith, *The Rise and Extension of Submarine Telegraphy,* London, 1867

Edward A Stoute, *Peeps Into Barbados History* (typescript), Barbados, 1976

Telegraph Construction & Maintenance Co Ltd. *The Telcon Story,* London 1950

Geoffrey Wilson, *The Old Telegraphs,* Chichester: Phillimore, 1976

Report on *The Construction of Submarine Telegraph Cables,* Parliamentary Papers, vol. LXII, 1860

Report of the Joint Committee appointed by the Lords of the Committee of the Privy Council for Trade and The Atlantic Telegraph Company to inquire into the *Construction of Submarine Telegraph Cables,* HMSO 1861

Parliamentary Blue Book, *The Establishment of Telegraph Communications in the Mediterranean and to India,* 1859

Parliamentary Blue Book, Select Committee of the House of Commons *Report on East India Communications* (428), 1866

Letters of Mrs (later Lady) John Pender, née Emma Denison, to her daughter Marion (Lady des Voeux) and others, 1875–1888 (typed transcripts), Cable and Wireless Archives, London [presented by Hon. Richard Denison-Pender who has no knowledge of the whereabouts of the hand-written originals]

Letters from the Superintendent Porthcurno Cable Station to Eastern Telegraph Head Office London 1876–1893 (typed transcripts), Cable and Wireless Archives, London

Souvenir Book of the Silver Jubilee of the Telegraph to the Far East Banquet and Fête, 1894, Eastern Telegraph and Eastern Extension companies, Cable and Wireless Archives, London

PART II (to 1938).

K C Baglehole, *Historical Notes on the Pacific Cable Board* (typescript), Cable and Wireless Archives, London

W J Baker, *A History of the Marconi Company,* London: Methuen, 1970

Sir Basil Blackett, *World Communications,* talk to Royal Empire Society, 1930

Charles Bright, *Imperial Telegraphic Communication,* London: P S King, 1911

Charles Bright (junr), *Imperial Communications,* talk to London Chamber of Commerce, 4 December 1902

F J Brown, *The Cable and Wireless Communications of the World,* London: Pitman, 1927

Cables and Wireless Ltd, *Annual Reports* (from 1930)

Cable and Wireless Ltd, *Annual Reports* (from 1934)

Cable and Wireless (Holding) Ltd, *Annual Reports* (from 1934)

The Canadian Annual Review 1904, pp 150 & 474–7 re Pacific Cable

Cocos Branch Data (typescript) December 1938, Cable and Wireless Archives, London

Frances Donaldson, *The Marconi Scandal,* London: Rupert Hart-Davis, 1962

Sir Sandford Fleming, *Memorandum* 'On the Pacific Cable and the Telegraph Service of the Empire', Coronation Conference, 1902

Keith Geddes, *G Marconi 1874–1937,* London: Science Museum HMSO, 1974

A Graves, *Zodiac,* 'Memories of Porthcurno Seventy Years Ago', 1971

K R Haigh, *Cable Ships and Submarine Cables,* London: Adlard Coles, 1968

Duff Hart-Davis, *Ascension, The Story of a South Atlantic Island,* London: Constable, 1972

W P Holly, *Marconi,* London: Constable, 1972

Imperial and International Communications Ltd, *Annual Reports* (from 1 July 1930)

Imperial and International Communications Ltd, *Imperial Wireless Telephony,* 1930

Imperial Wireless and Cable Conference, *Report,* London, HMSO (Cmd 3163), 1928

Imperial Wireless Telegraphy Committee 1919–1920, *Report,* London, HMSO (Cmd. 777), 1920

George Johnson (ed.) *The All-Red Line, the Annals and Aims of the Pacific Cable Project; the Problem of an Empire-Girdling State-owned Telegraph System,* Ottawa, James Hope, 1903

Robert V Kubicek, *The Administration of Imperialism: Joseph Chamberlain at the Colonial Office,* North Carolina: Duke University Press, 1969

Brigadier-General Sir Osborne Mance, *International Telecommunications* (pp viii & ix re 1919), Oxford University Press/Royal Institute of International Affairs, 1943

André Maurois, *Cecil Rhodes,* London: Collins, 1953

Sir Lewis Michell, *The Life and Times of the Right Honourable Cecil John Rhodes 1853–1902,* London: Edward Arnold, 1912

Pacific Cable Board, *Memorandum of Rules and Conditions of Employment,* January 1915

Quarterly Review, 'Imperial Telegraphs', April 1903

Harold C Rose, *The Military Wireless Station at Barbados* (typescript), Barbados, 1914

A V Ussher, *Tientsin Branch Data* (typescript) 1932, Cable and Wireless Archives, London

R N Vyvyan, *Wireless Over Thirty Years,* London: Routledge, 1933

Western Telegraph Company Ltd; Letter Books selected for preservation by Dr Platt and received from the WTC in Brazil in 1965, D M S Watson Library, University College, London

Edward Wilshaw, *Cable Communications and how to use them to the best advantage,* talk to Chartered Institute of Secretaries at Manchester, 1924

Blue Books and other Government Publications Concerning the Pacific Cable (3 vols), Cable and Wireless Archives, London

Inter-departmental Committee on Cable Communications of 1901, *Evidence,* London, HMSO (Cmd. 1118), 1902

Sir Campbell Stuart Files (41 items relating to Communications from 1912 to 1945) deposited by Sir Campbell's executors on his death in 1972 in the library of the Royal Commonwealth Society, London

PART III (to 1979)

Mrs H L N Ascough, *Saved by the Atomic Bomb* (typescript), 1941, Cable and Wireless Archives, Hong Kong

Andrew Boyle, *Only the Wind Will Listen, Reith of the BBC,* London: Hutchinson, 1972

Cable and Wireless Ltd, *World-Wide Communication,* 1954

Central Office of Information, *British and Commonwealth Telecommunications,* June 1963

Hugh Dalton, *High Tide and After, Memoirs 1945–60,* London: Frederick Muller, 1962

Charles Graves, *The Thin Red Lines,* London: Standard Art Book Co, 1946

R J Halsey, *Global Communication,* address to the Institution of Electrical Engineers, October 1961

R J Halsey, *Global Telecommunication,* Mitchell Memorial Lecture, 1965

Edgar Harcourt, 'Commonwealth Telecommunications after World War 2', *Transit,* May, June and July 1976

J G Alec Pierre, *Training at Barbados* (typescript), Cable and Wireless Archives, Barbados

J C W Reith, *Into the Wind,* London: Hodder & Stoughton, 1949

Michael Skinner, *History of the International Telephone Service Hong Kong* (manuscript), Cable and Wireless Archives, Hong Kong

Sir Campbell Stuart, *Opportunity Knocks Once,* London: Collins, 1952

Charles Stuart (ed.), *The Reith Diaries,* London: Collins, 1975

Submarine Cables Ltd., *First Transatlantic Telephone Cable,* 1956

Sir Edward Wilshaw, *Telecommunications and Commerce,* ad dress to the Conference of Federation of Chambers of Commerce of the British Empire, October 1945

Fifth Report from the Select Committee on Nationalised Industries 'Cable and Wireless Limited', London, HMSO (472), May 1976

Select Committee on the Cable and Wireless Bill, *Evidence,* 1946

Secret File, British Embassy Washington no G.285/1 'Communications: Peace Terms' Public Record Office ref FO 115 3571 [closed till 1972]

White Paper, *Cable and Wireless Ltd, Proposed Transfer to Public Ownership,* London, HMSO (Cmd. 6805), April 1946

OTHER SOURCES

The files of:

West Briton and Cornwall Advertiser (1870)

The Telegraphic Journal (1870s)

The Annual Register

The Times

Truth

Fairplay

Hansard

Zodiac house magazine (from 1906)

The Gazette, staff magazine of I&ICSA (from 1931 for a few years only)

INDEX

(f=footnote)

405